Pulsed-Field Gel Electrophoresis

Methods in Molecular Biology

John M. Walker, SERIES EDITOR

Methods in Molecular Biology • 12

Pulsed-Field Gel Electrophoresis

Protocols, Methods, and Theories

Edited by

Margit Burmeister

University of Michigan, Ann Arbor, MI

Levy Ulanovsky

Weizmann Institute of Science, Rehovot, Israel

Humana Press ✳ Totowa, New Jersey

© 1992 The Humana Press Inc.
999 Riverview Drive
Totowa, New Jersey 07512

WITHDRAWN

Printed in the United States of America

Library of Congress Cataloging in Publication Data
Main entry under title:

Methods in molecular biology.

Pulsed-field gel electrophoresis: protocols, methods, and theories / edited by
    Margit Burmeister, Levy Ulanovsky.
        p. cm. – (Methods in molecular biology ; v.12)
    Includes index.
    ISBN 0-89603-229-9
    1. Pulsed-field gel electrophoresis–Technique. 2, DNA–Analysis.
    I. Burmeister, Margit. II. Ulanovsky, Levy. III. Series.
    [DNLM: 1. Electrophoresis, Gel, Pulsed-Field–methods. 2. Genetic Techniques.
    W1 ME9616F v. 12/ / QU 25 P982]
    QP519.9.P84P85 1992
    574.8'8'028–dc20
    DLC
    for Library of Congress   92-1475
                                         CIP

Preface

Pulsed-field gel electrophoresis (PFGE) was invented barely 10 years ago, but has already become a major technique in molecular biology, with hundreds of applications. As is often the case with new techniques in their initial stages, most publications are outdated by the time they reach print. For this reason, a number of the authors of this book have effectively served as phone consultants on PFGE applications for several years. By now, many of the major methods have been tested and improved sufficiently that we feel a book of techniques will prove especially timely. During the preparation of *Pulsed-Field Gel Electrophoresis: Protocols, Methods, and Theories*, two other books reviewing PFGE were published. However, the present work is the first to incorporate the protocol style, the hallmark of the *Methods in Molecular Biology* series that has made that approach so successful in the past. In addition, many of the chapters include discussions on strategies helpful in planning experiments. The final part of the book contains theories and observations elucidating the complex behavior of DNA during PFGE.

For any given application of PFGE, the reader will usually proceed by assembling a protocol from more than one chapter (e.g., CHEF, preparation of agarose blocks, and mapping strategies). To help in this process, we have tried hard to provide an index and cross-references that allow easy access to all the methods. It may also be useful to read chapters not immediately related to the application in mind. For example, a user of CHEF may also find helpful hints in the chapters on FIGE, OFAGE, or TAFE. Certain methods, such as the preparation of yeast size markers, are done somewhat differently in different labs, and the reader will find several protocols from which to choose.

It is the inherent interdisciplinary character of PFGE that was most challenging to us. For the biologist, PFGE is a method that allows the analysis and separation of large DNA fragments or chromosomes. For the physicist, the behavior of DNA during PFGE presents a fascinating physical problem that happens to have great application potential. The present book gives some say to both groups. We feel that biologists and physicists interested in PFGE will benefit greatly from talking with each other. Biologists, one hopes, will be encouraged to read some of the theoretical work, even if it is harder to digest—there may well be practical applications from the more theoretical approaches. On the other hand, physicists may gain more insight into the mechanism of PFGE and appreciate where improvements are needed by reading through the experimental procedures and problems that biologists face.

We want to thank all those who helped tremendously, but whose names do not appear on the cover. The series editor, John Walker, shared with us his experience from previous volumes, read and co-edited many of the chapters, and helped us on every step on the way. The contributors struggled with great patience through the often painful editorial process. Rick Myers and the members of his lab in San Francisco suffered through Margit's endless phone conversations with great understanding and support. We also want to thank Hans Lehrach and the members of his lab when at the European Molecular Biology Laboratory in Heidelberg, Rick Myers and David Cox (University of California, San Francisco), and the members of their labs and Walter Gilbert (Harvard University, Cambridge) for encouragement and support.

Margit Burmeister
Levy Ulanovsky

Contents

Contents

Contributors

DENISE P. BARLOW • *Research Institute of Molecular Pathology, Vienna, Austria*

WILFRIED BAUTSCH • *Institut für Medizinische Mikrobiologie, Medizinsche Hochschule Hannover, Germany*

MARGIT BURMEISTER • *Department of Physiology, University of California at San Francisco, CA (Present address: Mental Health Research Institute, University of Michigan, Ann Arbor, MI)*

GEORGES F. CARLE • *Centre de Biochimie, Université de Nice, France*

SETTARA C. CHANDRASEKHARAPPA • *Department of Human Genetics, University of Michigan, Ann Arbor, MI*

FRANCIS S. COLLINS • *Department of Internal Medicine, Human Genetics, and Howard Hughes Medical Institute, University of Michigan, Ann Arbor, MI*

CRAIG A. COONEY • *Biology Department, Beckman Research Institute, City of Hope, Duarte, CA*

DIANE W. COX • *Research Institute, The Hospital for Sick Children, and Departments of Molecular and Medical Genetics, and Pediatrics, University of Toronto, Canada*

ANNE DOMINIQUE DÉFONTAINES • *Laboratoire de Physicochimie Théorique, ESPCI, Paris, France*

JOSHUA M. DEUTSCH • *Department of Physics, University of California, Santa Cruz, CA*

MARTIN W. GANAL • *Department of Plant Breeding and Biometry, Cornell University, Ithaca, NY*

KATHELEEN GARDINER • *Eleanor Roosevelt Institute for Cancer Research, Denver, CO*

KEITH M. GOTTESDIENER • *Department of Genetics and Development, Columbia University, New York*

GARY A. GRIESS • *Department of Biochemistry, University of Texas Health Science Center, San Antonio, TX*

STANLEY H. KORMAN • *Department of Genetics and Development, Columbia University, New York (Permanent address: Department of Pediatrics, Hadassah University Hospital, Jerusalem, Israel)*

ZOIA LARIN • *Genome Analysis Laboratory, Imperial Cancer Research Fund, Lincoln's Inn Fields, London, UK*

SYLVIE LE BLANCQ • *Department of Genetics and Development, Columbia University, New York*

HANS LEHRACH • *Genome Analysis Laboratory, Imperial Cancer Research Fund, Lincoln's Inn Fields, London, UK*

STEPHEN D. LEVENE • *Division of Biochemistry and Molecular Biology, Department of Molecular and Cell Biology, University of California, Berkeley, Berkeley, CA*

MELANIE M. MAHTANI • *Department of Genetics, Stanford University, Stanford, CA*

DOUGLAS A. MARCHUK • *Howard Hughes Medical Institute, University of Michigan, Ann Arbor, MI*

MICHAEL MCCLELLAND • *California Institute of Biological Research, La Jolla, CA*

ANTHONY P. MONACO • *Human Genetics Laboratory, Imperial Cancer Research Fund, Institute of Molecular Medicine, John Radcliffe Hospital, Headington, Oxford, UK*

ELENA T. MORENO • *Department of Biochemistry, University of Texas Health Science Center, San Antonio, TX*

JAAN NOOLANDI • *Xerox Research Centre of Canada, Mississauga, Ontario, Canada*

JOAN OVERHAUSER • *Department of Biochemistry and Molecular Bioogy, Thomas Jefferson University, Philadelphia, PA*

CATRIN PRITCHARD • *Department of Physiology, University of California at San Francisco, CA (Present address: Medical Research Council, Institute of Molecular Medicine, John Radcliffe Hospital, Headington, Oxford, UK)*

MARION S. RÖDER • *Department of Plant Breeding and Biometry, Cornell University, Ithaca, NY*

PHILIP SERWER • *Department of Biochemistry, The University of Texas Health Science Center, San Antonio, TX*

BRUNO W. S. SOBRAL • *California Institute of Biological Research, La Jolla, CA*

MARJATTA SON • *Department of Biochemistry, The University of Texas Health Science Center, San Antonio, TX*

NANCY C. STELLWAGEN • *Department of Biochemistry, University of Iowa, Iowa City, IA*

CHANTAL TURMEL • *Xerox Research Centre of Canada, Mississauga, Ontario, Canada*

LEX H. T. VAN DER PLOEG • *Department of Genetics and Development, Columbia University, New York*

JEAN LOUIS VIOVY • *Laboratoire de Physicochimie Théorique, ESPCI, Paris, France*

DOUGLAS VOLLRATH • *Whitehead Institute for Biomedical Research, Cambridge, MA*

ARMIN VOLZ • *Institut für Experimentelle Onkologie und Transplantationsmedizin, Universitätsklinikum Rudolf Virchow, Freie Universität Berlin, Germany*

MICHAEL A. WALTER • *Research Institute, The Hospital for Sick Children, and Departments of Molecular and Medical Genetics, and Pediatrics, University of Toronto, Canada (Present address: Laboratory of Human Genetics, Imperial Cancer Research Fund, Lincoln's Inn Field, London UK)*

PETER E. WARBURTON • *Department of Genetics, Stanford University, Stanford, CA and Department of Medical and Molecular Genetics, University of Toronto, Ontario, Canada*

ROBERT H. WATSON • *Department of Biochemistry, The University of Texas Health Science Center, San Antonio, TX*

MICHAEL WEIDEN • *Department of Genetics and Development, Columbia University, New York*

RACHEL WEVRICK • *Department of Genetics, Stanford University, Stanford, CA and Department of Medical and Molecular Genetics, University of Toronto, Ontario, Canada*

HUNTINGTON F. WILLARD • *Department of Genetics, Stanford University, Stanford, CA*

KUN-SHENG WU • *Department of Plant Breeding and Biometry, Cornell University, Ithaca, NY*

ANDREAS ZIEGLER • *Institut für Experimentelle Onkologie und Transplantationsmedizin, Universitätsklinikum Rudolf Virchow, Freie Universität Berlin, Germany*

PART I

ELECTROPHORETIC TECHNIQUES

CHAPTER 1

Field-Inversion Gel Electrophoresis

Georges F. Carle

1. Introduction

Among the techniques to separate large DNA fragments, field-inversion gel electrophoresis (FIGE, *1*) is probably the easiest to perform with a minimum of special equipment. Indeed, the only requirement besides a regular gel electrophoresis box and a power supply is a device enabling the periodic inversion of the electric field direction over the course of the experiment. This method was derived from experiments done on a modified orthogonal-field-alternation gel electrophoresis (OFAGE) apparatus *(2)* based on the observation that obtuse angles lead to a better separation. The widest angle being 180°, the four electrode pulsed-field gel electrophoresis (PFGE) system was reduced to a standard submarine gel electrophoresis box with only two electrodes. This simple configuration generates a highly uniform electric field across the gel, making the lane to lane comparison very easy (Fig. 1). Two basic electrophoretic modes can be used in order to achieve a net migration in this configuration:

- The same voltage (V) can be applied for a longer time (T) in the forward (T_F) than in the reverse direction (T_R). In general, the ratio T_F/T_R is kept constant over the course of the experiment to achieve maximum resolution; or

- A higher voltage can be applied in the forward (V_F) than in the reverse (V_R) direction but for the same amount of time ($T_F = T_R$).

From: *Methods in Molecular Biology, Vol. 12: Pulsed-Field Gel Electrophoresis*
Edited by: M. Burmeister and L. Ulanovsky

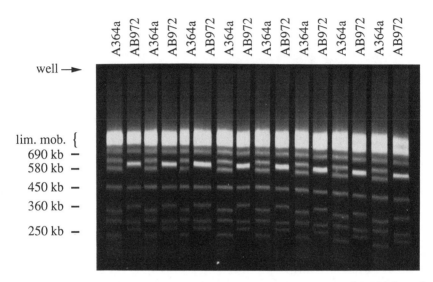

Fig. 1. Separation of the seven smallest chromosomal DNA bands of *S. cerevisiae* strains AB972 and A364a by FIGE. The other nine chromosomes comigrate in the limiting mobility bright band. The gel is 1% agarose, 0.5X TBE, and was run for 12 h at 9.5 V/cm with T_F increasing linearly from 10 s to 60 s and T_R kept constant at 5 s. The temperature was kept at 13°C.

Several pulse-time and voltage controllers to perform either of those electrophoretic modes are commercially available (DNASTAR PULSE™, MJ Research PPI-200™, Hoefer PC 750™, and so on), but can also be built by a local shop at cheaper costs *(1,3–6)*.

A unique feature of this technique when compared to the other PFGE methods is the ability to retain in the well a given size molecule, whereas smaller and larger ones will migrate in the gel (*see* Figs. 2B and 3). Therefore, a very high resolution of two molecules very close in molecular weight can be achieved by stopping the first one at the origin and letting the other one move. For a given size molecule, the switching frequency corresponding to the zero mobility has been referred to as the "resonance" point *(1,7)*. Indeed, when a constant switching regime is used over the course of the electrophoresis, the "resonance" effect leads to a sharp double-valued relationship between size and mobility, sometimes referred to as "band inversion" *(1)*: Where this effect is taking place, two DNA molecules of different size will have the same mobility. Although this method is very powerful when the sample size is already known, it can also be a source of

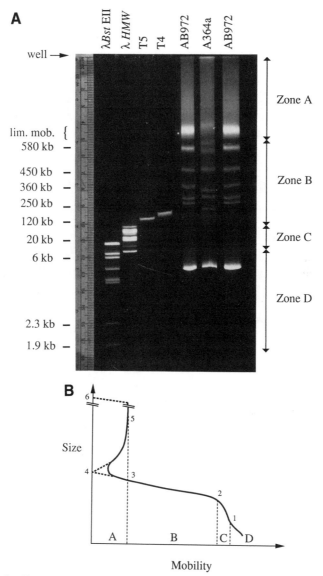

Fig. 2. **A**: Separation of linear DNA molecules ranging from 2 to 580 kbp. In this size range the relationship between size and mobility is a single-valued function. The four zones A–D described in Fig. 2B are shown on the right side.The experimental conditions are identical to the ones described in Fig. 1 except for the 1.5% agarose concentration. **B**: General representation of the doubled-valued relationship between size and mobility in FIGE. The four zones A–D correspond to the four area of the gel represented in Figs. 2A and 3. Point #4 represents the size corresponding to the "resonance" of this switching regime (*see* Introduction).

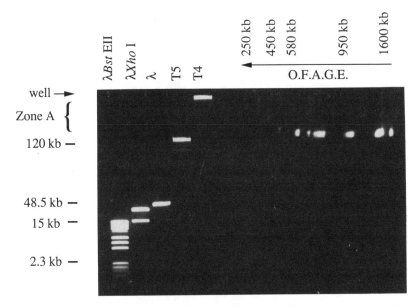

Fig. 3. Two-dimensional separation of DNA standards by OFAGE/FIGE illustrating the resonance effect. Yeast chromosomal bands were separated in a first dimension by OFAGE and in the second dimension by FIGE under the following conditions: 1% agarose, 0.5X TBE, $V_F = 14$ V/cm, $V_R = 9.5$ V/cm, and $T_F = T_R = 2$ s for 16 h. The temperature was kept at 13°C. The DNA molecules ranging from 170 kb (T4) to 250 kb (yeast chromosome I) display zero mobility and remain in the well.

some ambiguities in band sizing when all the parameters have not been set properly. One method to avoid this problem when analyzing unknown samples is to regularly increase the value of T_F and T_R during the course of the experiment (ramping mode). One such example is shown on Fig. 1, where T_F increases linearly from 10 s to 60 s, whereas T_R is kept constant at 5 s. Nevertheless, "ramping" the switching parameters is not always enough (8) to expect a monotonic function over the whole size range as the band inversion effect is tightly linked to other parameters, such as the voltage gradient.

As shown on Fig. 2A and B, four main zones can be distinguished on a size vs mobility plot of a FIGE gel: For low-mol-wt samples (up to point 1, zone D), the resolution is equivalent to what is achieved with a standard gel in continuous electrophoresis. Between points #1 and #2 (zone C), a compression zone of low resolution is generally present; as size increases, between points #2 and #3 (zone B), the mobility of the DNA molecules is inversely related to their size; therefore any band appearing in zone B can be sized without any ambiguity and with a maximum resolution. In the selection of new switching regime conditions, one should try to maximize zone B. Zone A is the more complex area of the gel: Any band appearing between points #3 and #5 can correspond to any of two sizes for a given mobility value. At a given size, and for some particular switching regime, the mobility value can even drop to zero at the "resonance" point #4 (*see* Fig. 3). Then, as size increases, the mobility becomes proportional to the DNA size until it reaches a plateau (point #5). This has been defined as the "limiting mobility" and its value is dependent on many parameters, such as the pulse frequency, voltage gradient, agarose concentration, and so on. Finally, one could try to extrapolate the curve for very large DNA molecules that are often found trapped in the well (point #6) unless special electrophoretic conditions are used *(9)*.

In summary, this method has some very attractive resolution power as long as the different zones of high resolution, compression, band inversion, and the like have been clearly defined with known size standards. Once this calibration has been done, FIGE can be very effective for screening large amounts of samples, like yeast artificial chromosomes *(10)*. However, one of the problems that limited the use of the method was the lack of band sharpness for samples over 700 kb, when compared with other PFGE methods. This feature appears to have been improved recently by using more complex switching programs (*11,12*, and Chapter 7).

At the present time, FIGE can be considered complementary to the other PFGE methods because it has some unique resolution features. However, as this method is more sensitive to many parameters than the other pulsed-field techniques, it appeared more complex to use and has been less practiced. Some electrophoretic conditions still need to be improved to reduce band broadening while achieving maximum resolution over a wider size range, but nevertheless it has proven very adequate for most sizing and screening purposes.

2. Materials

1. Agarose: purified regular agarose from major manufacturers (FMC, Bio-Rad, IBI) can be used for most experiments. Low-gelling agarose can be used if DNA needs to be recovered from the gel, but as in the case of regular gel electrophoresis, band sharpness is slightly decreased. The agarose concentration can vary between 0.5% to 1.8%.

2. 0.5X TBE buffer: 45 mM Tris base, 45 mM boric acid, 1.25 mM Na$_2$EDTA, unadjusted pH $\approx$ 8.2 has been used in all the experimental data reported in this chapter (*see* Note 1).

3. 10 mg/mL Ethidium bromide stock solution in water.

4. 6X loading dye: 15% Ficoll type 400 in H$_2$O, 0.25% bromophenol blue, 0.25% xylene cyanol.

5. Gel electrophoresis apparatus: A homemade (30-cm long/20-cm wide) submarine horizontal gel apparatus was used in this study but any standard horizontal or vertical model is adequate, from "mini" to "maxi" designs. Inlet and outlet manifolds should be present on the buffer tanks for recirculation and temperature control purposes. Independently of the apparatus shape, voltage gradient (V/cm) in the gel has to be measured accurately (*see* Note 2).

6. Power generator: The only requirement for the power supply is that it shouldn't be too sensitive to a charge drop, such as the one happening when the voltage is inverted. A standard 250 V/200 mA generator will be adequate for most gel configuration (*see* Note 3).

7. Timer and switching modules: Polarity inversion can be accomplished in many ways with very simple devices up to computer controlled power supplies *(1,3–6)*. We have used a homemade switching interface driven by a personal computer *(1)*, as well as the DNASTAR-PULSE™ controller, which is able to drive up to four independent gel boxes (*see* Note 4).

8. Pump and refrigerated water bath: In order to achieve reproducible results, it is important to keep the buffer concentration and temperature constant over the course of the experiment. We used heavy-duty peristaltic pumps (Masterflex™) equiped with industrial food-grade tubing (Norprene™) to isolate the pump from the electrically hot buffer. The cooling capacity of the refrigerated water bath should be chosen according to the number of gel boxes used. Most experiments were done at 13°C. **Warning:** The cooling unit should be equiped with an electrically isolated heat exchanger coil in order to avoid potential shocks. Avoid placing a FIGE setup in a cold room where electronic equipment is often subject to problems.

9. Voltage divider: This small electronic device is handy when a single power supply is used to achieve different voltage gradients in the forward and reverse direction ($V_F > V_R$). The basic circuitry is shown on Fig. 4. N.B.: The voltage (V_F or V_R) is always measured in parallel at the gel box level.

   a. *Switch position #1*: The current flows through the diode in the forward direction (V_F = voltage delivered by the power supply) but is blocked in the reverse direction and has to pass through a rheo-

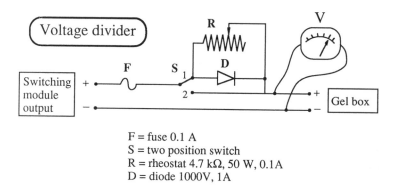

F = fuse 0.1 A
S = two position switch
R = rheostat 4.7 kΩ, 50 W, 0.1A
D = diode 1000V, 1A

Fig. 4. Voltage divider schematic.

stat ($V_R = V_F -$ voltage drop in the rheostat). The value of V_R is adjusted by changing the setting of the rheostat "R" while the field is being applied in the reverse direction.

b. *Switch position #2:* The voltage divider is bypassed, and the voltage applied to the gel box will be identical in the forward and reverse direction ($V_F = V_R$).

10. Voltmeter: A standard voltmeter (0–500V DC) is often needed to measure precisely the voltage gradient inside or at the edges of the gel. This measurement can easily be done for submarine horizontal gels using a small device made out of plexiglass with two platinum wires (0.4 mm ø, 3mm long) positioned 10 cm apart. The small platinum wires can be introduced into the gel while the calibration is being done, without causing any damage (*see* Note 2 and Fig. 5).

3. Method

3.1. Preliminary Considerations

When getting started, one should keep in mind a few points:

1. Electrophoretic conditions taken from the literature cannot be transposed directly to a particular gel apparatus, because some parameters have not always been reported or measured properly (such as voltage gradient, buffer temperature, and ramp increments). Each gel box model is somewhat unique, and it is important to test several parameters on your gel system in order to achieve the best reproducible results (*see* Note 5). One of the most efficient methods is to perform a set of two dimensional gel experiments using well-known size standards, such as multimers of lambda DNA, or yeast chromosomes from defined strains (AB972, YPH149, *S. pombe, see* Note 6).

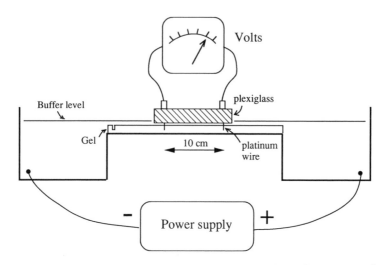

Fig. 5. Measurement of the voltage gradient for submarine gels.

2. In the majority of FIGE (as well as in the other PFGE methods), the precise limit of zone A (*see* Fig. 2B) is not always easy to determine if very large DNA size-standard molecules are not present (>1500 kb). Therefore, one should be very cautious about sizing unknown bands when using new electrophoretic conditions. The more reliable procedure to find out if a band is in zone A is based on the observation that, increasing of the forward switching time T_F (keeping $T_F/T_R = R_T$ constant) will affect larger DNA molecules by decreasing their mobility. Thus, if the mobility of a band increases or remains unchanged when T_F increases the band is likely to be located between points #1 and #4 (*see* Fig. 2B). But, on the other hand, if the mobility decreases when T_F increases, the band is probably located between points #4 and #5. By comparing band mobility behavior of standards as well as unknown samples, one can find out the relative size order and determine more appropriate switching conditions (ref. *13* and Note 7).

3.2. Apparatus Calibration

1. Prepare a 1% agarose solution in 05X TBE buffer in order to pour a 4–5-mm thick gel.
2. Level the gel tray to obtain a uniform thickness accross the gel. Cool the molten agarose to 65°C, and then pour the agarose over a glass or a plexiglass tray fitting in the gel box (*see* Note 8).
3. Let the gel solidify at room temperature for at least 15–20 min, and remove the comb.

4. Level the gel apparatus platform and fill the buffer tanks with 0.5X TBE buffer. Start buffer recirculation (0.2 L/min) and cooling in order to reach a temperature of 13–15°C inside the gel box.
5. Install the gel inside the electrophoresis box and cover it with 3–4 mm of buffer. Wait 20 min until a uniform temperature has been reached throughout the gel.
6. Connect the gel box to the power supply and using a voltmeter, perform a calibration curve of the voltage gradient measured inside or at the edges of the gel (V), as a function of the power supply output display. The distance (d) between the two probes of the voltmeter should be measured precisely. The ratio V/d corresponds to the voltage gradient in V/cm (*see* Notes 2 and 9).
7. If a voltage divider is going to be used in order to have the reverse voltage lower than the forward one, the same kind of calibration curve should be done. V_F and V_R have to be measured at the gel box level while adjusting the value of the rheostat "R" (*see* Section 2, Step 9).

3.3. Running Samples

1. Liquid or agarose-imbedded DNA samples preparation are described elsewhere (cf Chapter 8).
2. Repeat Steps 1–4 from Section 3.2. (apparatus calibration).
3. Samples in a solid form (block preparation) can be loaded at this time after having been equilibrated at room temperature in 0.5X TBE buffer (twice for 15 min). The blocks can be sealed in the well with molten agarose if they should remain in the gel later on for Southern transfer.
4. Install the gel apparatus on the platform and cover it with a defined amount of electrophoresis buffer (4–5 mm). Turn on cooling and buffer recirculation. Wait 20 min until a uniform temperature has been reached.
5. Turn the buffer recirculation off and load liquid DNA samples using standard loading dye.
6. Connect the gel box electrodes to the power supply (or to the switching module), and perform a prerun at (1 V/cm) for 10 min.
7. Switching parameters: Table 1 displays sets of parameters we have selected with our gel system (*see* also Note 7). Heller and Pohl *(3,14)* have tested systematically many switching regimes. As mentioned earlier, these conditions cannot always be used as such with one's own setup. Careful measurements and monitoring of the different parameters is recommended, and some adjustment might be required (like lowering the voltage gradient).

Table 1
FIGE Parameter Settings

Size range kb	V_f V/cm	V_r V/cm	T_f s	T_r s	Run time h	Example ref.
2–35	14	9.5	0.3	0.3	16	Fig.6
50–120	10.5	10.5	0.5	0.25	12	*1*
100–600	9.5	9.5	10→60	5	12	Fig. 1
2–600[a]	9.5	9.5	10→60	5	12	Fig. 2A
250–1600	9.5	9.5	9→60	3→20	20	*1*

[a]The agarose concentration was 1.5% instead of 1% for the other examples.

Switching conditions are very interdependent with the other electrophoretic parameters, and the references given below relative to particular parameters or to specific applications will be a good source of information:

- Nonlinear molecules *(15,16)*.
- Very large DNA *(9,13)*.
- Agarose concentration variation *(17)*.
- Voltage gradient variation *(3,17; see* Note 10).
- Switching ratio R_T *(3,17)*.
- Ramping *(14)*.
- Temperature *(18)*.
- Vertical apparatus *(19)*.
- Preparative gels *(20)*.
- Genome mapping *(21–26)*.

8. Staining: When the electrophoresis is over, turn the power off and remove the gel from the electrophoresis apparatus. Stain the gel with gentle shaking for 15 min in 500 mL of a 0.2 µg/mL ethidium bromide solution in deionized water.

9. Destain in 500 mL of deionized water for 30 min and take a picture under 300–320 nm UV light (*see* Note 11 and also Chapter 8).

10. Transfer and hybridization: Standard capillary Southern transfer and hybridization have been used successfully by us and other investigators (ref. *25; see* Note 12). However, Van Devanter and Von Hoff recently reported that acid depurination after FIGE reduces capillary transfer of large linear DNA molecules *(27)*. According to these authors, alkali transfer on nylon membranes without acid treatment seems to give the best results if multiple probings have to be performed (*see* Note 12).

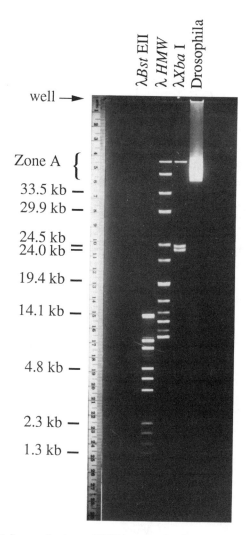

Fig. 6. High resolution of DNA samples between 1 to 34 kb by FIGE The 24.0/24.5 kb doublet of the λ*Xba* I digest is easily resolved under these electrophoretic conditions: 1% agarose (43 cm long), 0.5X TBE, $V_F = 14$ V/cm, $V_R = 9.5$ V/cm, and $T_F = T_R = 0.3$ s for 18 h. The temperature was kept at 14°C.

4. Notes

1. TAE buffer can also be used with appropriate recirculation *(16)*.
2. A good measurement of the voltage gradient is essential to extrapolate the results obtained with one gel box to another. It is inaccurate to estimate the voltage gradient inside the gel as being equal to the

ratio of the power supply readout over the distance between the electrodes. The voltage gradient should be determined with a voltmeter by a direct measurement at the edges or inside the gel itself (*see* Fig. 5): Once the gel is settled in the electrophoresis box and covered with buffer, a small device made out of plexiglass with two platinum wires set 10 cm apart can be placed in the gel parallel to the electric field direction. After the power is turned on, the reading on the voltmeter will give the voltage gradient in V/cm × 10. **Warning:** This measurement might require modifying temporarily some of the safety features of the commercially available gel boxes, causing a potential **electric hazard. Extreme caution** should be taken while performing this calibration and the safety interlocks should be put back in their original position once the measurements are done.

3. If a special power generator is going to be dedicated for either FIGE or other kind of PFGE, select a 500 V/400 mA model that gives more power flexibility. Old standard power supply are often a better choice, since newer models can be incompatible with sudden and repetitive load changes.

4. Depending on the potential number of users or how heavily the technique is going to be used, a modular system is a best buy. One gel box is rapidly not enough, and the pulse controller should have the flexibility of total independent parameters programming for each box.

5. During the test phase of your gel apparatus (and even later on), the following rules should be followed:

   - Pour always gels of identical thickness (3–4 mm) and keep the buffer level above the gel constant (4–5 mm).
   - Change systematically the electrophoretic buffer before each experiment in order to avoid nonreproducible results owing to variations in the pH and ionic strength.
   - Measure precisely the voltage gradient.
   - Monitor the buffer temperature during the course of the experiment.

6. As mentioned in the introduction, one would like to make sure of being preferably in zone B (*see* Fig. 2B) before estimating the size of a given band. Therefore, it is critical to have good size standards covering the whole range between 20 kb up to 6000 kb. This is easily done with the following DNAs (*see also* Chapter 8): < 50 kb bacteriophage lambda DNA cut with various restriction enzymes (*Hind* III, *Bst*EII, and the like), 10kb ladder (see Chapter 3), HMW markers (BRL) (*28*), 50–>500 kb "lambda ladders" (ref. *29* and Chapter 8),

200–1600 kb *Saccharomyces cerevisiae* chromosomes of well-characterized strains like AB972 *(30,31)* or YPH 149 *(32)*, 3500–5700 kb
Schizosaccharomyces pombe chromosomes *(33)*.

7. The double-valued relationship between mobility and size can be
overcome in most cases by ramping the forward (T_F) and the reverse
time (T_R) while keeping their ratio constant (R_T) *(1)*. Nonetheless,
this procedure is not always successful when applied directly to any
gel configuration or sets of switching conditions *(8)*: prior adjustment
of the voltage gradient might be required, and several ramping algorithm should be tested. Regarding the latter point, the ramping profile can be quite different depending on the method used: some
devices will interpolate linearly the values of T_F and T_R once the end
values and the total experimental run time have been programmed
(e.g., DNASTAR-PULSE™), whereas others will increase T_F and T_R
independently after each cycle with programmed values over the
course of *n* cycles (e.g., MJ Research PPI-200™). To a first approximation, we have derived the following set of equations to transform
switching parameters used with a linear ramping controller into ones
used with simple incremental type stepping controllers:

Number of cycles (Eq. 1): $\qquad$ $N_S = 2t / (T_{SF} + T_{EF} + T_{SR} + T_{ER})$

where N_S is number of cycles; t, duration of the electrophoresis (s);
T_{SF}, start forward time (s); T_{EF}, end forward time (s); T_{SR}, start reverse time (s); and T_{ER}, end reverse time (s).

Forward increment (Eq. 2): $\qquad$ $F_I = (T_{EF} - T_{SF}) / (N_S - 1)$

Reverse increment (Eq. 3): $\qquad$ $R_I = (T_{ER} - T_{SR}) / (N_S - 1)$

Example: Transformation of the switching conditions used in Fig. 1,
where T_{SF} is 10 s; T_{EF}, 60 s; T_{SR}, 5 s; T_{ER}, 5 s; and t, 12 h or 43200
therefore:

Number of cycles (Eq. 1): $\qquad$ $N_S = 2 \times 43200/(10 + 60 + 5 + 5)$
$\qquad\qquad\qquad\qquad\qquad\qquad$ $N_S = 1080$ cycles

Forward increment (Eq. 2): $\qquad$ $F_I = (60 - 10)/(1080 - 1)$
$\qquad\qquad\qquad\qquad\qquad\qquad$ $F_I = 0.046$ s

Reverse increment (Eq. 3): $\qquad$ $R_I = 0$ s

8. In order to prevent gel floatation if the electrophoresis buffer is
being recirculated, narrow Velcro™ strips (hook side up) can be
attached to the gel tray with silicone glue *(29)*.

9. The principle of FIGE allows one to use any kind of electrophoresis gel box, ranging from a "mini gel," to a "sequencing size" apparatus. The key feature is to be able to measure the voltage gradient inside the gel. Indeed, the transposition of any switching conditions from one gel box to another requires an identical voltage gradient. Failure to do so will lead to very different results. N.B. Independently of the platinum purity used in the electrophoresis apparatus, we have observed a significant reduction of the electrode diameter over time. Checking (and eventual replacement) of the electrodes should be done regularly.

10. Band trapping of large DNA molecules in FIGE is very sensitive to high electric fields, low ionic strength, and high temperature similar to what has been observed with OFAGE *(2)*. Band broadening in FIGE is frequently observed for fragments larger than 700 kb. Turmel et al. have reported recently that when high frequency reverse pulses are used to modulate the longer forward pulses, narrower bands can be observed and higher voltage gradients can be used (*11*, and Chapter 7).

11. In order to achieve the highest contrast and remove background fluorescence, destaining of the gel, with gentle shaking, should be done for 30 min to overnight.

12. Electroblotting has also been used successfully with fragments up to 500 kb *(34)*. An "in-gel" hybridization procedure after FIGE has been described in at least two publications *(35,36; see also* Chapter 22) but limits the procedure to a single probing experiment.

References

1. Carle, G. F., Frank, M., and Olson, M. V. (1986) Electrophoretic separations of large DNA molecules by periodic inversion of the electric field. *Science* **232**, 65–68.
2. Carle, G. F. and Olson, M. V. (1987) Orthogonal-field-alternation gel electrophoresis. *Methods Enzymol.* **155**, 468–482.
3. Heller, C. and Pohl, F. M. (1989) A systematic study of field inversion gel electrophoresis. *Nucleic Acids Res.* **17**, 5989–6003.
4. Roy, G., Wallenburg, J. C., and Chartrand, P. (1988) Inexpensive and simple set-up for field inversion gel electrophoresis. *Nucleic Acids Res.* **16**, 768.
5. Sor, F. (1988) A computer program allows the separation of a wide range of chromosome sizes by pulsed field gel electrophoresis. *Nucleic Acids Res.* **16**, 4853–4863.
6. Abbal, P. and Picou, C. (1990) A general purpose pulsed field controller *Electrophor.* **11**, 893,894.
7. Olson, M. V. (1989) Separation of large DNA molecules by pulsed-field gel electrophoresis. *J. Chromatogr.* **470**, 377–383.

8. Ellis,T. H. N., Cleary, W. G., Burcham, K. W. G., and Bowen, B. A. (1987) Ramped field inversion gel electrophoresis: A cautionary note. *Nucleic Acids Res.* **15**, 5489.

9. Turmel, C. and Lalande, M. (1988) Resolution of *Schizosaccharomyces pombe* by field inversion gel electrophoresis. *Nucleic Acids Res.* **16**, 4727.

10. Garza, D., Ajioka, J. W., Burke, D. T., and Hartl, D. L. (1989) Mapping the *Drosophila* genome with yeast artificial chromosomes. *Science* **246**, 641–646.

11. Turmel, C., Brassard, E., Slater, G. W., and Noolandi, J. (1990) Molecular detrapping and band narrowing with high frequency modulation of pulsed field electrophoresis *Nucleic Acids Res.* **18**, 569–575.

12. Turmel, C., Brassard, E., Forsyth, R., Hood, K., Slater, G. W., and Noolandi, J. (1990) High-resolution zero integrated field electrophoresis of DNA. *Current Communications in Cell and Molecular Biology: Electrophoresis of Large DNA Molecules: Theory and Applications,* Cold Spring Harbor Laboratories, Cold Spring Harbor, NY.

13. Lasker, B. A., Carle, G. F., Kobayashi, G. S., and Medoff, G. (1989) Comparison of the separation of *Candida albicans* chromosome-sized DNA by pulsed-field gel electrophoresis techniques. *Nucleic Acids Res.* **17**, 3783–3793.

14. Heller, C. and Pohl, F. M. (1990) Field inversion gel electrophoresis with different pulse time ramps. *Nucleic Acids Res.* **18**, 6299–6304.

15. Levene, S. D. and Zimm, B. H. (1987) Separations of open-circular DNA using pulsed-field electrophoresis. *Proc. Natl. Acad. Sci. USA* **84**, 4054–4057.

16. Sobral, B. W. S. and Atherly, A. G. (1989) Pulse time and agarose concentration affect the electrophoretic mobility of cccDNA during electrophoresis in CHEF and in FIGE. *Nucleic Acids Res.* **17**, 7359–7369.

17. Bostock, C. (1988) Parameters of field inversion gel electrophoresis for the analysis of pox virus genomes. *Nucleic Acids Res.* **16**, 4239–4252.

18. Olschwang, S. and Thomas, G. (1989) Temperature gradient increases FIGE resolution. *Nucleic Acids Res.* **17**, 2363.

19. Dawkins, H. J. S., Ferrier, D. J., and Spencer, T. L. (1987) Field inversion gel electrophoresis (FIGE) in vertical slabs as an improved method for large DNA separation. *Nucleic Acids Res.* **15**, 3634,3635.

20. Michiels, F., Burmeister, M., and Lehrach, H. (1987) Derivation of clones close to *met* by preparative field inversion gel electrophoresis. *Science* **236**, 1305–1308.

21. Daniels, D. L., Olson, C. H., Brumley, R., and Blattner, F. R. (1990) Field inversion gel electrophoresis applied to the rapid multi-enzyme restriction mapping of phage lambda clones. *Nucleic Acids Res.* **18**, 1312.

22. Woolf, T., Lai, E., Kronenberg, M., and Hood, L. (1988) Mapping genomic organization by field inversion and two-dimensional gel electrophoresis: Application to the murine T-cell receptor γ gene family *Nucleic Acids Res.* **16**, 3863–3875.

23. Gejman, P. V., Sitaram, N., Hsieh,W. T., Gelernter, J., and Gershon, E. S. (1988) The effects of field inversion electrophoresis on small DNA fragment mobility and its relevance to DNA polymorphism research. *Appl. Theor. Electrophor.* **1**, 29–34.

24. Den Dunnen, J. T., Bakker, E., Klein-Breteler, E. G., Pearson, P. L., and Van Ommen, G. J. B. (1987) Direct detection of more than 50% of the Duchenne muscular dystrophy mutations by field inversion gels. *Nature* **329**, 640–642.

25. Chen, J., Denton, M. J., Morgan, G., Pearn, J. H., and Mackinlay, A. G. (1988) The use of field inversion gel electrophoresis for deletion detection in Duchenne muscular dystrophy. *Am. J. Hum. Genet.* **42**, 777–780.

26. Bautsch, W. (1988) Rapid physical mapping of the *Mycoplasma mobile* genome by two-dimensional field inversion gel electrophoresis techniques. *Nucleic Acids Res.* **16**, 11461–11467.

27. Van Devanter, D. R. and Von Hoff, D. D. (1990) Acid depurination after field inversion agarose gel electrophoresis reduces transfer of large DNA molecules. *Appl. Theor. Electrophor.* **1**, 189–192.

28. Graham, M. Y., Otani, T., Boime, I., Olson, M. V., Carle, G. F., and Chaplin, D. (1987) Cosmid mapping of the human chorionic gonadotropin β subunit genes by field inversion gel electrophoresis. *Nucleic Acids Res.* **15**, 4437–4448.

29. Carle, G. F. and Olson, M. V. (1984) Separation of chromosomal DNA molecules from yeast by orthogonal-field-alternation gel electrophoresis. *Nucleic Acids Res.* **12**, 5647–5664.

30. Carle, G. F. and Olson, M. V. (1985) An electrophoretic karyotype for yeast. *Proc. Natl. Acad. Sci. USA* **85**, 3756–3760.

31. Link, A. J. and Olson, M. V. (1991) Physical map of *Saccharomyces cerevisiae* genome at 110-kilobase resolution. *Genetics* **127**, 681–698.

32. Vollrath, D., Davis, R. W., Connelly, C., and Hieter, P. (1988) Physical mapping of large DNA by chromosome fragmentation. *Proc. Natl. Acad. Sci. USA* **82**, 6027–6031.

33. Smith, C. L., Matsumoto, T., Niwa, O., Kico, S., Fan, J. B., Yanagida, M., and Cantor, C. R. (1987) An electrophoretic karyotype for *Schizosaccharomyces pombe* by pulsed field gel electrophoresis. *Nucleic Acids Res.* **15**, 4481–4488.

34. Satyanarayana, K., Hata, S., Devlin, P., Roncarolo, M. G., De Vries, J. E., Spits, H., Strominger, J. L., and Krangel, M. S. (1988) Genomic organisation of the human T-cell antigen-receptor α/δ locus. *Proc. Natl. Acad. Sci. USA* **85**, 8166–8170.

35. Son, M., Watson, R. H., and Serwer, P. (1990) In-gel hybridization of DNA separated by pulsed field agarose gel electrophoresis. *Nucleic Acids Res.* **18**, 3098.

36. Chou, H. S., Nelson, C. A., Godambe, S. A., Chaplin, D. D., and Loh, D. Y. (1987) Germline organization of the murine T cell receptor ß chain genes. *Science* **238**, 545–548.

CHAPTER 2

Resolving Multimegabase DNA Molecules Using Contour-Clamped Homogeneous Electric Fields (CHEF)

Douglas Vollrath

1. Introduction

Contour-clamped homogeneous electric field (CHEF) gel electrophoresis is a particular formulation of pulsed-field gel electrophoresis (PFGE), which uses an array of electrodes positioned around the gel (on a contour) and clamped to specific voltages to produce a nearly homogeneous electric field inside the contour *(1)*. The direction of the electric field is changed periodically, as with all pulsed-field techniques. In the case of CHEF, field reorientation is achieved electronically by changing the voltages (potentials) of the various electrodes in the array (*see* Fig. 1). Commercial CHEF devices currently employ a hexagonal electrode array, but other types of contours, such as circles or squares, if properly clamped, can also produce alternating homogeneous electric fields.

CHEF was developed for two primary reasons. The first pulsed-field machines employed inhomogeneous electric fields that caused the DNA molecules to migrate with curvalinear, arc-like, or even wave-like trajectories (*see* Chapter 4). Early workers suggested that the voltage gradient produced by these inhomogeneous fields might be required for the resolution of large DNA *(2)*. Thus, one motivation for

From: *Methods in Molecular Biology, Vol. 12: Pulsed-Field Gel Electrophoresis*
Edited by: M. Burmeister and L. Ulanovsky

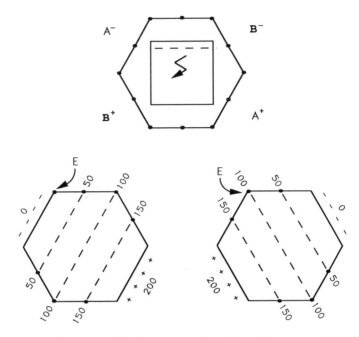

Fig. 1. The principle of contour-clamping. A hexagonal CHEF electrode array surrounding a gel is shown at the top. Black dots represent individual electrodes. The array contains two sets of driving electrodes (A and B) oriented 120° apart. An electric potential is periodically applied across each set for equal time intervals (the pulse time). The DNA molecules zigzag through the gel, reorienting with each change in the electric field, but the net direction of migration is straight. Nondriving electrodes are electronically clamped to intermediate potentials creating a homogeneous electric field inside of the contour, as illustrated in the lower portion of the figure. To the left, a potential difference of 200 (arbitrary units) has been applied across the A electrodes. Paired nondriving electrodes are clamped to the same potential and that potential varies linearly as a function of the perpendicular distance of the electrode pair from the driving electrodes. This creates a series of imaginary parallel isopotential field lines (dashed lines) and a homogeneous electric field. To the right, the potential difference has been switched to the B electrodes and the potentials of the other electrodes have been changed to create a homogeneous electric field in the new orientation. For example, the electrode marked (E) serves as a driving electrode with a value of 0 in one orientation, and as a nondriving electrode with a value of 100 in the other orientation.

inventing CHEF was theoretical, to test whether homogeneous electric fields could separate large DNA. A second motivation for the development of CHEF was practical. The nonlinear lane shapes made it difficult to compare samples across the gel and to use the gels preparatively to cut out a discrete size-fraction of DNA (*see* Chapters 16 and 21). This problem is eliminated in CHEF because homogeneous alternating electric fields produce straight lanes as in conventional electrophoresis.

One advantage of CHEF then is that it produces straight lanes. However, other techniques described in this volume can make the same claim. For example, field-inversion gel electrophoresis (FIGE) also produces straight lanes and is a simple and economical means of separating DNA of up to several hundred kilobase pairs (kbp) (*see* Chapter 1). However, above about 500 kbp, the bands of DNA become broad, compromising resolution. FIGE can also display the phenomenon of size/mobility inversion, in which large DNA molecules move with a much higher mobility than smaller molecules in the same sample *(3)*. These properties will obviously complicate analyses that require accurate sizing of DNA molecules. In CHEF, size/mobility inversion is much less pronounced, being limited to the region near the top of the gel (the exclusion limit), and the phenomenon can be easily identified by the lateral spreading of the DNA band *(4)*.

Rotating-field electrophoresis (ROFE) employs a single set of electrodes and rotates either the gel or the electrodes to effectively produce two homogeneous electric fields with different orientations (*see* Chapter 6). ROFE has the same resolution characteristics as CHEF, but has additional engineering problems associated with the need to precisely reorient the electric field by mechanical means. This can be especially problematic when short pulse times (a few seconds or less) are required.

Transverse-alternating-field electrophoresis (TAFE) produces straight lanes by suspending the gel vertically in a buffer tank with electrodes positioned parallel to the two large faces of the gel (*see* Chapter 5). The electric field produced by this arrangement is inhomogeneous, decreasing in strength from the top to the bottom of the gel. This results in a focusing effect whereby bands near the bottom are sharp and closely spaced. Although this can be useful for certain specialized applications, the unique electrode geometry constrains the thickness and overall size of the gel, limiting the usefulness of TAFE for preparative applications. In contrast, gels of many shapes and sizes can be accommodated within the CHEF electrode array.

Thus, CHEF compares favorably to the other formulations of PFGE that produce straight lanes. Its design retains the simplicity associated with conventional horizontal agarose electrophoresis, with no mechanical parts while providing excellent resolution characteristics. It has been successfully used to resolve molecules as small as 10 kbp (*see* Chapter 3 and ref. *5*) and as large as 10 Mb (*6*).

This chapter describes the use of CHEF to resolve molecules in the 2–10 Mb range. The major difficulty in separating molecules in this range has been that, as the size of the chromosomal DNA increases, the field strength used to resolve the molecules must be decreased to minimize trapping of the DNA as it moves through the agarose matrix. This necessarily leads to long periods of electrophoresis, more than a week for a 10-Mb molecule, making the determination of the optimal conditions for resolving a particular sample a time-consuming process. Recent advances involving the use of specialized agarose and insight into the behavior of these large molecules during electrophoresis show promise for shortening these times (Chapters 7 and 27 and ref. *4*). However, much experience has been gained using the original approach, and it represents a useful starting point for those new to the resolution of multimegabase molecules.

The window of resolution in CHEF electrophoresis is the result of an interplay of several factors. These include the electric field strength, the temperature, the agarose composition and concentration, the pulse time, and the angle between alternating electric fields (*1,7*). The results obtained with a given set of conditions are also affected by the particular electrophoresis apparatus used. In practice, when optimizing conditions for a particular sample, it is easiest to choose a single apparatus along with reasonable values for the temperature, agarose concentration, and field angle, and keep these constant from run to run. The pulse time or the field strength can then be varied until a suitable set of conditions has been found. Table 1 illustrates the application of this approach to the separation of fungal chromosomal DNA molecules, ranging in size from <1 to >10 Mb.

To successfully resolve multimegabase DNA molecules, one must first prepare them intact. It is useful to remember that a 3-Mb DNA molecule is approx 20 Å in diameter (2×10^{-6} mm), with a contour-length of about 1 mm, yielding an axial ratio of 500,000. To prepare such molecules, care must be taken to minimize shearing and enzy-

Table 1
CHEF Conditions for Separating Multimegabase Molecules

Range, Mb	Pulse, min	Volts/cm	Run, h	Agarose	Ref.
0.1–1	1	210/30	24	1%	7, Fig. 1B
1–1.5	4	120/30	54	1%	7, Fig. 2A
1.5–3	30	50/30	120	0.6%	7, Fig. 4A
3–6	60	40/30	130	0.6%	7, Fig. 4B
	60	40/30	137		6, Fig. 1A
8.5–10.5	140	30/30	168	0.6%	6, Fig. 2A
	180	30/30	192		6, Fig. 2D

All gels were run at 9°C on a custom-built hexagonal CHEF apparatus with vertical electrodes, an angle of 120° between alternating electric fields, and a 30-cm spacing between opposing faces of the hexagon.

matic degradation during all manipulations. Microorganisms have provided an abundant source of multimegabase molecules *(2, 6–9)*. The procedure for preparing chromosomal DNAs from these organisms is similar to that used for preparing large mammalian DNA (*see* Chapter 8) except that cell wall material is removed using the appropriate enzyme just prior to, or immediately after, embedding cells in low-melting-temperature agarose. This chapter contains a protocol for the preparation of chromosomal DNA molecules from the fission yeast *Schizosaccharomyces pombe.* This organism is easy to culture and contains three well-characterized linear chromosomal DNA molecules of about 3, 4.7, and 5.7 Mb in size *(10)*. These serve as useful size markers for multimegabase separations.

2. Materials

2.1. Materials
for S. pombe *DNA Preparation*

1. 200 mL YPD medium: 2% (w/v) Glucose, 2% (w/v) bactopeptone, 1% (w/v) yeast extract.
2. *S. pombe* strain 975 (strain 24811, American Type Culture Collection, Rockville, MD)
3. W buffer: 1.2M Sorbitol, 50 mM sodium citrate, 50 mM sodium phosphate, 40 mM sodium EDTA, pH to 5.6 with HCl.
4. S buffer: 1.2M Sorbitol, 50 mM sodium citrate, 50 mM sodium phosphate, pH to 5.6 with HCl.
5. Lysis buffer: 0.5M EDTA, pH 8, 1% (w/v) N-lauryl sarkosyl, 2 mg/mL proteinase K.

6. 1% Low-melting-temperature agarose in 125 mM EDTA, pH 8.
7. 100 mL TE 50: 10 mM Tris-HCL, 50 mM EDTA, pH 8.
8. Novozym SP234, Novo Nordisk Laboratories, Danbury, CT.

2.2. Materials for CHEF Gel

1. One CHEF apparatus with a 120° angle between the alternating electric fields, as well as a cooling unit and pump for buffer recirculation.
2. 0.5X TBE buffer, 1–2 L.
3. Agarose: Baker standard low electroendosmosis.
4. Glass slides.
5. Thin plastic ruler.
6. Comb, 1 mm × 10 cm.
7. Number 6 or 11 surgical scalpel.
8. Two small spatulas.
9. 1% Low-melting-temperature agarose in distilled water.
10. One thermometer.
11. One gel scoop.

3. Methods

3.1. Preparation of S. pombe Chromosomal DNA

1. Grow *S. pombe* strain 975 to an OD $_{550}$ of 0.3–0.4 at 30°C in 200 mL YPD medium (*see* Note 1).
2. Pellet the cells at 4000g for 10 min, aspirate the supernatant, and resuspend the pellet in 40 mL of W buffer plus 1% (v/v) β-mercaptoethanol. Incubate 10 min at room temperature (*see* Note 2). Spin the cells as before.
3. Resuspend the pellet to 5×10^8 cells/mL in S buffer plus 30 mM β-mercaptoethanol. Add 5 mg Novozym/1×10^8 cells (*see* Note 3). Incubate at 37°C. At 20-min intervals, examine a small aliquot of cells under a microscope to check for the formation of spheroplasts (*see* Note 4).
4. When at least 90% of the cells are spheroplasts (usually after 2 h), spin them at 300g for 5 min and gently resuspend the pellet in 10 mL of W buffer. Repeat for a total of three washes (*see* Note 5).
5. Resuspend the spheroplasts in W buffer at a concentration of 2.5×10^9 cells/mL. Add 1.67 mL of 1% low-melting-temperature agarose in 125 mM EDTA, pH 8, for every 1 mL of cells. The agarose should be cooled to 50°C and the cells warmed to 37°C before mixing. Pipet the solution gently to mix and allow it to solidify in a mold that has been placed on a bed of ice (*see* Note 6).

6. Place the solidified sample block in a volume of lysis buffer equal to that of the block and incubate at 50°C for 18–24 h (*see* Note 7).
7. Wash the block for 1 h each at 50°C at least four times in 5–10 vol of TE 50 to remove RNA and detergent/peptide complexes. Store the sample blocks in TE 50 at 4°C (*see* Note 8).

3.2. Running the CHEF Gel

1. Dissolve 0.36 g agarose in 60 mL 0.5X TBE buffer by heating. Replace any lost volume with distilled water. Mix and cool the molten agarose to 50°C in a water bath (*see* Note 9).
2. Assemble a 12 × 12 cm gel mold and position the comb 1 cm from, and parallel to, one edge of the gel mold (*see* Note 10). Raise the comb about 1 mm above the bottom of the mold using a thin ruler as a spacer (*see* Note 11).
3. Pour the molten agarose into the mold and allow it to solidify.
4. Carefully remove the comb and disassemble the gel mold (*see* Note 12).
5. Cut the sample blocks to slightly less than the dimensions of the wells. Place the agarose plug from which the sample is to be cut on a clean glass slide. Lay the comb flat and at one edge of the slide as a guide (*see* Note 13). With a spatula in one hand, gently pin the plug against the comb while cutting with a surgical scalpel held in the other hand (*see* Note 14).
6. Load the wells with the sample blocks of DNA. Place a spatula in the well and apply gentle pressure to the back to hold the well open. At the same time, use a second tool to fit the bottom of the sample block into the top of the well. With the second tool, apply pressure to the top of the sample block while withdrawing the spatula from the well. The block should slide in easily. Avoid trapping air bubbles (*see* Note 15).
7. Seal the blocks into the wells by overlaying them with molten 1% low-melting-temperature agarose (*see* Note 16). Allow 5–10 min for the agarose to gel.
8. Drain any existing buffer from the electrophoresis chamber and place the gel in it. Special care should be taken to support the area of the gel around the wells because this is the weakest part (*see* Note 17). Position the gel in the center of the electrode array with the line of wells parallel to the top and bottom faces of the hexagon (*see* Fig. 1). Secure the gel in the chamber so that it will not move during the electrophoresis (*see* Note 18).
9. Fill the chamber with 0.5X TBE buffer (without ethidium bromide) that has been chilled to 4°C. Depending on the particular apparatus, between 1 and 2 L is usually required. Be careful not to dislodge the

gel while adding the buffer. The final buffer level should be 2–3 mm above the surface of the gel (*see* Note 19).

10. Turn on the pump. Make sure that buffer flow is adequate and unrestricted by air bubbles (*see* Note 20).

11. Lay a thermometer in the buffer outside of the electrode contour and away from the buffer recirculation ports, so that it can be read without removing the lid of the electrophoresis chamber. Place the thermometer such that it measures the temperature of the buffer at the warmest point, as it exits the chamber.

12. When the buffer temperature has stabilized at 9°C, set the voltage and pulse time for the desired size range using the values in Table 1 as a guide (*see* Notes 21–23).

13. Turn on the power supply to begin the run. Observe the gas bubbles rising from the negative electrode. After one pulse time has elapsed, make sure that the field has switched by noting an increase in bubbles from the other bank of negative electrodes (*see* Note 24). Check the buffer level periodically during the run and replace lost volume with distilled water if necessary.

14. At the end of the run, carefully remove the gel with a gel scoop and stain it in 100 mL of 0.5 µg/mL ethidium bromide in water for 20 min at room temperature with gentle shaking. Remove the staining solution by suction and replace it with 100 mL of distilled water. Destain for 20 min. Photograph the gel under UV light with a ruler at its side. This facilitates comparison between different PFGE runs (*see* Note 25).

4. Notes

1. An OD_{550} of 0.3–0.4 equals a concentration of about 5×10^6 cells/mL.

2. This step is designed to reduce the disulfide bonds in the yeast cell wall to facilitate lysis. The 2-mercaptoethanol should be added to the S buffer just before use to ensure adequate reduction.

3. If the Novozym is added as a solid, be certain that it dissolves well. Alternatively, a stock of 50–75 mg Novozym/mL S buffer can be used.

4. Look at some normal *S. pombe* cells for comparison. They are usually cigar-shaped. Novozym digests the yeast cell wall resulting in a loss of the normal shape. Spheroplasts are round and will lyse in the absence of an osmotic-stabilizing agent, such as sorbitol. The rate of digestion can vary depending on the *S. pombe* strain, the growth conditions, and the enzyme preparation. Overdigestion will result in lysis, so the reaction should be carefully monitored. The extent of digestion can also be monitored by observing a decrease in turbidity when cells are diluted 1:50 in 1% *N*-lauryl sarkosyl.

5. Spheroplasts are fragile, treat them gently. Check the integrity of the cells by microscopy after the last wash is complete.

6. A simple mold can be constructed by spacing two stacks of glass slides parallel to one another and about 5 mm apart on top of a larger slide (e.g., a lantern slide) to form a rectangular trough. Seal the ends of the trough with tape. Excellent commercial molds are also available from, for example, Bio-Rad.

7. Avoid handling the sample blocks. This can introduce DNAase, which can degrade the large DNA molecules. Use spatulas that have been cleaned with ethanol instead.

8. If the sample block or storage solution becomes cloudy on cooling, the detergent is still present. Further washing is necessary. Blocks can be safely stored for several months. It may be useful to purchase *S. pombe* blocks as a positive control from, for example, Bio-Rad or FMC (Rockland, ME)

9. Any standard agarose, such as FMC SeaKem ME, will do. Low-melting-temperature agarose or agarose designed for speedier separation of chromosomal DNAs (Bio-Rad chromosomal grade agarose, FMC Fast Lane agarose, or Boehringer Mannheim PFGE agarose) may yield poor results using the conditions described here.

10. A variety of gel molds can be used, including both commercial and custom-built. Molds can be created by weighing down with lead pigs a Lucite frame of the type used to dry SDS-PAGE gels. The comb is supported using binder clips. If this approach is used, the gel can be cast directly in the electrophoresis chamber. Subsequent transport of the gel is then unnecessary.

11. The thickness of the comb is important in CHEF because there is no focusing effect, so the thickness of the chromosomal DNA bands will be at least that of the well. If there is sufficient DNA in the sample blocks, a 1-mm comb will give the best results. Raising the comb strengthens the thin portion of the gel near the wells, which helps to prevent cracking when the gel is transported.

12. Separation of the gel from the mold is aided by inserting a thin spatula between the two and passing it along the border of the gel.

13. The comb may be held in place by wetting it with water and pressing the wetted surface against the glass.

14. If the plug is not pinned, it tends to move while cutting, resulting in irregular cuts.

15. The spatula is basically acting like a shoehorn. If an air bubble does become trapped in the well, move the block to one side of the well and apply pressure to its top. If the blocks are cut smaller than the well, there will be room for the air to escape.

16. Sealing the wells helps prevent loss of the sample blocks during subsequent transport, staining, and destaining of the gel. It is convenient to keep the low-melting-temperature agarose in a 1.5-mL microfuge tube that can be heated to 65°C until molten.

17. Agarose gels (0.6%) are somewhat fragile and care should be taken in handling them. A thin rectangle of Plexiglass® with one bevelled edge is useful as a scoop for transporting gels. Alternatively, if a commercial casting apparatus is used the gel can be transported in it.

18. Because the buffer is recirculated continuously by a pump, there is a tendency for the gel to move from its original position, especially during the long periods of electrophoresis used to separate multimegabase molecules. Again, there are a variety of solutions to this problem. Casting the gel directly onto a sandblasted portion of the electrophoresis chamber is perhaps the simplest. Or, a small piece of Velcro® can be cast directly into the gel with its mate glued to the bottom of the chamber. Some commercial devices employ plastic stops placed at the corners of the gel and baffles outside the contour to reduce buffer flow over the gel.

19. In some of the older CHEF devices, the buffer level can influence the homogeneity of the electric field. More recent devices are less sensitive to this.

20. Some pumps require priming. If a variable speed pump is used, turn it to high speed initially to expel all the air from the system. The flow rate can then be reduced to the minimum necessary to maintain the buffer at 9°C during electrophoresis. This rate will vary depending on the efficiency of the cooling unit. A reasonable rate is 1 L/min. If the buffer has not been pre-chilled, air bubbles will form in the buffer as it cools.

21. To obtain the results listed in Table 1, it is best to measure the actual field strength (V/cm) in the gel or buffer using a voltmeter. A less accurate alternative is to measure the distance between two opposing faces of the hexagon and divide that distance (in cm) into the nominal voltage appearing on the power supply. When the latter method is used, several runs at varying field strengths may be required to achieve optimal results.

22. Ramping the pulse time means changing its value incrementally during the course of the run. Most commercial devices have the capability to program a ramped pulse time. Generally, ramping results in more even resolution of molecules over a broader size range than can be achieved using a single pulse value. A good rule of thumb is that, if the pulse time changes linearly during the electrophoresis,

the average pulse time can be compared to the values given in Table 1. Alternatively, if linear ramping is not possible, stepping the pulse time through several discrete values during the run will broaden the range of resolution.

23. It may be possible to shorten the run time by altering the field strength and pulse time using the recently described formula of Gunderson and Chu *(4)*:

$$W = (E)^{1.4} \times T_P$$

where W is a function describing the "window" of molecular sizes that are resolved, E is the field strength, and T_P is the pulse time. Given a particular set of conditions that resolve a certain size range, if E is increased and T_p is decreased such that W remains constant, a similar range of molecules should be resolved, but over a run time that is decreased in proportion to the increase in field strength. It may thus be possible to achieve good separation in a shorter time. However, band smearing at higher values of E is still observed, possibly as a result of electrostatic interactions between the DNA and agarose matrix.

24. The long electrophoresis times used to resolve multimegabase molecules mean that the run may be subject to power failures or surges in line voltage that lead to loss of switching. These kinds of failures can be minimized by confirming the switching at least once a day and by interposing a surge suppressor between the wall outlet and switching box. Some commercial devices have built-in surge suppressors.

25. If the hoped-for results are not achieved, troubleshooting will be necessary. First, determine whether the problem lies in the chromosomal DNA preparation or in the electrophoresis. Purchase or borrow *Saccharomyces cerevisiae* chromosomal DNAs of good quality and separate them in the 1.5–3 Mb range as outlined in Table 1. Include *S. pombe* samples in several lanes. The *S. pombe* DNA should produce a single unresolved band positioned near the well (*see* Fig. 4A in ref. *7*). If it does not, the *S. pombe* DNA may be degraded.

After the integrity of *S. pombe* molecules has been established, choose conditions that will resolve the 3–6 Mb range. If the bands appear smeary and indistinct, the DNA molecules are probably moving too rapidly through the gel. All of the parameters mentioned in the Introduction can affect the rate of migration, including the field strength, the temperature, the agarose composition and concentration, and the pulse time. Of these, the field strength is the most likely culprit because, for best results, it must be measured directly with a voltmeter but rarely is. Try lowering the power supply voltage by one-third and increasing the run time proportionally. If improvement is observed, optimize the separation by adjusting the pulse time.

If the lanes do not run straight there may be a problem with the potential at one of the electrodes. Carefully inspect each of the electrodes to ensure that none is broken or corroded. When possible, exchange compatible components (such as the switching box) with an apparatus that is working to localize the problem. If the problem lies in the circuitry, consult an electrician or a company field representative. Crooked lanes can also be caused by objects that distort the electric field, such as items used to hold the gel in place, or the thermometer. Move or remove such items to determine whether they are the source of the problem.

References

1. Chu, G., Vollrath, D., and Davis, R. W. (1986) Separation of large DNA molecules by contour-clamped homogeneous electric fields. *Science* **234,** 1582–1585.
2. Schwartz, D. C. and Cantor, C. R. (1984) Separation of yeast chromosome-sized DNAs by pulsed field gradient gel electrophoresis. *Cell* **37,** 67–75.
3. Carle, G. F., Frank, M., and Olson, M. V. (1986) Electrophoretic separations of large DNA molecules by periodic inversion of the electric field. *Science* **232,** 65–68.
4. Gunderson, K. and Chu, G. (1991) Pulsed-field electrophoresis of multimegabase-sized DNA. *Mol. Cell. Biol.* **11,** 3348–3354.
5. Kuspa, A., Vollrath, D., Cheng, Y., and Kaiser, D. (1989) Physical mapping of the *Myxococcus xanthus* genome by random cloning in yeast artificial chromosomes. *Proc. Natl. Acad. Sci. USA* **86,** 8917–8921.
6. Orbach, M. J., Vollrath, D., Davis, R. W., and Yanofsky, C. (1988) An electrophoretic karyotype of *Neurospora crassa. Mol. Cell. Biol.* **8,** 1469–1473.
7. Vollrath, D. and Davis, R. W. (1987) Resolution of DNA molecules greater than 5 megabases by contour-clamped homogeneous electric fields. *Nucleic Acids Res.* **15,** 7865–7876.
8. Cox, E. C., Vocke, C. D., Walter, S., Gregg, K. Y., and Bain, E. S. (1990) Electrophoretic karyotype for *Dictyostelium discoideum. Proc. Natl. Acad. Sci. USA* **87,** 8247–8251.
9. Brody, H. and Carbon, J. (1989) Electrophoretic karyotype of *Aspergillus nidulans. Proc. Natl. Acad. Sci. USA* **86,** 6260–6263.
10. Fan, J. B., Chikashige, Y., Smith, C. L., Niwa, O., Yanagida, M., and Cantor, C. R. (1989) Construction of a Not I restriction map of the fission yeast *Schizosaccharomyces pombe* genome. *Nucleic Acids Res.* **17,** 2801–2818.

CHAPTER 3

Separation and Size Determination of DNA over a 10–200 kbp Range

Craig A. Cooney

1. Introduction

Some types of pulsed-field gel electrophoresis (PFGE), including CHEF (contour-clamped homogeneous electric field), give straight lanes and excellent separations of a wide range of DNA sizes on one gel (ref. *1* and Chapters 1,2,5–7). Here conditions are described for the separation of DNA in the 10–200-kbp size range by CHEF. CHEF resolution in this range exceeds that of many commonly used size standards. Production and use of a size standard giving a 10-kbp ladder in the 10–200-kbp range *(2)* is described and examples of separations are shown.

Good separation over a wide range of DNA sizes is achieved by ramping the pulse times through a range that includes the optimum pulse times for all DNAs to be separated. Ranges of short pulse times are used here because the DNA sizes are just above or in the range of conventional agarose gel electrophoresis.

CHEF even offers advantages when separations are in the range of conventional electrophoresis. These advantages include separation of larger DNAs (e.g., 60 from 70 kbp) on the same gel as DNAs of just a few kbp (e.g., 2 from 3 kbp). Because CHEF gels for this size range are typically 1% agarose, they give much sharper bands with small DNA and are much easier to handle than conventional gels (e.g., of ~0.4% agarose).

From: *Methods in Molecular Biology, Vol. 12: Pulsed-Field Gel Electrophoresis*
Edited by: M. Burmeister and L. Ulanovsky
Copyright © 1992 The Humana Press Inc., Totowa, NJ

2. Materials

2.1. Materials for Size Standard

1. Lambda DNA: 0.5 µg/µL in 10 mM Tris-HCl, pH 8.0, 1 mM EDTA.
2. Ligase buffer (5X): 250 mM Tris-HCl, pH 7.6, 50 mM MgCl$_2$, 5 mM dithiothreitol, 25% (w/v) polyethylene glycol-8000.
3. ATP: 0.1M Na$_2$ATP.
4. T4 DNA ligase.
5. Restriction enzymes: *Apa*I and *Nae*I.
6. Buffers for restriction enzymes.
    a. 10X for *Apa*I: 500 mM potassium acetate, 200 mM Tris-acetate, 100 mM magnesium acetate, 10 mM dithiothreitol, pH 7.9,
    b. 10X for *Nae*I: 100 mM Bis-Tris Propane-HCl, 100 mM MgCl$_2$, 10 mM dithiothreitol, pH 7.0. *(See* Note 1.)
7. EDTA: 0.5M Na$_2$EDTA adjusted to pH 7.5 with NaOH.
8. SDS: 10% w/v in H$_2$O.

2.2. Materials for CHEF

1. Tris-borate-EDTA (TBE) buffer (0.5X): 45 mM Tris-borate, 1 mM Na$_2$EDTA, pH ~8.3.
2. Agarose (e.g., FMC Bioproducts [Rockland, ME] Genetic Technology Grade).
3. Tracking dye mixture: 70% glycerol, 3 mg/mL bromophenol blue, 0.5X TBE, 1% SDS, 10 mM EDTA.
4. CHEF electrophoresis system (e.g., Bio-Rad [Richmond, CA] CHEF-DR II with chiller).

3. Methods

3.1. Production of the Size Standard

Lambda ladder is partially cleaved at regular intervals to produce intermediate bands. The restriction enzymes *Apa*I and *Nae*I are used to cleave lambda ladder because of the spacing of their sites (10 and 20 kbp respectively from one end of each lambda monomer) and because they each have only one site per lambda monomer (Fig. 1, ref. *2* and *3*, pp. 100–101). *See* Notes 2 and 3.

3.1.1. Ligation of Lambda Monomers

1. Add in a microcentrifuge tube: 200 µL of lambda DNA (100 µg), 80 µL of ligase buffer (5X), and 96 µL of H$_2$O. Mix and heat to 65°C for 15 min (*see* Note 4).

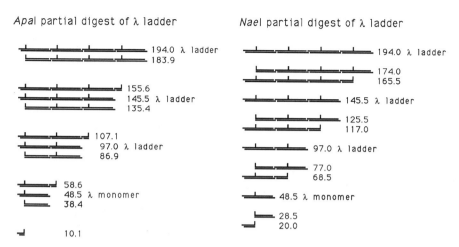

Apal partial digest of λ ladder

194.0 λ ladder
183.9

155.6
145.5 λ ladder
135.4

107.1
97.0 λ ladder
86.9

58.6
48.5 λ monomer
38.4

10.1

Nael partial digest of λ ladder

194.0 λ ladder

174.0
165.5

145.5 λ ladder

125.5
117.0

97.0 λ ladder

77.0
68.5

48.5 λ monomer

28.5
20.0

Fig. 1. Maps of lambda ladder partially digested with *Apa*I and partially digested with *Nae*I. Intact lambda ladder rungs are marked "λ ladder" and the sizes of these and their partial digestion products are given in kbp. Vertical lines mark the respective restriction sites. When there are two cuts in a lambda polymer, fragments with the same sizes as the lambda ladder are produced (maps not shown). This insures that the lambda ladder size fragments are always more numerous (and more intense on a gel) than the nearby fragments between ladder rungs. As shown in Fig. 2, a combination of the *Apa*I and *Nae*I partial digests gives a series of fragments of known size that are ~10 kbp apart.

2. Cool to ~16°C, spin down, and add 4 µL of 0.1*M* ATP and 800 cohesive-end ligation U or ~12 Weiss U *(3, p. 51)* of T4 DNA ligase. Mix gently, spin down, and incubate at ~16°C for 6 h. From this point on the solution is very viscous.
3. Add 4 µL of 0.1*M* ATP and 400 cohesive-end ligation U of T4 DNA ligase. Mix very gently, spin down, and incubate at ~16°C overnight.
4. Store at 4°C and test 3 to 5 µL by PFGE to check extent of ligation. Partial digests are made from the lambda ladder once suitable ligation is achieved (e.g., visible signal from the tetramers or pentamers).

This is a typical protocol. The principle consideration is that lambda DNA be significantly ligated in a solution that (once diluted) is compatible with the restriction digests to be described herein. In order not to shear the large DNA use wide bore pipet tips to transfer the lambda ladder. Minimize steps and handling to keep DNA intact. *See* Note 5.

3.1.2. ApaI *Partial Restriction Digest*
of Lambda Ladders

1. Add in a microcentrifuge tube 100 μL of lambda ladder as described earlier (~24 μg), 60 μL of 10X *Apa*I buffer, and 430 μL of H$_2$O. Mix very gently and spin down.
2. Add 1 U *Apa*I (to accurately measure 1 U of *Apa*I, dilute a small amount of *Apa*I in 1X *Apa*I digestion buffer at 0–4°C just before use). Mix very gently, spin down, then incubate at 37°C for 30 min. (*See* Note 6).

3.1.3. NaeI *Partial Restriction Digest*
of Lambda Ladders

1. Add in a microcentrifuge tube 100 μL of lambda ladder (~24 μg), 60 μL of 10X *Nae*I buffer and 430 μL of H$_2$O. Mix very gently and spin down.
2. Add 20 U of *Nae*I. Mix very gently, spin down, then incubate at 37°C for 30 min. (*See* Notes 6 and 7).

Stop the reactions with EDTA and store at 4°C. Check the extent of digestion by PFGE of 15–30 μL of each digest. If digestion gives satisfactory bands between lambda ladder rungs, then SDS can be added (e.g., to ~0.5% w/v) to stop enzyme activity and EDTA can be added in excess (e.g., to 20 m*M*). Mix the *Apa*I and *Nae*I partial digests in equal ratio to make the finished size standard.

3.2. Separation by CHEF

1. Cast a 1.0% agarose gel with 0.5X TBE buffer. Fill the CHEF system with 0.5X TBE buffer and cool to ~14°C. For overall CHEF set-up, *see* Chapter 2 and specific instructions for your system.

   Be careful with any electrophoresis system that the power supply is off and the lines disconnected when handling the gel or the buffer in the electrophoresis chamber. Be certain that the electrophoresis chamber and any buffer recirculation system is free of leaks.
2. Mount the gel in the electrophoresis chamber.
3. Load first any DNAs in agarose plugs because loading these is least affected by moving the gel. Next, load any DNAs in agarose beads. Lastly load any liquid samples including the marker by mixing them gently with ~0.2 vol of the tracking dye mixture and then pipetting with a wide bore pipet tip with the open end of the tip just inside the well. (*See* Note 8.)
4. The system can be "pre-run" for ~1 h, usually through the ramp of pulse times that will be used in the run. This assures that the DNA is in the gel for the entire ramp of pulses. (*See* Note 9.)

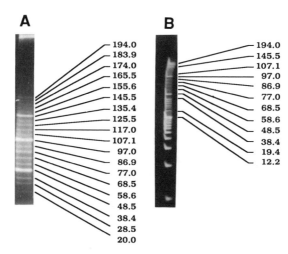

Fig. 2. **(A)** An equal mixture of *Apa*I and *Nae*I partial digests of lambda ladder were electrophoresed with a linear ramp of pulse times between 2 and 15 s for 24 h. **(B)** An equal mixture of *Apa*I and *Nae*I partial digests of lambda ladder, high mol wt marker (BRL) and 1 kbp ladder (BRL) were electrophoresed with a linear ramp of pulse times between 0.1 and 8 s for 14 h. In both **(A)** and **(B)** the gels are 1% agarose (FMC Bioproducts Genetic Technology Grade), run at 200 V and 14°C in 0.5X TBE in a Bio-Rad CHEF-DR II system and visualized with ethidium bromide stain. The sizes of selected bands are indicated in kbp.

5. Electrophorese with a ramp of short pulse times. Figure 2 gives two examples. Figure 2A shows the result of 2–15 s pulses at 200 V (6 V/cm) for separation emphasis in the 20–200 kbp range and Fig. 2B shows the result of 0.1–8 s pulses at 200 V (6 V/cm) to emphasize separation up to ~100 kbp. A ramp of 0.1–12 s with this system will emphasize separation up to ~150 kbp (not shown). *(See* Note 10.)

4. Notes
4.1. Size Standard

1. Most manufacturers supply digestion buffers at 10X concentration with their restriction enzymes. Using the buffer supplied or recommended by the manufacturer usually facilitates prediction of an enzyme's activity.

2. Other size ladders could be produced using different restriction enzymes. Note that restriction enzymes cutting lambda monomer more than once will produce complex patterns with lambda ladder and that this is compounded because some sites on the monomer are more readily cut than others.

3. Other methods to produce markers in this range include ligation of small DNAs to produce a ladder with the resolution of the monomer (10 kbp, ref. *4* and 5 kbp, ref. *5*) or digestion of bacterial genomes with infrequent cutting restriction enzymes (e.g., ref. *4*). The marker described in this chapter has the advantage that it is regularly spaced (~10 kbp) with more intense bands at regular intervals (~50 kbp), which make sizes easy to locate.

4. The 65°C heating (Step 1) is designed to reduce or eliminate any stray nucleases that may be present before the long incubations.

5. The same sizes described earlier can be covered by partial digestion of lambda ladder in agarose blocks with *Apa*I or *Nae*I and by electrophoresing these *Apa*I and *Nae*I partial digests side by side on a gel. Lambda ladder in agarose blocks often includes twentymers (~1Mb). As a result, *Apa*I and *Nae*I partial digests of lambda ladder in agarose can also cover size ranges significantly higher than 10–200 kbp (Cooney, unpublished data). See Chapter 8 for digestion of DNA in agarose blocks.

6. The exact amounts of restriction enzymes needed for partial digests may vary with factors such as enzyme manufacturer, enzyme lot, and extent of ladder ligation. It is an excellent idea to assay your enzyme on lambda monomer before digesting the lambda ladder. The lambda DNA and enzymes used to develop these protocols were from New England Biolabs (Beverly, MA).

7. *Nae*I cuts its site in lambda DNA very slowly and unit activity (based on digestion of adenovirus 2 DNA) will give only a partial digest with lambda DNA (*3*, p. 135).

4.2. CHEF

8. When working with liquid samples it is often useful to place a piece of GelBond film (FMC Bioproducts) underneath the area around the wells and pour the gel over this. The bottom of the comb rests on the GelBond when the gel is solidifying. Gel bond is recommended with liquid samples to prevent their leaking through the bottom of the wells. Even with beads and plugs it may be desirable to keep solutions in which they were restriction digested with them in the well. This would help prevent loss of small DNA fragments *(6)*.

9. If buffer circulation washes samples from the wells, circulation can be turned off for the pre-run while the DNA enters the gel. In the Bio-Rad CHEF-DR II system the gel can be mounted in the electrophoresis chamber with four gel stops to prevent it from moving during the pre-run. Once the buffer circulation is started, remove the two gel stops at the top of the gel (leave the two at the bottom of the gel in place).

10. Pulse times are guidelines and need to be adapted to the particular system in use. For example, good separation in the 10–250 kbp range with a hexagonal array "homemade" CHEF system (based on ref. 7) can be achieved with a 1–25 s ramp at 160 V (~5.8 V/cm) for 24 h (not shown).

Acknowledgments

The methods development and the results presented were largely done at the University of California at Davis with the support of NIH and DOE grants to Dr. E. Morton Bradbury. This work at the City of Hope is supported by an NIH grant to Dr. Arthur D. Riggs. I thank Ronald Goto and Dr. Marcia Miller at the City of Hope for use of and help with their CHEF, which is supported with NIH Biomedical Research Support Grant #SO7RR05471. I thank Jodie Galbraith of U. C. Davis for helpful discussions and Dr. James Spencer at the City of Hope for helpful discussions and helpful comments on the manuscript.

References

1. Dawkins, H. J. S. (1989) Large DNA separation using field alternation agar gel electrophoresis. *J. Chrom.* **492**, 615–639.
2. Cooney, C. A., Galbraith, J. L., and Bradbury, E. M. (1989) A regularly spaced DNA size standard with 10 kbp resolution for pulsed field gel electrophoresis. *Nucl. Acids Res.* **17**, 5412.
3. New England Biolabs Catalog 1990-91, Beverly, MA.
4. Hanlon, D. J., Smardon, A. M., and Lane, M. J. (1989) Plasmid multimers as high resolution molecular weight standards for pulsed field gel electrophoresis. *Nucl. Acids Res.* **17**, 5413.
5. New Pulsed Field Size Standard for 5–100kb DNAs (1990) Bio-Radiations No. 75 p. 7, Bio-Rad Laboratories, Richmond, CA.
6. Fritz, R. B. and Musich, P. R. (1990) Unexpected loss of genomic DNA from agarose gel plugs. *BioTechniques* **9**, 542–550.
7. Chu, G., Vollrath, D., and Davis, R. W. (1986) Separation of large DNA molecules by contour-clamped homogeneous electric fields. *Science* **234**, 1582–1585.

CHAPTER 4

PFGE Using Double-Inhomogeneous Fields or Orthogonal Field-Alternating Gel Electrophoresis (OFAGE)

Margit Burmeister

1. Introduction

The methods discussed in this chapter were the first pulsed-field gel electrophoresis (PFGE) methods developed. As a curiosity, in the first publications, PFGE stood for pulsed-field *gradient* gel electrophoresis *(1)*. However, later it was recognized that field gradients or inhomogeneous fields were not necessary to separate large DNA fragments, and nowadays, PFGE is generally used to mean pulsed-field gel electrophoresis, i.e., all kinds of methods involving switching (pulsing) of fields to separate large DNAs. The two configurations discussed here are the original PFGE version with point electrodes and a later version using continuous electrodes *(2)* (*see* Fig. 1). In the latter publication, the abbreviation OFAGE was introduced, which stands for orthogonal field-alternating gel electrophoresis. Since both versions create approximately orthogonal fields, OFAGE is now often, and will be in this chapter, used for either version. Both versions result in double inhomogeneous fields, and their electrode configurations are shown in Fig. 1. (Initially, also a configuration with only one inhomogeneous field was used, but that will not be discussed here, since it was not used in later studies *[3]*). As in all PFGE techniques, two different fields are generated by switching the voltage between two different sets

From: *Methods in Molecular Biology, Vol. 12: Pulsed-Field Gel Electrophoresis*
Edited by: M. Burmeister and L. Ulanovsky
Copyright © 1992 The Humana Press Inc., Totowa, NJ

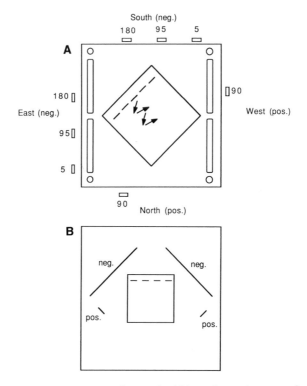

Fig. 1. Set-up of point-electrode (**A**) and continuous electrode (**B**) configuration. The point electrode configuration is based on the LKB-Pharmacia apparatus. In order to compare home-made apparatus: The gel is 20×20 cm^2, and the box itself is 33×33 cm^2. The positions of the electrodes are given in centimeters, and can be read along the sides of the box.

of electrodes. Each time one configuration is on, that is called a pulse, thus pulsed field.

The configuration of electrodes found empirically when PFGE and OFAGE were invented is not easily understood, although field strength and resolution can be modeled *(3)*. The electrodes are positioned in such a way that the DNA migration between different pulses is at an angle of larger than 90° and less than 180°. However, later experiments have shown that an angle of 180° and homogenous field, as in field inversion gel electrophoresis (FIGE, *see* Chapter 1) or one dimensional pulsed-field gel electrophoresis (ODPFGE, *see* Chapter

7), will separate large DNA as well, and that inhomogeneity of fields is not necessary for achieving separation of large DNA fragments. Because of the inhomogeneity of the fields in OFAGE, the angle of electric forces for a DNA molecule located in the slot of the gel is different than that in the middle of the gel. Altogether, this results in a complex DNA migration pattern, in which lanes form sigmoid patterns, except for the very middle lane of the gel (*see* Fig. 2). Because of this complexity, the majority of research using PFGE has switched to later PFGE versions, in which the DNA migrates more or less in straight lanes (*see* Chapters 1–3 and 5–7). However, some research is still done using OFAGE-like systems, and there are several advantages. Some specialized applications may give different results in different systems, such as migration of circular DNA in OFAGE *(4–7)* or in FIGE *(8)*, and it may be advantageous to use double inhomogeneous fields for such special applications.

However, for the use of PFGE in the separation of double-stranded, linear DNA molecules, the most important advantage is the increased resolution of bands. As can be seen in Fig. 3, where a complex pattern of bands is generated by hybridization to a repetitive probe, the inhomogeneity of the field widens each lane during a run, and that results in thinner, sharper bands at the bottom of a gel compared to the top part. This phenomenon has been called *self-sharpening effect (3)*. These sharper bands help to resolve many more bands than is possible with other systems. Unfortunately, the self-sharpening effect is directly correlated to the curvature of the gel, i.e., it is not possible to have sharper bands as well as straight lanes. In certain types of experiments one might wish to accept the curved lanes in order to increase the resolution, such as when repetitive probes like human Alu or alphoid repeat probes (*see* Fig. 3) are hybridized to cell lines containing a single human chromosome or fractions thereof. Other applications that might need high resolutions are comparisons of different strains of bacteria, yeast, or protozoa, where sometimes more than 20 chromosomes can be resolved (*see*, e.g., Chapter 15). On the other hand, curved lanes make it harder to compare lanes, except when they are immediately adjacent. Size markers in every other lane will solve that problem *(9)*, but then fewer samples can be analyzed in each gel.

Currently, the only commercially available apparatus using the double inhomogeneous configuration is available from LKB-Pharmacia (Piscataway, NJ). Since the continuous electrode arrangement has the

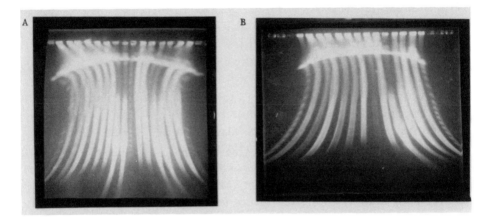

Fig. 2. Effect of comb position. Varies digests of mammalian DNA are shown on gels run at 180 V (LKB-Pharmacia unit), with electrode positions as described in the text, and switching times of 70 s for 44 h. In **A,** a lambda ladder of multimers of 48 kb are run in the central lane, in **B** in the two outside lanes. The comb was positioned about 1 cm from the start of the gel in **A,** and about 5 cm from the start in **B.** Note that by moving the comb further into the middle of the gel, the sigmoid pattern seen in **A** can be avoided.

drawback of requiring a large bench space for a relatively small gel (*see* Fig. 1) and is to my knowledge not commercially available, I focus on problems encountered with the LKB-Pharmacia point electrode system. Detailed instructions to build an apparatus with continuous electrodes are given in ref. *10,* which also discusses many aspects of working with OFAGE. LKB-Pharmacia provides an extremely useful and extensive manual. This chapter can by no means be as extensive as the manual, and users are strongly advised to read the manufacturer's manual before using the LKB-Pharmacia apparatus. It is useful even if you are using a home-built apparatus. Reference *9* also contains an appendix with pictures, in which the results of common mistakes are shown. Several other chapters of this book, especially Chapters 1 and 2, discuss the effect of varying conditions such as switching time, agarose concentration, and voltage. Since many of the general trends described in the chapters on CHEF and FIGE will also be true for inhomogeneous fields, OFAGE users are advised to read those chapters as well. Here I mainly focus on special problems not encountered in other PFGE methods.

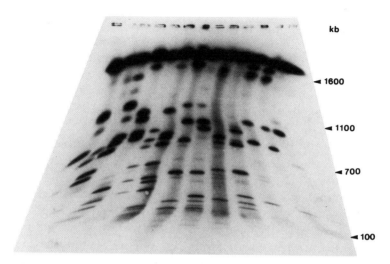

kb

◀ 1600

◀ 1100

◀ 700

◀ 100

Fig. 3. Self-sharpening effect and lane-spreading in OFAGE. DNA from a variety of unrelated and related people was digested with the restriction enzyme *Bam*HI, separated by OFAGE in an LKB-Pharmacia apparatus as described in this chapter with a switching time of 5 min for 3.5 d, and of 2 min for 1.5 d. Waterbath was set at 14°C, the voltage was 150 V. The DNA was blotted onto Nylon membranes, and hybridized to the probe L1.26, which recognizes the alphoid repeats of human chromosomes 13 and 21 *(14)*. These polymorphisms were also described in ref. *15*. Note that the smaller the fragments, the sharper they are.

2. Materials

1. LKB-Pharmacia PFGE set-up for double-inhomogeneous fields.
2. 0.5X TBE (*see* Note 1).
3. 1% LE-agarose in 0.5X TBE.
4. Agarose blocks or agarose beads containing the DNA samples.
5. Gel comb with slots as large or slightly larger than the agarose blocks. Plastic frame to hold the comb for pouring the gel.
6. 1% Low-melting point (LMP) agarose in 0.5X TBE.

3. Method

3.1. Setting Up the Apparatus

1. Start precooling about 3 L of 0.5X TBE (Note 1) in the apparatus (*see* Note 2). Put the cooling unit to 12°C, or whatever temperature you have determined best for your separation (*see* below).

2. Put negative electrodes (black) at positions 5, 95, and 180 on the South and East side of the box, and positive electrodes (orange) at position 90 at the North and West sides of the box (*see* Note 3).

3. Melt LE agarose, and cool until hand-warm (≈55°C). Put the gel frame on an even surface in the position shown in Fig. 1. Place the rubber frame and the gel comb as shown in Fig. 1 (*see* Note 4). The comb should be 1–2 in. (3–5 cm) from the start of the gel (*see* Note 5). Pour the agarose (about 200 mL) into the frame. The comb might bend a little bit from the heat, but will be straight when the agarose has cooled down.

4. Carefully remove the comb, and load the DNA blocks while the gel is still on the bench. If the blocks are slightly smaller than the slots, they will easily drop to the bottom of the slot. If they are sealing the block air-tight, push a syringe with a fine-gage needle behind the slot, and suck out the air. The block will slide into the slot from the vacuum that you created (*see* Note 6). You can also load liquid DNA samples into the dry gel. In both cases, seal the slots with 0.5–1% liquified LMP agarose gel in 0.5X TBE.

5. Once the sealing agarose is firm, put the gel into the chamber into the position shown in Fig. 1.

6. Program the appropriate switching times and voltage. If you have no experience, follow initially the manufacturer's recommended conditions. To further optimize your running conditions, consider the rules in Section 3.2. and in Chapters 1–6.

7. Start the run. Watch bubbles forming at the electrodes, and look at the amperage for at least one full cycle (both directions). There should not be more than about 10 mA difference between the two directions (*see* Note 7).

8. After the run, remove the gel, switch off cooling, empty the buffer from the box. If you are not running another gel immediately afterwards, wash the box once in distilled water (deionized water is often too acidic and should not be used), and empty again, tilting the box, using the pump to remove most of the liquid. Switch off the pump (*see* Note 8).

3.2. Parameters Affecting OFAGE

1. The *position of electrodes* affects mainly the shape of the lanes and the sharpness of bands. Generally, the further toward the corner the positive electrode is positioned, the more curvature and band-sharpening occurs. Putting the electrode close to the middle of the gel results in straight lanes, but in nearly no resolution *(3)*. Increasing the num-

ber of negative electrodes can help in getting straighter gels. If the negative electrodes are put too far into the corners (positions >200), the DNA will first move toward the middle of the gel before spreading outwards, i.e., a C-shape rather than an L-shape for the right lanes (*see* Fig. 2). Generally, one should determine a good compromise of electrode positions for the purpose, and not change the positions for every run.

2. The effect of the *agarose concentration* is important *(11)*. Although one can speed up long runs by using 0.7% rather than 1% agarose, this does decrease the resolution and broadens the bandwidth. If you do want to speed up the run, some companies now offer agarose that will be stiff enough at 0.4% concentration to work with. Concentrations of agarose above about 1.2% seem to slow down the migration without a significant increase of resolution, and are thus not recommended. Concentration between 0.9 and 1.2% seem to be optimal.

3. *Temperature* effects PFGE more than other electrophoresis *(11)*, and it is therefore very important to keep a constant temperature. The opinions over what temperature is optimal differ widely, from around 5°C to about 35°C *(12)*. Whatever is preferred, a system that keeps the temperature constant is important. It is good to stick to one temperature that your apparatus is capable of keeping reliably. The main parameters you want to change are voltage and switching time, and changing the set temperature would introduce another variable. Generally, a higher temperature has a similar effect as an increased switching time: Everything else being constant, larger DNAs will be separated at higher temperatures and the total running time will be shortened. It may, however, increase the band widths and thus affect the resolution. This is the reason why in most publications the preferred temperatures are somewhat below ambient, i.e., 10–15°C.

4. The effects of the *switching time* on the separation of different DNA fragment sizes are discussed in other chapters of this book (*see* Chapters 1–3), as well as in the LKB instruction manual. When referring to published tables to choose the correct condition for the size range of interest, be aware that temperature, voltage, and agarose concentrations should be similar in order to get comparable results. The dependence of the mol wt from the distance migrated is not simple in OFAGE *(13)*. As in CHEF (*see* Chapter 2), there are basically three zones of separation, from bottom to top: a first, main region of resolution, a region of higher resolution (by about a factor of two), and the limiting mobility zone. The region of higher resolution can be seen in Fig. 2 as a region where the distance between individual lambda

marker bands is increased. Thus, the best resolution in OFAGE is actually achieved in the region shortly below the limiting mobility. If you are interested in high resolution separation of one particular fragment size, it will be optimal to choose conditions under which those fragments will be in the region just before the limiting mobility. This may require some trial and error. In contrast, for separations of fragments covering a large size range, it is useful to combine several switching times, also called ramping. For such overviews, it may be best to start with the long switching times, and end the run with the shorter switching times (e.g., a separation of 100–6,000 kb can be achieved by using successively switching times of 1 h for 7 d, 30 min for 4 d, 15 min for 2 d, and 5 min for 10 h). However, by using several different switching times on the same gel, the three zones mentioned earlier can no longer be discriminated.

5. The *voltage* used for separations up to about 1 Mb is generally about 8–10 V/cm (measured from electrode to electrode) (*see* Note 9). Above that size range, the voltage has to be lowered because of trapping effects (*see* Chapter 7). To separate DNA fragments between 0.5 and 2 Mb, about 5 V/cm seems to be sufficient, and for fragments larger than about 2 Mb, one has to go down to 1–3 V/cm. This, of course, increases the running time for separation of large fragments (*see* Note 10).

4. Notes

1. I have generally used standard 0.5X TBE. 1X TBE would generate too high an amperage. To have the advantage of 1X TBE buffering capacity, but lower amperage, you can reduce the EDTA content by fivefold relative to standard TBE *(9)*. Lower buffer capacities generally tend to broaden the bands observed.

2. Many cooling units (but not the one supplied by LKB) contain both a heater and a cooler switch. Switch the heater on, even if your desired temperature is way below ambient. The way these units work is that they cool pretty much constantly, and the regulation of the temperature is through heating. Running them in a "cooling only" mode results at best in smeary bands owing to the variation in the temperature, and at worst can freeze the whole bath, although you had set it to a temperature well above 0°C.

3. The effect of different positioning of the electrode is very complex. They are the main determinant of the degree of inhomogeneity of the fields, and therefore how curved or straight the lanes will run.

Other suggested positions are at about 11, 44, and 77% of the length (at about position 36, 145, and 250) for the negative and 22% (about 73) for the positive electrodes *(9)*.

4. It is important to orient the comb relative to the two slits in the frame as shown, because the slits let the buffer be pumped through. If you inadvertently have poured the gel at 90° from the position shown in Fig. 1, the easiest solution is to redo it. If you only realize the mistake when the precious samples are already loaded, detach the gel from the frame, turn it 90°, and then run the gel with a heavy glass plate put on top of the gel to prevent it from floating around.

5. By putting the comb further toward the middle of the gel at the electrode positions given, the DNA will move only outwards. If the comb is put at the very start of the gel, the DNA will migrate toward the middle of the gel first, and then outwards (*see* Fig. 2). If that happens, it is often hard to align a hybridizing fragment with the correct slot. Even crossing of bands can occur under some circumstances.

6. Do not try to push the block into the slot by force. If you poke into the block, introduce air bubbles, or hurt the block in any way, you will see imperfections such as streaks in the lanes.

7. If there is a difference in current between N/S and E/W runs, level the box again, which will usually solve the problem. If that does not help, consider that after several months of continuous running, the electrodes tend to wear out, usually the single positive electrode first. The platinum wire gets thin, resulting in a different amperage in N/S than E/W direction. This will lead to samples running slightly to one side. Check whether changing the electrodes between N/S and E/W will change the amperage accordingly. If yes, replace the electrodes. Since this is a frequent problem, you should have an electrode replacement set at hand. The life-time, under constant high-voltage running conditions, is about 1 mo for the positive and 10 mo for a negative electrode *(9)*.

8. In our experience, after the electrodes, the pump is one of the most frequently replaced spare parts of the LKB-Pharmacia box. Running dry, it will burn out and has to be replaced. These costs can easily be avoided by always switching off the pump after emptying the buffer. To our experience, no matter how big the warning note on the machine, someone will forget to switch it off one day, and it is useful to have a spare pump at hand.

9. Unfortunately, many publications give an absolute voltage rather than V/cm. Remember that your box may have a different geometry from the one in publications. Try to calculate the correct voltage rather

than taking the value straight from a publication in which a different equipment was used.

10. If your switching times result in a gel without any limiting mobility zone, and with a faint smear in the first part of the gel, the voltage used was too high for the size of molecules separated.

Acknowledgments

Most of the work described here is based on my work performed while I was in the laboratories of Hans Lehrach (European Molecular Biology Laboratory, Heidelberg, now at Imperial Cancer Research Fund, London) and of Richard M. Myers (University of California, San Francisco). I thank them and the members of their labs for discussions, encouragement, and support.

References

1. Schwartz, D. C. and Cantor, C. R. (1984) Separation of yeast chromosome-sized DNAs by pulsed field gradient gel electrophoresis. *Cell* **37,** 67–75.
2. Carle, G. F. and Olson, M. V. (1984) Separation of chromosomal DNA molecules from yeast by orthogonal-field-alternation gel electrophoresis. *Nucleic Acids Res.* **12,** 5647–5664.
3. Cantor, C. R., Gaal, A., and Smith, C. L. (1988) High-resolution separation and accurate size determination in pulsed-field gel electrophoresis of DNA. 3. Effect of electrical field shape. *Biochemistry* **27,** 9216–9221.
4. Hightower, R. C. and Santi, D. V. (1989) Migration properties of circular DNAs using orthogonal-field-alternation gel electrophoresis. *Electrophoresis* **10,** 283–290.
5. Beverley, S. M. (1988) Characterization of the 'unusual' mobility of large circular DNAs in pulsed field-gradient electrophoresis. *Nucleic Acids Res.* **16,** 925–939.
6. Beverley, S. M. (1989) Estimation of circular DNA size using gamma-irradiation and pulsed-field gel electrophoresis. *Anal. Biochem.* **177,** 110–114.
7. Mathew, M. K., Hui, C. F., Smith, C. L., and Cantor, C. R. (1988) High-resolution separation and accurate size determination in pulsed-field gel electrophoresis of DNA. 4. Influence of DNA topology. *Biochemistry* **27,** 9222–9226.
8. Levene, S. D., and Zimm, B. H. (1987) Separations of open-circular DNA using pulsed-field electrophoresis. *Proc. Natl. Acad. Sci. USA* **84,** 4054–4057.
9. Smith, C. L., Klco, S. R., and Cantor, C. R. (1988) Pulsed-field gel electrophoresis, in *Genome Analysis, a Practical Approach* (K. E. Davies, ed.), IRL Press, Oxford, pp. 41–72.
10. Van Ommen, G. J. B. and Verkerk, J. M. H. (1986) Restriction analysis of chromosomal DNA in a size range up to two million base pairs by pulse field gradient electrophoresis, in *Human Genetic Diseases: A Practical Approach* (K. E. Davies, ed.), IRL Press, Oxford, pp. 113–133.

11. Mathew, M. K., Smith, C. L., and Cantor, C. R. (1988) High-resolution separation and accurate size determination in pulsed-field gel electrophoresis of DNA. 1. DNA size standards and the effect of agarose and temperature. *Biochemistry* **27**, 9204–9210.

12. Snell, R. G., and Wilkins, R. J. (1986) Separation of chromosomal DNA molecules from C.albicans by pulsed field gel electrophoresis. *Nucleic Acids Res.* **14**, 4401–4406.

13. Mathew, M. K., Smith, C. L., and Cantor, C. R. (1988) High-resolution separation and accurate size determination in pulsed-field gel electrophoresis of DNA. 2. Effect of pulse time and electric field strength and implications for models of the separation process. *Biochemistry* **27**, 9210–9216.

14. Devilee, P., Slagboom, P., Cornelisse, C. J., and Pearson, P. L. (1986) Sequence heterogeneity within the human alphoid repetitive DNA family. *Nucleic Acids Res.* **14**, 2059–2073.

15. Marçais, B., Bellis, M., Gerard, A., Pages, M., Boublik, Y., and Roizes, G. (1991) Structural organization and polymorphism of the alpha satellite DNA sequences of chromosomes 13 and 21 as revealed by pulse field gel electrophoresis *Hum. Genet.* **86**, 311–316.

CHAPTER 5

Transverse Alternating-Field Electrophoresis

Katheleen Gardiner

1. Introduction

Transverse alternating-field electrophoresis (TAFE) refers to a pulsed-field system that uses a vertical gel and a simple electrode geometry. A schematic of the apparatus is shown in Fig. 1 *(1)*. The electrophoresis tank is a large plexiglass box, in which the gel stands vertically, supported at each side by two thin plexiglass slots and by the buoyancy of the buffer. The electrodes, represented by dots in the figure, are wires stretched across the width of the box (in commercial designs, the electrodes are wired to removable plexiglass inserts). During electrophoresis, the DNA is alternately moved downward and to the left, when the A electrodes are activated, and downward and to the right, when the B electrodes are activated. Note that the angle between the A and B fields is not constant down the length of the gel. At the wells, it is 115°, but is much greater at increasing distances from the wells. As the angle increases, the downward component of the field decreases.

The placement of the electrodes parallel to the gel faces generates fields that are equivalent across all lanes of the gel. Therefore, in the TAFE system, DNA runs absolutely straight in all lanes, free of even the moderate distortions seen in other systems. Additional advantages to the TAFE system are: (1) band sharpening, because the downward component of the field decreases down the gel; (2) good resolution

From: *Methods in Molecular Biology, Vol. 12: Pulsed-Field Gel Electrophoresis*
Edited by: M. Burmeister and L. Ulanovsky
Copyright © 1992 The Humana Press Inc., Totowa, NJ

Gardiner

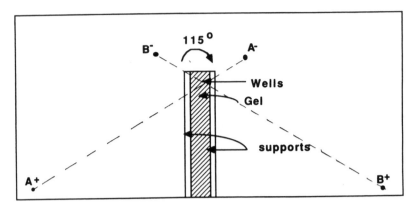

Fig. 1. Schematic representation of the TAFE system. The gel is held vertically in a slot in the center of the tank, and is supported largely by the buoyancy of the buffer. The electrode pairs A (+ and –) and B (+ and –), shown as points, are wires spanning the width of the tank, parallel to the gel faces. The angle between the fields generated by the A and B electrodes is 115° at the wells. Reprinted by permission from *Nature* vol. 331, p. 371. Copyright, 1988, Macmillan Magazines Inc.

and very sharp bands obtained for all size classes <1600 kbp, in relatively short electrophoresis times (<18 h); and (3) useful resolution of fragments in the Mb range in 3 d. The gels are shorter than many (10 cm) and the angle between fields is not readily altered. However, neither of these latter features have hampered any practical applications.

As with all pulsed-field systems (discussed elsewhere in this volume), factors, such as voltage, pulse time, temperature, buffer composition, and so on, can affect the size range of molecules optimally separated. Again, as in other pulsed-field systems, if all other conditions are held constant, the pulse time can be used to optimize separation of the size class of interest. However, because of the unique geometry of the TAFE system, the pulse times and voltages required for resolution of a particular size class can not always be directly inferred from those used in other types of apparatus. This chapter discusses four basic procedures that contain features unique to the TAFE system: building the gel, sample concentrations and loading, choosing electrophoresis conditions, and staining and transferring the gel. Included in staining are discussions of what should be seen in the stained gel, and what to do if this is not seen. (For discussions on the preparation of DNA in agarose, the general effects of various electrophoresis parameters [temperature, voltage, and so on], and the causes and effects of trapping, *see* other chapters in this volume.)

2. Materials

1. 20X PF/EB buffer: $0.2M$ Tris-HCl, 10 mM EDTA (free acid, not Na salt), pH 8.3 (approx) with glacial acetic acid (*see* Note 1).
2. Agarose.
3. Glass plate or tray (approx 20 × 20 cm), large enough for transporting the gel with the comb in.
4. TAFE pulsed-field electrophoresis chamber and electrodes (Beckman,* or homemade), with gel former (plastic base, glass plate, comb, and comb holder).
5. Power supply, with constant current capability, 150 mA at 250 V (*see* Note 2).
6. Refrigerated water bath filled with 50% antifreeze/H_2O, and set at –5°C to –10°C (*see* Note 3).
7. Pump setup to recirculate electrophoresis buffer through refrigeration unit.
8. Pulse timer and switch (Beckman, or other).
9. DNA size markers: Lambda *Hind*III digests, lambda ladder, *S.cerevisiae* chromosomes, and *S.pombe* chromosomes, all in agarose blocks (Beckman or homemade) (*see* Note 4).

3. Methods

3.1. Building the Gel

1. For resolution of fragments <1600 kbp, heat 0.48 g agarose in 60 mL 1X PF/EB + 3 mL H_2O in a microwave (to give a 0.8% gel). Boil until completely dissolved. Cover and equilibrate in 50°C H_2O bath. For resolution of fragments >1600, use 0.33 g agarose, to give a 0.55% gel (2).
2. To assemble gel former, slide the glass plate onto the plastic base. Position the comb in the holder over the plastic base at the open end. Adjust the height of the comb so that the teeth are slightly below the center of the thickness of the gel. The aim is to position the teeth so that the plugs (approx 3-mm thick) will be centered in the thickness of the gel. Remove comb (*see* Note 5).
3. Tape across the open end of the plastic base.

---

*Beckman Instruments now provides a Geneline II version of the TAFE system. The design is essentially the same as that described here, except that the gel is 14 cm × 15 cm. This is sufficient for 20 lanes and can separate the DNA over a greater length. Running conditions are: constant current of 300 mA, with an initial voltage of 320 V. Pulse times should be chosen 30–50% longer than those shown later in Table 1. Run lengths can be increased from 18–24 or 36 h if desired.

4. Lay the assembled gel former flat on the tray. Use a Pasteur pipet and seal the edges of the glass plate with about 2 mL of agarose.

5. When the seal has set, prop the open end of the gel former on a pipet (for ease of pouring the gel). With a 25 mL pipet, quickly pipet 50 mL of agarose between the plastic base and plate. Avoid trapping bubbles in the gel (most will rise quite quickly to the open end). Immediately lay the gel flat, place the comb in the gel with the teeth against the edge of the glass plate, and fill the area between the teeth and the tape with agarose. Return the remaining agarose to the 50°C bath.

6. When agarose has set, chill the gel at 4°C for 20 min.

7. While gel is chilling, fill the electrophoresis tank with 3.5 L 1X PF/EB buffer (for 0.8% gels; for 0.55% gels, use 2X PF/EB), and start recirculation to chill buffer to approx 4°C. The buffer level must reach a height of 2.5 cm above the negative eletrodes.

3.2. Loading the Gel

1. Remove comb and tape.

2. With a razor blade, trim agarose from the top edge of the glass plate, and, using slight pressure on the gel, slide the plate upward and off. Do not try to pry the plate off the gel; this can cause it to tear.

3. Using a sterile spatula, load plugs into the wells. Include markers that span the size class of interest, e.g., use lambda digested with *Hind*III and lambda ladders, for 10–300 kbp; lambda ladder and *S.cerevisiae* chromosomes for 200–1600 kbp; both *S.cerevisiae* and *S. pombe* chromosomes for the Mb range. With a pipet tip, push the plugs down to the bottom and the back of the wells, to ensure that all plugs are centered in the thickness of the gel and at the same level. Be careful not to puncture or tear the plugs (*see* Note 6).

4. Seal plugs in the well using the remaining agarose at 50°C. Allow seal to set for 2–3 min.

5. Run a razor blade around the edges of the gel to separate it from the base. Slide the gel upward off the base and onto your hand (left hand works best for right handed people). Place the bottom of the gel in the center slots of the electrophoresis tank, and slowly let the gel slide from your hand down into the tank. Be sure it remains within the center slot (*see* Note 7).

6. Place the plexiglass bar on top of the gel, to prevent the gel from rising up during the run; place the electrode inserts into the buffer chamber and slide together. Connect leads to chamber and power supply.

3.3. Electrophoresis Conditions

1. Set power supply to constant current. For resolution of fragments <1600 kbp, set voltage range to >250 V, and current range to >150 mA. For >1600 kbp, use >80 V, and >80 mA (*see* Note 8).
2. Use Table 1 to select the switch time to optimize resolution of the fragment size of interest. Set timer for run length (*see* Note 9).
3. For <1600-kbp resolution, set the voltage to 250 V; current should be approx 150 mA, and temperature between 5 and 15°C. For >1600 kbp, start at 50 V; the current should be 50 mA (*see* Note 10).

3.4. Staining and Transfer

1. Stain the gel in 2-L running buffer plus 150 µL of 10 mg/mL ethidium bromide for 30 min at room temperature. Destain twice for 30 min in 1 L of distilled water (*see* Note 11).
2. Visualize and photograph DNA in UV light. The number of yeast and lambda bands visible will depend on the pulse time as described in Table 1 (*see* Note 12). Undigested mammalian DNA should be retained within the wells. The appearance of digested DNA will also vary with the pulse time. For examples of resolution at 20-s and 20-min pulse times, *see* Figs. 2 and 3 (*see also* Note 13).
3. In the 20-s gel, three regions with different separation characteristics are seen. Region A is a faintly staining area immediately below the wells. Although some DNA is present here, no ethidium (or hybridization) bands are seen. Most likely, the DNA here is too small to remain in the wells, but is too large to resolve, owing to the effects of trapping (*see* Chapters 7, 23, 26, and 27 for discussions of the causes and effects of trapping). Region B is a densely staining broad band, often called a compression zone. Here, molecules that are small enough to have escaped trapping but still too large to resolve are comigrating. All that can be inferred about their size is that it is between approx 600 and 1600 kbp. Region C shows distinct bands in the marker lanes, and a uniform smear in the mammalian digests. It is in this area that good resolution occurs, as can be inferred from the sharp and well-spaced marker bands. These three regions are seen at all pulse times, although the size class they contain varies. The compression zone (B) becomes fainter with increasing pulse times, as large DNA fragments move into region C, and at a 60-s pulse time, may be almost nonexistent.
4. The 20- plus 30-min pulse times resolve fragments >1600 kbp. There is generally a faint compression zone nearer the wells. In the mammalian DNA, note that fragments <1600 kbp do not resolve, and run

Gardiner

Table 1

Electrophoresis Conditions

Size class kbp	Pulse time	Initial voltage V	Current mA	Run length h	Linear range kb
2–150	4 s	250	150	12	30–150
10–400	15 s	250	150	14	50–350
200–600	30 s	250	150	16	100–600
50–1600	60 s	250	150	18	200–600
					600–1100
1600–	30 min	50	50	10	
5800	plus 30 min	80	80	10	
	plus 20 min	80	80	50	
200–	same as 1600–5800,				
5800	plus 60 s	250	150	10	

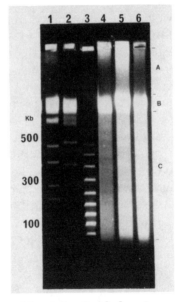

Fig. 2. Ethidium bromide-stained gel showing separation of fragments <600 Kbp. Electrophoresis conditions were: 18 h, at 150 mA (initial voltage, 250 V; final, 220 V), with a 20-s pulse time; temperature, 15°C. Lanes 1 and 2: chromosomes from two different strains of *S. cerevisiae*; lane 3, lambda ladder; lanes 4–6, mammalian DNAs digested with the rare-cutter, *BssH*II. Approximate sizes are given to the left of the figure in kbp. A, B, and C (at the right of the figure) indicate three regions of the gel with different separation characteristics.

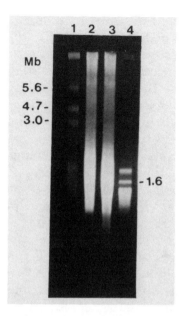

Fig. 3. Ethidium bromide-stained gel showing separation of fragments >1600 kbp. Electrophoresis conditions: 10 h, 50 mA (50 V), 30-min pulse time; plus, 10 h, 50 mA (50 V), 20-min pulse time; plus 50 h, 80 mA (80 V), 20-min pulse time. Lane 1, chromosomes from the yeast *S. pombe;* lanes 2 and 3, mammalian DNA digested with *Not*I and *Bss*HII; lane 4, chromosomes from the yeast *S. cerevisiae.* Fragment sizes are given in Mb.

as a second lower mol-wt compression zone, seen most clearly in the *S. cerevisiae* marker lane. This can be eliminated by including a fourth stage with a 60-s pulse time, at 250 V, for 10 h. Doing so, however, tends to sacrifice the best resolution of all size classes.

5. Soak the gel for approx 5 min in 200 mL of 25 m*M* HCl, and transfer in 20X SSC by standard procedures (*see* Notes 14 and 15).

4. Notes

1. The free acid form of EDTA is used to keep the buffer at low ionic strength. Use of the more common trisodium salt will cause excessively high currents, low voltages, and intolerable heat generation. This is a characteristic of the TAFE system and is caused by the three-dimensional positioning of the electrodes and the large buffer volumes. Excessive current can also be caused by too much acetic acid. Addition of 5 mL acid/L produces acceptable buffer, although the pH may be as high as 8.6.

2. Ideally, gels can be run under constant voltage. However, small changes in the geometry, as a result of warping of the tank or electrode inserts, can cause the voltage drop across the gel in one direction to be greater than that in the opposite direction. If this difference is great enough, the DNA, especially the smaller molecules, will migrate out one face of the gel. Running the gels under constant current can compensate for these small distortions, and should be used as a precaution.

3. It is possible to use this system at room temperature (if lower voltages are used), or to run the gels in a cold room. The drawback is that the bands become less sharp at higher temperatures, and that reproducibility (in terms of run length and optimum pulse time) is less easy to control. Note that more heat is generated in the TAFE system because of the larger buffer volumes and the electrode–gel orientation.

4. Many other commercially available markers are too concentrated to produce sharp bands in the TAFE system. Recommended concentrations for yeast markers are 5×10^7 cells/mL in agarose, using 1/8-inch id tygon tubing as plug molds. A 1/2-cm plug is 40 μL and contains 0.25 μg of DNA. Lambda ladders are made at 15 μg/mL in 1/16-inch id tygon tubing, where a 1/2-cm plug is 10 μL and contains 150 ng of DNA. At least 10–12 bands should be visible. Mammalian plugs are best made at approx 7×10^6 cells/mL in 1/8-inch id tubing. Higher concentrations result in lower mobilities of the DNA, and therefore anomalously high estimations of fragment sizes. (*see* Chapters 8 or 9 for details of DNA preparation, *see also* Chapter 18).

5. It is important to have the DNA loaded precisely in the center of the gel because of the geometry of the system. DNA loaded off center will be pulled more and more off center with each successive switch, and may eventually escape out one face of the gel.

6. The integrity of the plugs is important. Physical damage to the agarose translates into mechanical shearing of the DNA and can result in smearing of the DNA down the gel tracks.

7. Check the placement of the gel in the slot by looking through the electrophoresis chamber from the side. The gel faces should be flat. If the gel appears bowed, the box may have warped causing the gel to bulge on one side. Remove the gel from the chamber, return it to the base of the gel former, and with a razor blade, trim 1–2 mm from one side. This will usually correct the problem. Failure to do this may cause loss of DNA from the gel in some lanes.

8. Separation of molecules >1600 kbp (e.g., the chromosomes of *S. pombe*) require somewhat different conditions in order to prevent trap-

ping (as discussed elsewhere in this volume): a 0.55% agarose gel in 2X PF/EB, 2X PF/EB running buffer, temperatures of 4–8°C, and three stages (*see* Table 1).

9. The conditions given in Table 1 are intended as a guide. They work well, but are not unique. Lower voltages, with corresponding longer run times, can be as effective, but note that this will require increases in the pulse time to effect the same size class resolution. Higher voltages are not recommended because of the heating effects of higher currents. Other pulse times (e.g., 7 s, 20 s, 40 s, and so on) are effective for ranges intermediate between those shown. Reliable size determinations require the correct choice of pulse time. Whereas a 60-s pulse time will resolve fragments of 300 kbp, use of a 15-s or 30-s pulse time clearly will give greater accuracy. Longer electrophoresis times do not generally give better separations, and will decrease the resolution of the smaller molecules. Near the bottom of the gel the downward component of the field is very small, and therefore the molecules begin to stack up, losing separation.

10. The voltages given are for the initial voltage. In the early hours of the run, the buffer will heat up, decreasing resistance. Because the gels are run at constant current, the voltage will therefore decrease to compensate.

11. Best contrast between DNA and background in the gel is achieved when the ethidium bromide concentration is very low and the staining buffer volume is large. Best staining is obtained when a tray is designated only for staining pulsed-field gels.

12. The number of bands seen in the *S. cerevisiae* lanes will also depend on the strain of yeast used, because of the variability of individual chromosome sizes. For example, *see* Fig. 2, lanes 1 and 2.

13. Several problems can be diagnosed from the ethidium bromide-stained gel:

    a. Material in the wells but no DNA visible in the lanes: Most likely, the switch is not working, and DNA has run out one face of the gel.

    b. DNA appears faint in the center lanes: The gel may be bowed, causing DNA to escape out one face of the gel (*see* Note 8).

    c. Chromosomal bands of *S. cerevisiae* decrease in intensity down the gel or smallest bands missing entirely: After viewing the DNA through one face of the gel, view it from one side, through the thickness of the gel. The DNA should be centered within the thickness of the gel. If instead the path of the DNA slants upward or downward toward one face, it suggests that some DNA may have been lost. Check the position of the wells, and that the DNA plugs were loaded dead center (*see* Note 5). If this is fine, then the tank or the electrode inserts may have warped, causing asymmetry in the electrode placements relative to the gel. This sometimes happens, in particular when the tank is in constant use, or after use at temperatures much exceeding 23°C. Allow the tank and inserts to dry at room temperature for 2–3 d.

d. At a 60-s pulse time, largest (>1100 kbp) *S. cerevisiae* chromosomes not visible: Caused by trapping, the irreversible entanglement of larger DNA fragments in multiple pores of the agarose matrix, from which they cannot be extricated. This would most likely be attributable to either the temperature or the voltage being too high (*see also* Chapter 7).

e. Marker bands are fuzzy and diffuse: Check the temperature of the buffer after the gel has been running for 2–3 h. This should be <18°C. Above approx 23°C, the higher the temperature, the broader the bands will be. Also, check the concentration of cells (and therefore DNA) in the plugs (*see* Note 3).

f. Under appropriate conditions, chromosomes from *S. pombe* not visible: The separation of fragments of this size range is more sensitive to the electrophoresis conditions than are smaller fragments (<1600 kbp). Small (10%) decreases from the voltages given in Table 1 can cause the two largest bands to migrate close together in a broad diffuse band; small increases may cause trapping and result in no bands appearing. For reproducible results in resolving this size class of molecules, it is particularly important to carefully control the voltage, current, temperature and buffer composition.

14. Often, the quality of a pulsed-field gel is best judged by the quality of the hybridization results. Weak hybridization signals can be caused by excessive depurination prior to transfer. Strongest hybridization signals have been obtained when the HCl treatment is sufficient to result in the transfer of 80–90% of the DNA, as estimated by restaining the gel after transfer is complete (it is necessary to destain the gel thoroughly, preferably overnight, to get an accurate impression). Complete transfer of the DNA generally lowers the strength of the hybridization signal. It is best to titrate the length of the HCl treatment, and choose conditions that optimize the hybridization signal, not the amount of transfer. For transfer of Mb-sized fragments, slightly longer acid treatment times may be required.

15. Capillary transfers, either neutral or alkaline Southern, work with good uniformity for all size classes of molecules. The electroblotting procedure that can be used with TAFE does not save much time in handling, and requires different conditions (voltages and time) for fragments in different size ranges. For example, single conditions do not give uniform transfer and/or binding of all sizes from a 60-s gel.

References

1. Gardiner, K. and Patterson, D. (1989) Transverse alternating field electrophoresis and applications to mammalian genome mapping. *Electrophoresis J.* **10,** 296–301.
2. Gardiner, K. (1991) Pulsed field gel electrophoresis. *J. Anal. Chem.* **63,** 658–665.

CHAPTER 6

Rotating Field Gel Electrophoresis (ROFE)

Andreas Ziegler and Armin Volz

1. Introduction

Rotating field gel electrophoresis (ROFE) *(1–4)* is an alternative to orthogonal field alternating (OFAGE) *(5)*, transverse alternating field (TAFE) *(6)*, field inversion (FIGE) *(7)*, or related pulsed-field gel electrophoretic procedures for the separation of very large nucleic acid molecules. The latter methods all involve the electronic switching between two electric fields oriented at an obtuse angle (or 180° in FIGE) toward each other. In ROFE, the anodes and cathodes are carried by a rotor that can be turned around the stationary gel in which the separation takes place. The electric field, generated between two main electrodes and stabilized with electronically regulated additional sets of electrodes, can be reoriented after predetermined intervals (Fig. 1). In principle, this leads to a separation of DNA molecules in the gel analogous to that obtainable with devices using purely electronic switching.

ROFE is carried out using the ROTAPHOR apparatus and offers a number of distinct advantages over other designs. The large gel size (20 × 20 cm), combined with relatively small dimensions of the apparatus (area 35 × 40 cm, height 24 cm), permits the comparison of up to 50 different samples in the same gel. Several variables can be freely chosen, and in many cases varied automatically during an individual

From: *Methods in Molecular Biology, Vol. 12: Pulsed-Field Gel Electrophoresis*
Edited by: M. Burmeister and L. Ulanovsky

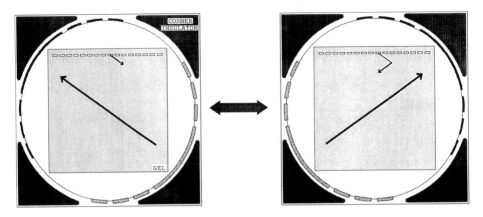

Fig. 1. Principle of rotating field gel electrophoresis. Cathode
(■■) and Anode (▭▭) are part of a movable rotor (—). The direction of the
electric field (long arrows) depends on the position of the rotor relative to
the gel in which the separation takes place. The movement of a DNA molecule out of a sample slot is schematically depicted (short arrows). The
field angle in this example is 110°.

run (Table 1): duration of electrophoresis (0–1000 h), pulse time (1–
9999 s), angle between field orientations (0–255°, 180° for FIGE mode),
field strength (0–8.5 V/cm), and buffer temperature. A single thermostat can be employed to cool several ROTAPHOR instruments, each
with its individually controlled buffer temperature. Since the gel is
horizontally positioned and stationary, the latter feature being in contrast to the "waltzer" technique in which the gel is turned *(8)*, it is
possible to employ very low agarose concentrations. Furthermore, the
time needed to reorient the rotor and, consequently, the electric field,
can be short in comparison with the waltzer technique, in which the
torque exerted on the gel limits the angular velocity. The reliability of
ROFE as performed in the ROTAPHOR even extends to the fact that
electricity dropouts do not affect the performance, since electrophoresis will be continued correctly after the predetermined buffer temperature has been reached again.

The versatility of rotating field gel electrophoresis is shown in Figs.
2 and 3, which depict typical separations of various DNA molecules.
Depending on the electrophoresis conditions, ROFE can separate short
(0.5 kbp) or very long (10,000 kbp) nucleic acid molecules with excellent resolution. It is also possible to separate molecules in two dimensions, simply by turning the gel tablet by 90° after the first run and

Table 1
Conditions for Separating DNA Molecules Falling into a Particular Size Range[a]

DNA size, kbps	Duration, h	Interval, s	Angle, °	Voltage, V	Temp., °C	Agarose, % w/v	Buffer	Resolution
1–100	18	4→2 lin	120→110 lin	110→90 lin	12	0.8	Loening	very high
0.2–350	18	40→3 lin	120→130 lin	200→140 lin	12	1.3	TBE	high
0.6–440	11	30→3 lin	120→95 log	200 const	14	1.2	TBE	average
3–1000	37	5→90 lin	110→125 log	120→180 log	13	1.0	TBE	high
100–2000	20	85→10 log	105→95 log	240 const	15	0.8	TBE	average
100–2000	36	100→20 lin	100→125 log	200→180 lin	13	1.0	TBE	high
100–2000	74	10→125 lin	100→125 log	120→180 log	11	1.0	TBE	very high
500–6000	130	5000→500 log	97→110 lin	46→49 lin	12	1.0	TBE	high
	80	500→50 lin	110→115 log	48→59 log	12	1.0		
1000–6000	48	4000→2000 log	95→105 lin	48 const	13	0.7	TBE	low
1000–6000	96	4500→800 log	95→110 lin	48 const	13	0.7	TBE	average
1000–9000	200	9000→500 lin	110→98 log	49→46 log	10	0.8	TBE	very high

[a]The term "resolution" refers to the distances by which individual DNA molecules (e.g., two yeast chromosomes) are separated.

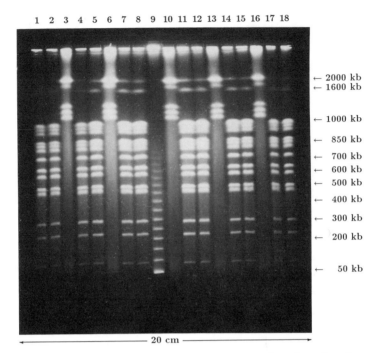

Fig. 2. Separation by ROFE of various DNA molecules: *Saccharomyces cerevisiae* YPH 149 (lanes 1,2,4,5,7,8,11,12,14,15,17,18), *Schwanniomyces occidentalis* CBS819 (3,6,10,13,16), λ concatemers *(9)*. Electrophoresis parameters: duration, 45 h; interval, 150–20 s lin.; field angle, 130°–115° log.; field strength, 6.71–5.67 V/cm log.; temperature, 14°C; 1% agarose, 0.035*M* TBE. These conditions are suitable to separate molecules in the 10–1500 kbp range.

starting a new electrophoresis experiment, possibly under totally different separating conditions. This feature, for example, can be employed to separate closed circular from linear DNA molecules (*see* Note 15).

2. Materials

1. TBE buffer stock: 900 m*M* Tris, 900 m*M* boric acid, 20 m*M* EDTA. Dissolve 108 g Tris and 55 g boric acid in 900 mL distilled water, add 40 mL 0.5*M* EDTA, pH 9, and adjust to 1 L. It is not necessary to adjust pH. Dilute this stock solution 1:40 for use in ROFE. Store at 4°C and do not use solution longer than 10 d.
2. Loening buffer stock: 360 m*M* Tris, 300 m*M* NaH₂PO₄, 10 m*M* EDTA. Dissolve 43.6 g Tris, 41.4 g NaH₂PO₄ • 1 H₂O, and 3.72 g EDTA in

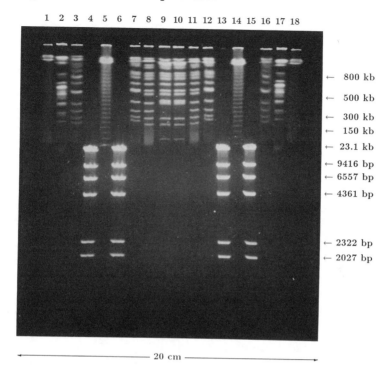

Fig. 3. Separation by ROFE of various DNA molecules: *Schwanniomyces occidentalis* CBS819 (lanes 1,18), *Leishmania mexicana* (2,17), *Saccharomyces cerevisiae* Way 5-4A (3,16), *S. cerevisiae* YNN 281 (7,12), *S. cerevisiae* YNN 295 (8,11), *S. cerevisiae* YPH 149 (9,10), *Hind III*-cut λ (4,6,13,15), λ concatemers (5,14). Electrophoresis parameters: duration, 22 h; interval, 100–10 s log.; field angle, 120–130° log.; field strength, 5.0–5.7 V/cm log.; temperature, 12°C; 1% agarose, 0.025*M* TBE. Molecules larger than about 1000 kbp are not resolved under these conditions, which are best suited for molecules 1–800 kbp long.

900 mL distilled water and adjust to 1 L. It is not necessary to adjust pH. Dilute this stock solution 1:20 for use in ROFE (*see* Note 1).

3. Agarose, high-grade, for gel electrophoresis (e.g., Seakem GTG or Fastlane).

4. 5X sample buffer: 15% w/v Ficoll MW 400,000, 0.5% xylenexyanol, 0.5% bromphenol blue (*see* Note 3).

5. Agarose inserts with various embedded size standard DNA molecules and sample DNA (*see* Chapters 7 and 8).

6. Staining Solution: Ethidium bromide, 10 mg/mL in distilled water. Dilute 1:10,000 before use. *(Handle with extreme care and wear gloves!* **Ethidium bromide is highly carcinogenic.**)

7. ROFE electrophoresis chamber "ROTAPHOR" with control unit (Geiger, Tübingen, Germany).
8. Power supply (0–250 V, 300 mA), best with remote control facility (e.g., type P24 from Biometra, Göttingen, Germany).
9. Laboratory cooler or thermostat.
10. UV illuminator with a wavelength of 305 nm.
11. Microwave oven or heating plate.
12. Magnetic stirrer.

3. Methods

3.1. Separation of DNA Molecules

1. Dissolve agarose of desired concentration in 350 mL buffer (20 × 20 cm gel). *See* Table 1 and Note 1 for buffer type and concentration suitable for a particular separation problem. Heat in a covered beaker to 100°C, then place the beaker in a 55°C water bath, or stir with a magnetic stirrer for about 10 min at room temperature.
2. Mount the gel casting frame on the gel support plate and seal the gap in-between with some agarose solution using a Pasteur pipet (the plastic sealing sometimes leaks with hot agarose).
3. Place the gel support plate on a horizontal surface (check with spirit level). Stir the agarose solution to be sure it is homogeneous, pour the gel (keep 5 mL at 55°C) and mount the comb (up to 50 slots can be formed). Remove air bubbles from the surface and from the holes that form the "feet" of the gel with a pipet. Allow the gel to set for 30 min (*see* Note 2).
4. While the gel is setting, pour 2.3 L of the desired buffer (Table 1) into the electrophoresis chamber and start the buffer pump and the thermostatic circulator. Select the running temperature (*see* Table 1 and Note 13) at the control unit. Make sure that the instrument is positioned horizontally (*see* Note 6).
5. Carefully remove the comb, separate the gel with a scalpel from the casting frame, and lift the latter, without detaching the gel from the support plate.
6. Cut the agarose inserts with your samples and the size standard to a suitable size (*see* Note 8) and insert them (*see* Note 9) with a spatula into the slots (for separation of DNA in solution, *see* Notes 3 and 4 and skip step 6). Make sure that the inserts attach to the front wall (direction of electrophoresis) of the slot and finally fill the slots with the remaining hot agarose from step 3.

7. At the edges of the gel and at the slots, surface tension tends to create unevenness of the gel surface. Remove this spare agarose with a scalpel to allow a well-balanced buffer flow (*see* Note 5).
8. Mount the corner insulators (press them down until they attach firmly to the gel support plate) and place the plate into the electrophoresis chamber (*see* Note 7). Lower the plate carefully until it rests firmly on the inner edges of the tank. Make sure that the slots are at the correct side of the apparatus, i.e., near the electrical connectors of the chamber (but *see* Note 15!).
9. If DNA in solution is used, wait 10 min to equilibrate the temperature of buffer and gel. Then stop the buffer pump and pipet the samples carefully into the slots (*see* Notes 3, 4, and 9). With samples in agarose inserts, skip point 9.
10. Close the lid and enter the desired parameters (*see* Table 1 and Notes 10–14) into the controller and start electrophoresis (*see* Note 4).

3.2. Gel Handling After Electrophoresis

1. When electrophoresis is finished, open lid, remove corner insulators, and lift the plate with the gel out of the chamber. Cut off the "feet" of the gel with a scalpel and slide gel carefully into a staining tray. Add 1 L staining solution and incubate 30 min, while slowly agitating the tray. Destain another 30 min in 1 L distilled water.
2. Transfer gel to the transilluminator and take a photograph for documentation.

4. Notes

1. Loening buffer is not stable with high voltages and especially over prolonged electrophoresis durations, but DNA molecules of 1–50 kbp are really well resolved.
2. It is possible to store gels made with TBE buffer at 4°C for about 1 wk. If more than one gel support plate is available, the gel supply for the next week may be produced in advance.
3. DNA smaller than 150 kbp can be applied in solution. Add 0.2-vol sample buffer and mix very carefully to avoid shearing. Sample buffer made with Ficoll results in sharper DNA bands than sample buffer with glycerol.
4. Apply DNA samples in solution to the slots only with the buffer pump switched off, otherwise the samples may be washed out by the buffer circulation. When xylenexyanol has entered the gel following the beginning of electrophoresis, switch on buffer pump again. During

electrophoresis without buffer circulation, buffer temperature at the gel will increase. The temperature detector is located near the cooling coil and comes into contact with this warmer buffer only after restarting the buffer pump.

5. If the gel is not properly covered with buffer 10–15 mm high, the field will not be homogeneous, and the outer lanes may then be curved toward the outside.

6. The electrophoresis chamber must be placed horizontally, because a field gradient will develop otherwise. This may also result in distortion of the outer lanes.

7. Do not forget to mount the corner insulators. If the corners of the chamber are filled with buffer, the shape of the electric field becomes angle-dependent. The outer lanes are then bent toward the middle of the gel.

8. Apply about 5–10 μg digested genomic DNA to one slot. The more DNA you apply, the slower it will move. If you apply more than the amount suggested, you may obtain overloading effects which will make a comparison of different lanes impossible.

9. If you want to load DNA of very different sizes, or a very different amount of DNA in neighboring lanes, it is advisable to keep one or two lanes free in-between. In PFGE, the electric field vector is not parallel to the lanes but intersects them. The charge of DNA appears to cause a microinhomogeneity of the electric field that influences the velocity of DNA in the neighboring lanes.

10. DNA molecules larger than 2000 kbp will only migrate as a defined band when a low field strength is applied. A 6 Mbp molecule, e.g., reveals a nice band at 1.7 V/cm, only a faint band at 2 V/cm, and no band at all at 3 V/cm after an electrophoresis for 5 d. With Fastlane agarose (Seakem), slightly higher voltages can be applied.

11. Field angles of 95–160° (and 180° in FIGE mode) can be used to separate DNAs larger than 100 kbp. With angles more acute than 95°, there is no resolution; at 95° DNA molecules move fast but do not produce sharp bands; at 160° bands are very sharp but migration is slow, and with angles more obtuse than 160°, DNA molecules will hardly move at all. A good compromise is to employ a fixed angle of about 100° or an alteration of the angle from 100 to 130° during the run.

12. The interval between reorientations of the electric field is of fundamental importance to achieve separations. Refer to Table 1 for suggestions.

13. At higher temperatures all bands move faster than at lower temperatures, but the resolution (distance between individual bands) is unchanged.

14. It is advisable to change the running buffer every 5 d when a very long electrophoresis experiment is carried out. Care must be taken that the gel is not detached from the gel support plate when it is lifted out of the chamber.

15. To separate closed circular from linear DNA, use a parameter list of Table 1 but change angles according to the following formula:

$$\text{new angle} = 360° - \text{old angle}$$

The new angle may maximally obtain a value of 255°. Choose the option "Continue electrophoresis while the rotor is moving" at the controller and run electrophoresis with the slots at the **opposite** side of the electrical connectors. After this first dimension, turn the gel support plate by 90° such that the lane with the separated DNA molecules is now near the electrical connectors and append an electrophoresis with the following parameters:

Duration, 2 h
Interval, 500 s
Angle, 1°
Voltage, 200 V
Temperature, 12°C.

Closed circular DNA moves down its lane at the first dimension but stops moving at the second dimension, whereas linear DNA leaves the original lane at the second dimension.

References

1. Ziegler, A., Geiger, K. H., Ragoussis, J., and Szalay, G. (1987) A new electrophoresis apparatus for separating very large DNA molecules. *J. Clin. Chem. Clin. Biochem.* **25,** 578,579.

2. Ragoussis, J., Bloemer, K., Pohla, H., Messer, G., Weiss, E. H., and Ziegler, A. (1989) A physical map including a new class I gene (cda12) of the human major histocompatibility complex (A2/B13 haplotype) derived from a monosomy 6 mutant cell line. *Genomics* **4,** 301–308.

3. Weichhold, G. M., Klobeck, H. G., Ohnheiser, R., Combriato, G., and Zachau, H. G. (1990) Megabase inversions in the human genome as physiological events. *Nature* **347,** 90–92.

4. US Patent number 4,995,957. Device and method for the electrophoretic separation of macromolecules.

5. Schwartz, D. C. and Cantor, C. R. (1984) Separation of yeast chromosome-sized DNAs by pulsed field gradient gel electrophoresis. *Cell* **37,** 67–75.

6. Gardiner, K., Laas, W., and Patterson, D. (1986) Fractionation of large mammalian DNA restriction fragments using vertical pulsed-field gradient gel electrophoresis. *Somat. Cell. Mol. Genet.* **12,** 185–195.

7. Carle, G. F., Frank, M., and Olsen, M. V. (1986) Electrophoretic separations of large DNA molecules by periodic inversion of the electric field. *Science* **232,** 65–68.

8. Southern, E. M., Anand, R., Brown, W. R., and Fletcher, D. S. (1987) A model for the separation of large DNA molecules by crossed field gel electrophoresis. *Nucleic Acids Res.* **15,** 5925–5943.

CHAPTER 7

Preparation, Manipulation, and Pulse Strategy for One-Dimensional Pulsed-Field Gel Electrophoresis (ODPFGE)

Jaan Noolandi and Chantal Turmel

1. Introduction

The preparation and manipulation of very large DNA molecules for pulsed-field gel electrophoresis (PFGE) requires more care than normally used for smaller molecules. The general protocol used is the preparation of DNA directly in solid agarose blocks (plugs) or beads. Intact cells are encapsidated in agarose (plugs or beads) and are treated with different combinations of enzymes and detergent to remove cell walls, membranes, RNA, proteins, and other materials in order to obtain naked DNA. This is possible because detergent and enzymes can diffuse by Brownian motion through the agarose pores of the plug, whereas the large pieces of DNA cannot and remain sequestered inside the plug. Here, we give detailed protocols of preparation for yeast and mammalian DNA. In addition, we describe the power supplies, switchers, gels, and gel trays required for one-dimensional pulsed-field gel electrophoresis (ODPFGE), as well as the pulse strategies used for the separations.

We describe here three distinct modes of running ODPFGE. In zero-integrated-field electrophoresis (ZIFE), which is used for the 10–2000-kbp range, the product of field strength and pulse time is approximately the same for backward and forward pulses, although

From: *Methods in Molecular Biology, Vol. 12: Pulsed-Field Gel Electrophoresis*
Edited by: M. Burmeister and L. Ulanovsky
Copyright © 1992 The Humana Press Inc., Totowa, NJ

the actual field strengths and pulse times are different. This regime avoids band inversion effects in a preselected size window, but at the high end of the size range results in a fairly slow migration of the molecules. A similar pulse shape, but with a positive bias in the forward direction, is used in the size range above 2000 kbp. Generally, for megabase-sized molecules, extremely short, high voltage spikes in the reverse direction are inserted into the long pulses to avoid the trapping commonly observed for large DNA molecules at high field strengths. Computer algorithms are available that integrate all three approaches.

2. Materials

2.1. Materials for Yeast
Chromosomes Preparation

1. Bacto peptone.
2. Yeast extract.
3. Dextrose.
4. $5M$ NaOH.
5. $0.5M$ EDTA, pH 8.0.
6. Low melting-point (LMP) agarose (BRL, Gaithersburg, MD): Make 1% in appropriate buffer (distilled water, CPE, or PBS), boil or autoclave to melt, then keep at 42°C until used.
7. Zymolyase 100 T (ICN Biomedical).
8. Silicone tubing (id 3/32 inch Cole Parmer 6411-63).
9. $1M$ Tris-HCl, pH 7.5.
10. β-Mercaptoethanol.
11. Sarcosyl (N-lauroylsarcosine sodium salt from Sigma).
12. Proteinase K.
13. Dithiothreithol (DTT).
14. Sorbitol.
15. CPE solution: $0.04M$ citric acid, $0.12M$ sodium phosphate, $0.02M$ EDTA.
16. CPES solution: CPE, $1.2M$ sorbitol, 5 mM DTT.
17. Novozym SP234 (Novo Biolab, Danbury, CT).
18. YEPD: 20 g of Bacto peptone, 10 g of yeast extract, 20 g of dextrose, dissolved in 1 L of water, adjusted to pH 7.4 with $5M$ NaOH and autoclaved.
19. PBS (Ca–Mg-free).
20. TE: 10 mM Tris-HCl, pH 7.6, 1 mM EDTA, pH 8.0.

21. PMSF (phenylmethyl sulfonate fluoride): Make 40 mg/mL stock in isopropanol, fresh before use.
22. *C. shehatae* yeast chromosomes [from American Type Culture Collection (ATCC 34887)].
23. Yeast cell lysis (YCL) buffer: $0.5M$ EDTA, pH 8.0, $0.01M$ Tris-HCl, pH 7.5, 7.5% β-mercaptoethanol.
24. SPE buffer: 1% Sarcosyl, 2 mg/mL proteinase K, $0.5M$ EDTA, pH 8.0.
25. Pre-lysis buffer: $0.01M$ Tris-HCl, pH 7, $0.05M$ EDTA, $0.01M$ dithiothreithol (DTT).

2.2. Buffers, Gels, Gel Trays, Power Supplies for ODPFGE

1. 1X TBE: 89 mM Tris borate, 89 mM boric acid, 2 mM EDTA, pH 8.3.
2. Agarose (type NA Pharmacia, Uppsala, Sweden or, Ultra Pure DNA agarose, Chromosomal Grade, Bio-Rad laboratories, or Fast Lane agarose from FMC Bio Products or Agarose Electrophoresis Grade from ICN) (*see* Note 1).
3. 10 mg/mL of Ethidium bromide dissolved in water.
4. BRL gel box model H-l (Bethesda Research Laboratory), with 41 cm between platinum electrodes. Other gel boxes with the same separation between the electrodes and the same buffer capacity (~2.5 L) could also be used.
5. Combs to produce gel slots 2-mm long × 6.4-mm wide.
6. Two direct current 250 V power supplies with an intervalometer (Sound Scientific, Seattle, WA) can be used to generate asymmetric voltage pulses in the "forward" and "reverse" directions by switching between the power supplies. High-frequency secondary pulses can be inserted into the primary voltage pulses using additional programmable timer controllers. Alternatively, a power supply that operates two gel trays independently, generating both primary and secondary pulses, and including a library of ROM cards for preprogrammed windows of size resolution, can be used (Q-Life, Kingston, Ontario, Canada; Xerox Research Center, Mississauga, Ontario, Canada).

3. Methods
3.1. Yeast Cells Culture

1. Inoculate 100 mL of YEPD sterile media with 0.1 mL of liquid culture stock or with one colony.
2. Incubate at 30°C for 24 to 48 h with slow agitation (*see* Note 2).

3.2. S. cerevisiae and C. shehatae *Yeast* *Chromosomes Preparation*

For 100 mL of culture suspension in stationary phase.

1. Pour cell culture suspension into 50-mL plastic tubes.
2. Pellet cells by centrifugation at $300g$ for 5 min.
3. Discard supernate and fill tubes with $0.5M$ EDTA. Suspend the pellet gently.
4. Repeat Steps 2 and 3 two times.
5. After last wash, suspend the cells in 6–10 mL of $0.5M$ EDTA.
6. Add equal volume of sterile 1.0% LMP agarose in distilled water at 42°C.
7. Add Zymolyase 100 T to a final concentration of 1 mg/mL and mix the suspension.
8. Pipet into autoclaved silicone tubing. Clamp both ends of tubing.
9. Place tubing at 4°C for 5–10 min to allow agarose to set.
10. Unclamp ends and allow agarose to slide out onto a sheet of Saran Wrap®. Gentle pressure must be applied to one end of the tubing.
11. Line up agarose tube with a ruler and cut into 0.5-cm long plugs.
12. Place plugs into a 50-mL conical tube containing 5 vol of YCL buffer.
13. Incubate 16 h at 37°C.
14. Pour off solution and fill tube with $0.5M$ EDTA. Wash at room temperature with gentle agitation for 30 min. Repeat three times.
15. After the third wash, add 5 vol of SPE buffer.
16. Incubate at 50°C for 48 h.
17. Wash plugs three times 30 min in $0.5M$ EDTA. Store for extended period at 4°C. Samples are stable for at least 6 mo.

3.3. S. pombe *Yeast Chromosomes Preparation*

For 100 mL of culture suspension in stationary phase.

1. Pour cell culture suspension into 50-mL plastic tubes.
2. Pellet cells by centrifugation at $300g$ for 5 min.
3. Discard supernate and fill tubes with $0.5M$ EDTA. Suspend the pellet gently.
4. Repeat Steps 2 and 3 two times.
5. After last wash, suspend cells in 20 mL of pre-lysis buffer.
6. Incubate for 15 min at 30°C with gentle agitation.
7. Pellet the cells and suspend in 20 mL of CPE solution.
8. Pellet and suspend the cells in 10 mL of CPES solution.
9. Keep the cells at 37°C for 10 min.
10. Dissolve 40 mg of Novozym in 10 mL of 1% LMP agarose (dissolved in CPE and maintained at 42°C).

11. Add 10 mL of Novozym suspension to 6 mL of cell suspension.
12. Pipet into autoclaved silicone tubing. Clamp both ends of tubing.
13. Place tubing at 4°C for 10 min.
14. Unclamp ends and allow agarose to slide out onto a sheet of Saran Wrap®. Pressure must be applied to one end of the tubing.
15. Line up agarose tubes with a ruler and cut into 0.5-cm long plugs.
16. Plugs are then placed in a 50-mL conical tube containing 5 vol of CPE. Incubate for 2 h at 30°C.
17. Rinse plugs three times 30 min with gentle agitation in 0.5M EDTA.
18. Incubate for 48 h at 50°C in 5 vol of SPE buffer.
19. Rinse plugs three times 30 min in 0.5M EDTA.
20. Store at 4°C. Samples are stable for at least 6 mo.

3.4. Preparation and Restriction Enzyme Digestion of DNA in Agarose Plugs

3.4.1. DNA Preparation (in Agarose Plugs)

This protocol has been used for lymphoblasts and fibroblast (subconfluent cultures) and for mouse spleen and lymph node suspensions.

1. Pellet cells and resuspend at appropriate concentration (*see* Note 3) in Ca–Mg-free PBS.
2. Add equal vol of sterile 1.0% LMP agarose (dissolved in Ca–Mg-free PBS) and maintained at 42°C after melting by autoclaving or boiling.
3. Pipet into autoclaved silicone tubing.
4. Place tubing at 4°C for 5–10 min to allow agarose to set.
5. Unclamp ends and allow agarose to slide out onto a sheet of Saran Wrap®. (Occasionally pressure must be applied to one end of tubing.)
6. Line up agarose tube with a ruler and cut into 1-cm long plugs.
7 Place 1-cm plugs in 50-mL conical tube containing 5 vol of cell lysis buffer. Incubate at 50°C for 24–48 h.
8. Pour off EDTA solution and fill tube with sterile TE (10 mM Tris-HCl, pH 7.5, 1 mM EDTA, pH 8.0). Allow plugs to settle and pour off TE. Repeat rinse three times.
9. Fill tube with TE + 0.04 mg/mL PMSF (phenylmethyl sulfonyl fluoride, (Boehringer Mannheim). After a 30-min incubation at 50°C, repeat the PMSF treatment and incubation.
10. Plugs can be digested at this point or stored for extended periods of time at 4°C in 0.5M EDTA.

3.4.2. Restriction Enzyme Digestion of DNA in Agarose Plugs

1. Rinse plugs that have been stored in EDTA twice in TE for 30 min at room temperature.
2. The 1-cm plugs are suspended in 250-μL restriction enzyme buffer (as recommended by the manufacturer) containing 5 mM spermidine (Sigma) and 0.5 mg/mL bovine serum albumin.
3. Incubate at the appropriate temperature for 2 h with 20 U of restriction enzyme.
4. Add 75 L of 0.5M EDTA to each reaction mixture to stop the reaction.

3.5. Separating DNA by ODPFGE

Standard ODPFGE is carried out using 0.8% agarose gel in 1X TBE.

1. Pour 0.8% agarose gel by dissolving 2.4 g of agarose in 300 mL of 1X TBE (*see* Note 4).
2. Cut plugs to a length of 0.5 cm and insert them into gel slots using spatula (*see* Note 5).
3. Immerse the gel in 2.5 L of 1X TBE in BRL gel box model H-1.
4. Run the gel with appropriate ODPFGE conditions (*see* Section 3.7 and Note 6).
5. Stain the DNA by washing the gel in a 1X TBE solution containing 1 μg/mL of ethidium bromide for 1 h followed by a destaining wash in 1X TBE for 30 min.
6. Photograph the gel under UV light.

3.6. Southern Blot Hybridization

We use the protocol described in detail in Lalande et al. *(1),* except that the DNA probes are labeled using the random primer method of Feinberg and Vogelstein *(2).* Many other methods will work as well (*see also* Chapter 8).

3.7. ODPFGE Pulse Strategy

Figure 1, panel A, shows the limiting mobility for a broad range of DNA sizes obtained with continuous-field electrophoresis. Panel B shows the almost miraculous effect of pulsed fields in resolving the same sample shown in panel A into bands corresponding to different sizes. Figure 2 displays the plots of distance of migration (drift) vs molecular weight for the separations shown in Fig. 1. These pulses were chosen to give an almost linear relationship on the semilog plot shown in Fig. 2, and are given in Table 1. A large number of computer

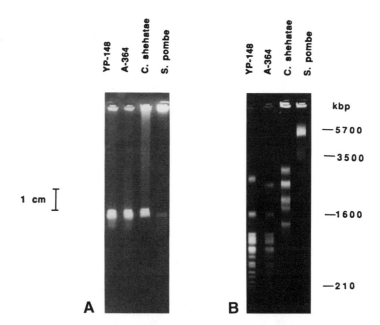

Fig. 1. Continuous-field electrophoresis **(A)** was carried out at E = 2.65 V/cm for 12 h. One-dimensional pulsed-field gel electrophoresis (ZIFE and ODPFGE) **(B)** was carried out by applying a complex pulse train (Table 1) for a total duration of 100 h. In both cases, the electrophoresis was performed in a BRL model H-1 gel box using 0.8% agarose gel (NA Pharmacia) in 1X TBE buffer.

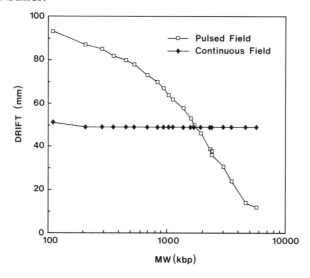

Fig. 2. Plots of distance of migration for continuous- and pulsed-field electrophoresis experiments shown in Fig. 1.

Table 1
Pulse Conditions for Separation
Shown in Fig. 1B

t_+ s	t_- s	E_+ V/cm	E_- V/cm	Duration h
400	160	1.93	0.96	0.16
500	200	1.88	0.94	0.39
650	260	1.85	0.93	9.10
800	320	1.83	0.91	0.62
950	380	1.83	0.91	13.67
1250	500	1.71	0.85	19.44
1600	640	1.58	0.79	6.84
2000	800	1.46	0.73	14.00
2300	920	1.39	0.70	2.68
2500	1000	1.37	0.68	1.94
3000	1200	1.27	0.63	2.33
60	84	3.41	1.07	4.32
70	98	3.41	1.07	8.45
80	112	3.41	1.07	2.29
115	161	3.41	1.07	0.46
130	182	3.41	1.07	0.69
160	224	3.41	1.07	11.63

algorithms to specify different pulse regimes for different separations can be created to suit the individual user. Some are available on commercial ROM cards and others can be obtained on request (*see* Section 2.2.). The pulse regimes are based on primary data discussed below (and shown later in Figs. 4, 5, 7, and 8).

Two main types of pulse shapes are used. The first, corresponding to the technique called ZIFE (zero integrated field electrophoresis) is shown in Fig. 3. The pulse shape has four parameters, E_+ (forward electric field), E_- (reverse electric field), t_+ (duration of forward pulse), t_- (duration of reverse pulse). Equivalently, one can use the four parameters $R_E = E_+/E_-$, $R_t = t_+/t_-$, E_+, t_+ since it is more convenient to program two ratios and two absolute values rather than four absolute values. As shown in the schematic below the pulse in Fig. 3, this shape affects the dynamics of different-sized fragments in different ways, as we explain below.

First we have already noted that pulsed electric fields must be used because continuous electric fields do not separate large DNA fragments (*see* Figs. 1 and 2). The obvious question is—Why run the frag-

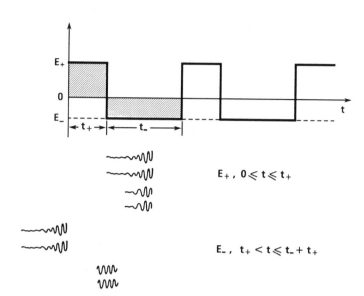

Fig. 3. Zero-integrated-field (ZIFE) pulse. This shape allows the proper size-migration distance relationship to be maintained, and gives rise to sharp drop in the mobility as a function of size for fixed pulse parameters. The reverse time, t_-, is proportional to the relaxation time of the largest molecule, which still has a net forward displacement during the cycle, as shown.

ments backward with a reverse field, since this merely lengthens the time of the experiment? The answer is that there is no choice, since a lowering of the electric field in the forward direction for part of the time (instead of using a reverse field) does not significantly increase the resolution and sometimes causes band inversion (the larger fragments respond more slowly to field changes than the smaller fragments and continue to move at the same speed for a longer time than the smaller fragments, thus overtaking the smaller fragments), and turning off the field for some period of time will result in relaxation times that are too long to be of any practical value. Hence, we are left with the only option of reorienting the molecule to a new conformation in the backward direction at a rate that is size-dependent. Incidentally, pulsed-field systems that reorient the molecule in a different direction (at a different angle than 180°) also pay the added penalty of a reorientation time, and the total separation time is just as long or longer (depending on the field intensity) as for a one-dimensional electric field system in order to achieve comparable resolution.

Returning to Fig. 3, we now calculate the displacements of a small fragment (L_{small}) and a large fragment (L_{big}) during a zero-integrated-field pulse, to show why this is a useful pulse strategy. The mobility (or velocity per unit electric field) of a DNA fragment is denoted by μ (i.e., $\mu = v/E$). This is a quantity that depends in a complicated way on the gel pore size, electric field, fragment length, and buffer composition. Assuming that the buffer composition and gel concentration are fixed, we focus only on the length and electric field dependence of the mobility, which fortunately we do not have to calculate explicitly, but merely indicate the functional dependence by μ (E, L) for the purpose of qualitatively analyzing the dynamic response of this quantity.

The displacement of a fragment during a time t is simply vt, and since $v = \mu E$ (according to the definition of mobility given above) the displacement is μEt. The small fragment, during the forward pulse t_+, therefore moves forward a distance μ (E_+, L_{small}) E_+t_+, where E_+, L_{small} in the brackets indicate that we are using the mobility of the small fragment evaluated at the field strength E_+. The large fragment, during the same forward pulse t_+, moves forward a distance μ (E_+, L_{big}) E_+t_+. During the reverse pulse the small fragment moves backward a displacement $-\mu E_- t_-$, but we can use the value of the mobility for the new reverse field E_- only if the time t_- is long enough that the small fragment has adjusted its mobility to the new electric field. If this is the case, the backward displacement is $-\mu$ (E_-, L_{small}) $E_- t_-$, and the net displacement over the whole cycle, including the forward and reverse displacements is μ (E_+, L_{small}) $E_+t_+ - \mu$ (E_-, L_{small}) $E_- t_-$.

For the large fragment the situation is different. The backward displacement is still $-\mu E_- t_-$, but if the time of the reverse pulse is not long enough to allow the large fragment to adjust to the new reverse field then we must write for the backward displacement $-\mu(E_+, L_{big})$ $E_- t_-$, because the large fragment simply moves backward with its original mobility, which is why E_+ instead of E_- is written inside the bracket. The net displacement of the large fragment is then μ (E_+, L_{big}) $E_+t_+ - \mu$ (E_+, L_{big}) $E_- t_-$. Notice that since the mobility in the forward and backward directions is the same for the large fragment, we can arrange that the forward and backward displacements cancel for the large fragment simply by setting $E_+t_+ = E_- t_-$. This is a good choice because it does not lead to a cancellation of the displacement of the small fragment, which now becomes $[\mu(E_+, L_{small}) - \mu(E_-, L_{small})]\, E_+ t_+$.

We obtain this result by substituting $E_- t_- = E_+ t_+$ in the earlier expression for the net displacement of the small fragment. Note that here the forward mobility depends on E_+, but the backward mobility depends on E_-, since the small fragment is able to adjust to changing field conditions more rapidly than the large fragment. Since μ (E_+, L_{small}) > μ (E_-, L_{small}), because the magnitude of E_+ is larger than the magnitude of E_-, the small fragment has moved forward during the cycle, whereas the large fragment has returned to its original position, leading to the desired result, which is a sharp drop in the mobility as a function of size, as shown in Figs. 4 and 5.

In practice we often use the condition $E_+ t_+ \geq E_- t_-$ to speed up the separation, although the average field over the entire cycle is still lower than in other techniques, and we still call this zero-integrated field. When $E_+ t_+ \geq E_- t_-$, a repeat of the above discussion of the dynamics of fragments of different lengths in response to field changes leads to the conclusion that band inversion always takes place at the high mol wt end, as shown in the figure. This does not cause any problems if it is recognized that one should not continue to use the same pulse parameters over too wide a size range. A set of pulse parameters can always be chosen that eliminates band inversion in a preselected size range. The idea then is to control the band inversion, so that it does not affect separations in the desired range, and not to waste effort in trying to eliminate this effect universally, which is not possible because of the physics of the process.

Above about 2 Mbp it is advantageous to switch from ZIFE to the forward-biased pulse shape, shown in Fig. 6, in order to speed up the time of separation *(4)*. Figure 7 shows that this strategy works because the entire mobility curve can be made to shift substantially by systematically increasing the pulse time so that the limiting mobility can be positioned within a preselected size range. Figure 8 shows the data for very long pulses, which can be used to separate DNA fragments up to 6 Mbp. As mentioned earlier, data such as shown in Figs. 4 and 5 and Figs. 7 and 8 can be used to develop a complex pulse code for a desired separation (by choosing a piece of one mobility curve for a certain time, followed by a piece of another curve for another length of time, and so on) by designing the appropriate computer algorithm; however, simpler requirements can be met by using the data given in the figures.

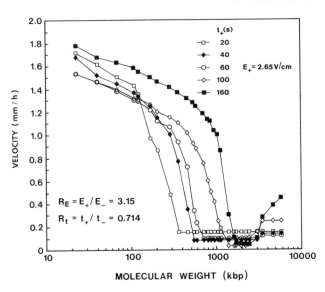

Fig. 4. Measurements of DNA velocities as a function of size for near-ZIFE pulse conditions *(see text)* with 0.8% agarose gel, 1X TBE buffer, and a model H-1 BRL gel box.

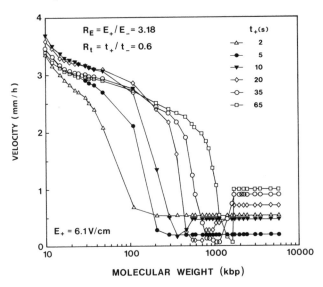

Fig. 5. High field measurements of DNA velocities as a function of size for near-ZIFE pulse conditions, using buffer recirculation and cooling to maintain a constant temperature of 20°C. Other experimental conditions are the same as for Fig. 4.

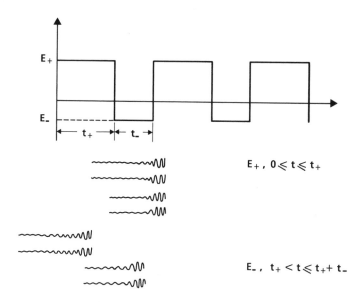

Fig. 6. Forward-biased pulse shape, with $t_- < t_+$, and $E_- < E_+$, which is effective in separating DNA molecules larger than approx 2 Mbp.

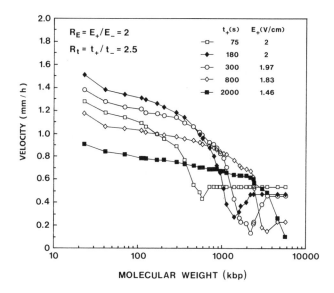

Fig. 7. Forward-biased measurements of DNA velocities as a function of size for various pulses, using experimental conditions noted in Fig. 4.

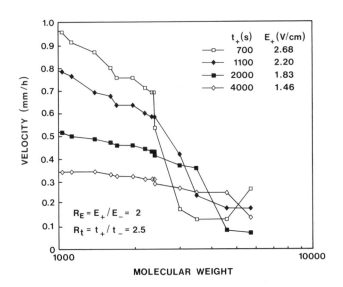

Fig. 8. Forward-biased measurements of DNA velocities as a function of size above 1000 kbp, for indicated pulses, with 0.10 field reversal after every second of both forward and backward long pulses. The other experimental conditions are indicated in Fig. 4.

3.8. Effect of Gel Pore Size and Static Electric Field on Migration of Multimegabase Molecules

We saw previously (Fig. 1A) that mixtures of DNA molecules larger than the average agarose pore size comigrate in a constant electric field, resulting in a limiting of mobility. The molecular size-independent gel pattern is related to the alignment of the molecules in the direction of the electric field. In order to establish the dependence of this size-independent migration on key parameters, we carried out a series of experiments in which the agarose concentration and the electric field intensity were varied. *S. cerevisiae* (200–2000 kbp), *C. shehatae* (1000–3000 kbp), *S. pombe* (3000–6000), and human chromosomes (>50,000 kbp) were chosen to represent broad classes of mol wt distributions.

Figure 9 shows the effect of agarose concentration on the migration of the three classes of molecules. The three panels (from left to right) correspond to concentrations of 0.2, 0.8, and 1.2% agarose, respectively. A static electric field (32 V/41 cm) was applied for 65 h. In the 0.2% agarose gel, the three different types of yeast chromo-

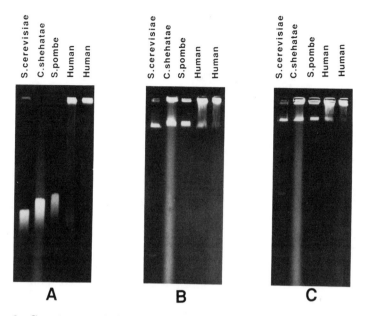

Fig. 9. Continuous-field (32 V/41 cm) electrophoresis was carried out on 0.2% (**A**), 0.8% (**B**), and 1.2% (**C**) agarose gels (NA Pharmacia) for 65 h. Same gel boxes and buffer were used as for Fig. 1.

somes enter the gel and are separated from the loading well, though not from each other. The separation is lost when the concentration of agarose is increased, as shown in the other two panels of Fig. 9. In all three panels, there is no evidence for the entrance of intact human chromosomes into the gel. In particular in a static electric field, even a low agarose concentration by itself does not facilitate the entrance of intact human chromosomes. When the field is increased (42, 62, and 82 V/41 cm) for 0.2% agarose concentration, the mobilities of the two groups of yeast molecules (*S. cerevisiae* and *S. pombe*) reach a common limiting value (plateau) at 62 V/41 cm and 82 V/41 cm, as shown in Fig. 10.

The above results are consistent with earlier observations on smaller molecules (2–48 kbp) for different agarose concentrations and field intensities *(5)*. We previously identified three electrophoresis regimes for DNA in agarose by varying the gel concentration and the static electric field strength: the Ogston regime (small DNA fragments in large pores of agarose), the reptation regime without DNA stretching (small pores of agarose and weak electric fields) and a reptation regime with molecular stretching (small pores of agarose, strong elec-

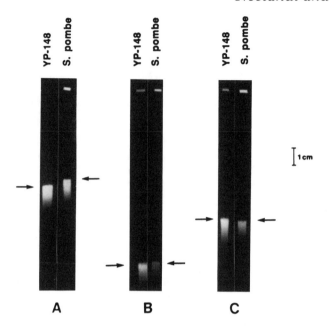

Fig. 10. Continuous-field electrophoresis of 42 V/41 cm (**A**), 62 V/41 cm
(**B**), and 82 V/41 cm (**C**) was applied for 50, 40, and 20 h, respectively. The
gel concentration was 0.2% agarose in 1X TBE buffer. The arrows indicate
the lead positions of the bands.

tric fields, and/or large DNA fragments). The references for the theory and the analysis of experimental data using some of these models is
given in Chapter 27.

The results shown in Fig. 9 with large molecules, low agarose concentrations, and weak fields are consistent with the reptation-without-
stretching regime, in which continuous field mobilities are proportional
to the inverse sizes of the molecules. At higher electric fields, the
results shown in Fig. 10 are consistent with the reptation-with-stretch-
ing regime, in which continuous field mobilities are independent of
molecular size.

3.9. Trapping of Megabase Molecules

It is now well known that the electrophoresis of multimegabase
DNA fragments is very sensitive to the intensity of the electric field.
The electric-field strength in PFGE follows the paradigm for medica-
tion—more does not necessarily give better results. When subjected
to high electric fields, large molecules stay trapped in the well or show

a nonspecific mobility represented by a diffuse smear. PFGE users have sometimes thought that the smeared pattern of separation or entrapment of molecules at the loading site was associated with a small amount of nuclease activity affecting the quality of the sample, or possibly to a nonlinear structural conformation resulting in topological entanglements in the well, or to shear breakage, or to an incomplete lysis of the crude cellular extracts so that the DNA was not completely separated from cellular proteins. Although these may all be contributing factors to poor separations, it is just as likely that the electric-field strength chosen was too high. As shown in Chapter 10, however, chicken metaphase but not interphase microchromosomes can enter the gel with the right pulse conditions, so that the sample preparation protocol is also very important.

Recently, we demonstrated that trapping and/or nonspecific mobilities are directly related to the field intensity *(6)*. We observed trapping for molecules as small as 1 Mbp in a continuous field of 10 V/cm. The trapping had some finite lifetime as evidenced by the extremely faint and diffuse smears that would spread in the gel for longer times of electrophoresis. We showed that the trapping effect could be reduced by using high-frequency intermittent fields or high-frequency field inversion. These observations suggested that the mechanisms involved in trapping act over a short-length scale of a few pore sizes, where high-frequency modulation of the electric field could shake the molecules out of awkward conformations in the presence of obstacles. We also showed that the high-frequency pulses could be used to modulate longer pulses without appreciably altering the relative separation of the bands *(6)*, allowing the use of higher fields with little band broadening.

In the following, we show examples of the trapping phenomenon for multimegabase DNA fragments. Figure 11 shows the effect of low-frequency pulses at different voltages on the entrance into the gel and the migration of multimegabase molecules. The five lanes in each of the four panels contain the same samples as for Fig. 5; *S. cerevisiae* YP 148, *C. shehatae*, *S. pombe*, and two lanes of intact human chromosomes. Panel A shows the result of applying 10,000 s forward pulses (32 V/41 cm) with 4000 s reverse pulses (–15 V/41 cm) for a duration of 65 h. Panel B shows the effect of applying 10,000 s (52 V/41 cm) forward pulses with 4000 s reverse pulses (–23 V/41 cm) for 44 h after and in addition to the pulsing conditions for panel A, and panel C shows the

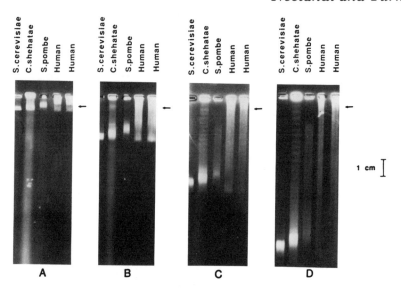

Fig. 11. The low-frequency pulses were applied (as described in the text) on 0.8% agarose gel in 1X TBE buffer using a model H-1 BRL gel box. The arrows indicate the maximal entrance of intact human chromosomes. In panel **B**, the two lanes containing human chromosomes are slightly overexposed, making the front of the chromosomal smear difficult to see.

results after an additional 40 h of 10,000 s forward pulses (72 V/41 cm) with 4000 s reverse pulses (–36 V/41 cm). Finally, panel D shows the effect of applying an additional 10,000 s (82 V/41 cm) forward pulses with 4000 s reverse pulses (–41 V/41 cm) for 40 h. The low-frequency pulses appear as a continuous field for molecules smaller than 6 Mbp since no separation is obtained for this size range. The low-frequency and low-field pulses have an effect on the entrance of intact human chromosomes into the gel. As noted earlier, when a constant field is used (*see* Fig. 9A, 0.8% agarose gel) we do not observe any entrance of intact human chromosomes.

The third lane in all four panels shows the effect of the electric field on the *S. pombe* band. At 32 V, the band is very sharp, whereas at 52 V, it starts to become diffuse. At 72 and 82 V, the band degenerates into a faint smear. This effect is also apparent for the *C. shehatae* lane, in which the sample had a higher DNA concentration than the others. As the field was increased in steps in the four panels, moving from left to right, the *C. shehatae* lane shows the same broadening/smearing as the *S. pombe* lane. However, an artificial band pattern also appears, which is related to the switching of the electric field (ten bands appear after ten field reversals in panel C). *This is possibly the result of a field-stimulated emission of a lump of DNA from the highly concentrated DNA in the plug each time the field is reversed.*

The field-induced band broadening does not take place for the *S. cerevisaie* YP 148 sample (first lane), which is consistent with earlier observations that a field of 2 V/cm does not result in the trapping of molecules in the size range of 100–1000 kbp in a 0.8% agarose gel. Intact human chromosomes have a limited penetration (0.5–0.8 cm from the well) into the gel at 32 V/41 cm (panel A) after 64 h. Longer electrophoresis times of 100 and 150 h (data not shown) do not increase this penetration. The limited penetration does not increase at higher field strengths (panels B, C, and D) and is related to the difficulty of leaving the plug, whereas the trapping of the yeast chromosomes at higher electric fields takes place after the entrance into the gel at the initially low field. The trapping in the gel has a finite lifetime resulting in the formation of faint and wide bands. From this set of experiments, we can deduce that the size of the molecules trapped in the gel is approximately inversely proportional to the magnitude of the electric field intensity.

The molecular trapping effect may be a result of local conformational obstructions since it can be eliminated or reduced by the application of high-frequency pulses. Figure 12 shows a series of continuous-field experiments in which a high-frequency 0.05 s field reversal (the same field strength in the backward direction) was added after each second of electrophoresis to reduce specifically the trapping of *S. pombe* (lane 2) at higher voltage. In panel A, a field of 32 V/41 cm was run for 65 h followed in addition by a 40 h run at 52 V/41 cm (panel B), continuing with an additional 72 V/41 cm run for 28 h (panel C), and adding a 82 V/41 cm run for 20 h (panel D). The bands are not as diffuse as in Fig. 11 (without the high-frequency component) but artificial bands are again produced each time the field is increased in the same way as with field reversal (Figs. 11, C and D). Three bands appear in Fig. 12D (human chromosomes, lane 3) after three steps of increasing the field. Three faint bands are also apparent in lane 2 (*S. pombe*). The artificial band formation was also observed when the field was decreased or turned off for a period of time during the run.

In summary we have been discussing two effects here. First is the formation of artificial bands by field-stimulated emission of DNA from the plug. Second is the elimination of trapping in the gel (this

Noolandi and Turmel

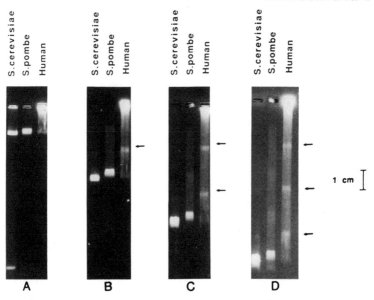

Fig. 12. The pulse parameters were applied as described in the text on
0.8% agarose gel with 1X TBE buffer in a model H-1 BRL gel box. The
arrows indicate the positions of the artificial bands.

includes the artificial bands) by high-frequency pulses. In Fig. 12 the
artificial bands shown in Fig. 11 are gone for the *C. shehatae* lanes, as a
result of the superimposed high-frequency pulsing. However, the high-
frequency pulsing does not increase the entrance of human intact
chromosomes at the very low field (panel A), nor does it eliminate the
formation of artificial bands. Their entrance was more efficient in the
same field at very low pulsing frequency (Fig. 11A). One possible reason for this is that the high-frequency pulses (voltage spikes) used to
detrap the *S. pombe* (3–6 Mbp) are not adequate to produce the same
effect on molecules in the size range of 50–200 Mbp. In fact, the optimum high-frequency pulses are specific to the size range of molecules
to be detrapped and the intensity of the field, as discussed below.

As the pulse durations needed for the separation of multimegabase-
size molecules are very long, the separation that we would expect for a
given pulse condition is often lost because of trapping during the constant field part of the long pulse cycle. One of the difficult problems in
the separation of large molecules is that the effects of the long pulses
necessary to move the entire molecule are not entirely independent
of the effects of the short pulses necessary to clear obstructions and/
or conformational entrapments. Inserting short pulses into the long

pulses is the only available approach to this problem at present, and the choice of the parameters for both types of pulses is critical for obtaining a good separation.

Figures 13 and 14 show a series of experiments relating the different aspects discussed earlier. In order to eliminate the possibility that band broadening could arise from difficulties encountered by long molecules entering the gel, we submitted the DNA to an electrophoresis prerun. In Fig. 13A, an electrophoresis preexperiment is shown, in which pulses (30 s forward, 82 V/41 cm; 10 s reverse, –41 V/41 cm) were used for 4 h to facilitate entry of DNA into the gel for a distance of 3–4 mm from the well, without any resolution of the chromosomes into bands. The electrophoresis prerun was followed by separating pulses where the forward (1600 s at 82 V/41 cm) and backward pulses (640 s at –41 V/41 cm) were kept constant. These pulses were chosen to obtain separations between 3 to 6 Mbp. The modulation of the pulses is the only variable parameter between Figs. 13B and C. Figure 13B shows the result of the standard pulse conditions (without modulation) applied for 65 h. Chromosomes XII (2.4–2.6 Mbp) of the two *S. cerevisiae* strains are resolved as well as the *C. shehatae* chromosomes of 2.38 and 3 Mbp. The 3-Mbp band is very diffuse. However, the three chromosomes of *S. pombe* are totally trapped and no clear separation is seen. Figure 13C shows the results obtained with the same low-frequency pulses, including high-frequency 0.1 s reverse pulses and 0.16 s zero voltage pulses after every second of the long pulses both in the forward and backward directions. These hybrid pulse conditions narrow the width of the 3.5-Mbp band of *S. pombe.* The 4.6 and 5.7-Mbp bands, however, are still diffuse.

Figures 14A and B show the results obtained with the same low-frequency pulses as above except that the reverse field intensity was –26 V/41 cm instead of –41 V/41 cm, and a more complex high-frequency pulse spectrum was used. After every second of the forward pulse a 0.1 s reverse pulse, followed by a 0.1 s zero-voltage pulse, then a 0.1 s reverse pulse, followed by another 0.1 s zero-voltage pulse was used, and after every second of the reverse pulse a 0.1 s zero-voltage pulse was inserted. These conditions give rise to a good separation of the three *S. pombe* chromosomes, both with (Fig. 14A) and without (Fig. 14B) an electrophoresis prerun. In fact, the electrophoresis prerun does not sharpen the bands. The trapping affecting the 3–5 Mbp bands under these conditions is not related to the difficulty of entering the gel. The band diffusion takes place during the separa-

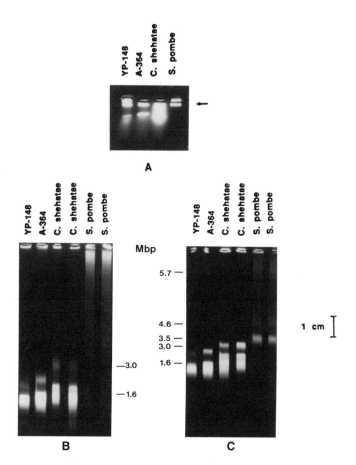

Fig. 13. Panel **A** shows the result of an electrophoresis prerun as described in the text. The arrows indicate the positions of the chromosomal bands. The smear observed in the *C. shehatae* track is caused by RNA (treatment of the gel with RNase enzyme eliminated the smear). Panels **B** and **C** were obtained as described in the text. In both cases, 0.8% agarose gels were run in a model H-1 BRL gel box in 1X TBE buffer.

tion in the gel. This result eliminates the possibility that trapping is only associated with DNA breakage by shearing or to an incomplete lysis of the crude cellular extracts.

In the next section, we show examples of *S. pombe* separation obtained without high-frequency modulation of the low-frequency pulses. With these conditions, smaller fields are used to obtain similar separations as obtained with higher low-frequency fields with high-frequency modulation.

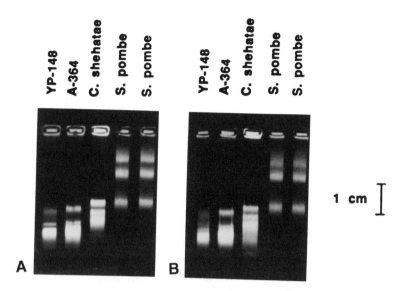

Fig. 14. Panels **A** and **B** show separations of the three *S. pombe* chromosomes obtained with pulsed electric fields described in the text for a total time duration of 65 h. Panel **A** was submitted to an electrophoresis prerun as described in Fig. 12A, whereas in panel **B** the separation pulses were applied without a prerun. A gel of 0.8% agarose in 1X TBE buffer was used in a model H-1 BRL gel box.

3.10. Examples Showing the Methodology of ODPFGE

As discussed earlier, the method known as field-inversion gel electrophoresis (FIGE, *see* Chapter 1) is often associated with large band inversion limiting this technique to relatively smaller fragments (50–1500 kbp) *(7)*. The use of ramped pulses can in some cases restore a monotonic size–distance relationship *(8)* for a given size range, but it is still arbitrary and can produce errors in the interpretation of the results. Another unhelpful phenomenon associated with FIGE is the migration of large molecules (>1000 kbp) in broad bands.

Recently, it was demonstrated that band inversion and band broadening also occur in crossed-field gel electrophoresis *(9)*. In this section, we show examples of multimegabases separation obtained by one-dimensional pulsed-field electrophoresis (ODPFGE). This overview of different separations will show that ODPFGE can be used effectively for multimegabase molecules, in some cases with better results than for crossed-field systems.

3.10.1. Separation of Candida shehatae
Yeast Chromosomes

Although many pulsed-field electrophoresis systems can now separate megabase (Mbp) DNA molecules, few markers can be used in the 1–3-Mbp range. *S. cerevisiae* yeasts possess only two chromosomes in the 1.5–3-Mbp range, whereas the smallest *S. pombe* chromosome is 3.5 Mbp in size *(10)*. *C. shehatae* has chromosomes in the 1–3-Mbp range, but they have not been previously identified. We used one-dimensional pulsed fields to study the genome of *C. shehatae*. In Fig. 15, panel B, we show a ZIFE separation that produces three bands in the 1–2-Mbp range, and two more between 2 and 3 Mbp. Figure 15, panel A, where we used experimental conditions that can separate chicken microchromosomes, indicates that there are also three chromosomes in the 2–3-Mbp size range. Analyzing the results of ten such separations, and interpolating the sizes of the six *C. shehatae* chromosomes, we obtained the following sizes (standard deviations correspond to the number of data points given in parenthesis): I: 3.00 ± 0.10 Mbp (3); II: 2.38 ± 0.04 Mbp (2); III: 2.33 ± 0.04 Mbp (2); IV: 1.96 ± 0.08 Mbp (6); V: 1.70 ± 0.09 Mbp (6); VI: 1.39 ± 0.05 Mbp (9). The total genome of *C. shehatae* is thus 12.76 ± 0.40 Mbp, divided into six chromosomes, including a narrow doublet at 2.33–2.38 Mbp.

3.10.2. Candida albicans *Yeast Chromosomes*

We estimated the sizes of the *C. albicans* chromosomes based on 20 different ZIFE pulsed-field separations. The results indicate that *C. albicans* has eight chromosomes, in agreement with Lasker et al. *(11)*. These authors also published estimated sizes for the four smallest chromosomes. In our experiments we used ZIFE to stretch the 1–3.5-Mbp region. The gel picture (Fig. 16) shows clearly that *C. albicans* has eight chromosomes. The estimated sizes are (in Mbp): 1.11 ± 0.05; 1.21 ± 0.06; 1.36 ± 0.07; 1.85 ± 0.07; 2.10 ± 0.10; 2.37 ± 0.03; 2.71 ± 0.05; 3.17 ± 0.12 for a total genome size of 15.88 ± .55 Mbp, where the errors are the standard deviations only.

3.10.3. Separation of S. Pombe *Yeast Chromosomes*

The sizes of the three *S. pombe* chromosomes have been determined by adding the sizes of *Not* I fragments *(10)*. These chromosomes are currently used as size markers in the 3–6-Mbp range. Their separation usually requires the use of small electric fields and long-time durations.

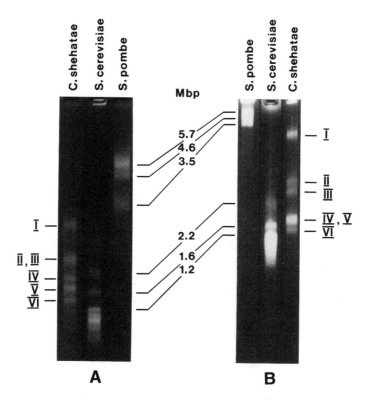

Fig. 15. Separation of *C. shehatae* chromosomes: **(A)** separation carried out for 187 h as described in Fig. 2, Chapter 10, the line indicated by
triangles, **(B)** with 22 h each of the following pulses: (i) 350 s at 2.65 V/cm,
then 490 s at –0.84 V/cm; (ii) 600 s at 2.0 V/cm, then 840 s at –0.65 V/cm;
(iii) 800 s at 2.0 V/cm, then 1120 s at –0.65 V/cm.

Recently, an effort has been made to shorten the time of separation for the *S. pombe* chromosomes. The CHEF mapper decreases the
switching angle to obtain a separation in 24 h *(12)*. Other systems use
the high-frequency modulation of low-frequency pulses to reach the
same level of performance. In both cases, the three chromosomes are
resolved over approx 0.5–1 cm, which does not make the separation
very useful for genomic mapping and intact chromosomal karyotyping, since a resolution of only 200–500 kbp/mm is obtained. At present,
it seems prudent to avoid short-duration runs unless only the separation of the three *S. pombe* chromosomes is needed.

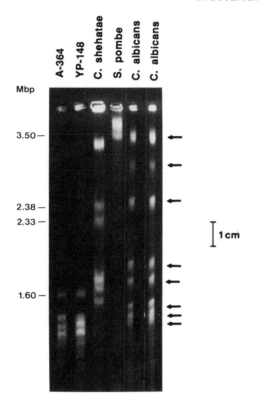

Fig. 16. Separation of the six *C. shehatae* and eight *C. albicans* chromosomes by ZIFE on a single gel. Three different pulses were used, each for 48 h: (i) 350 s, 82 V/31 cm forward and 490 s, 26 V/31 cm reverse; (ii) 600 s, 62 V/31 cm forward and 840 s, 20 V/31 cm reverse; (iii) 800 s, 62 V/31 cm forward and 1120 s, 20 V/31 cm reverse. High-frequency modulation, consisting of 0.ls of field inversion after every second, added to the low-frequency ZIFE pulses during the entire separation. A BRL gel box was used, along with 0.8% agarose gel and 1X TBE buffer.

Figures 17A, B, C, and D show four different separations of the three *S. pombe* chromosomes. Panels A and B were obtained after 36 and 45 h runs, respectively, using an ODPFGE pulse strategy with $R_E = 2.0$ and $R_t = 2.5$. Three pulse conditions with $t_+ = 2100$ s and $E_+ = 59$ V/41 cm, $t_+ = 2200$ s and $E_+ = 59$ V/41 cm, $t_+ = 2300$ s and $E_+ = 59$ V/41 cm were applied for 12 h each (panel A) and 15 h each (panel B) without modulation of the low-frequency pulses. In panel B, the three chromosomes are separated over 2 cm, which gives a separation of approx 100 kbp/mm. Panel C shows the result obtained with a single ODPFGE

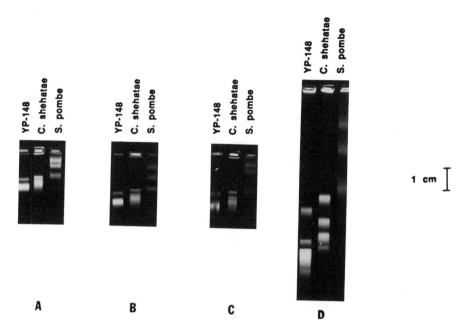

Fig. 17. Separation of the three *S. pombe* chromosomes with different pulse conditions as described in the text. The electrophoresis was carried out in a model H-1 BRL gel box in 1X TBE buffer on 0.8% agarose gel. Panels **A, B,** and **D** were run at room temperature, whereas panel **C** was run while recirculating and cooling the buffer to maintain a temperature of 20°C. The pulse conditions for the separation shown in panel **D** are given in Table 2.

condition (800 s at 140 V/41 cm followed by 320 s at –70 V/41 cm), where modulation of the pulse (0.025 s of reverse field followed by 0.3 s of zero field repeated four times at every second of the long pulse) was added for a total run duration of 40 h. In this case, the separation of the three chromosomes is also obtained on more than 2 cm but in 40 h, which is an improvement compared to the panel B. Panel D shows a separation in the 2–6-Mbp range. A complex pulse train (Table 2) was applied for a total duration of 130 h and a separation of 50 kbp/mm is obtained for the 1.4–5.7-Mbp size range.

3.10.4. Genomic Digest

For reasons mentioned earlier (band inversion and band broadening), one-dimensional pulsed-field electrophoresis is presently rarely used for analyzing complex mammalian genomes. After fragmenta-

Table 2
Pulse Conditions for Separation Shown in Fig. 17D

Pulse duration, s	Electrical field, E_+ V/cm	Total time of application[a], h
250	2	3.11
500	1.90	4.67
650	1.88	4.04
800	1.85	4.98
950	1.83	5.91
1100	1.76	6.84
1250	1.71	3.89
1400	1.66	4.56
1600	1.58	4.98
2000	1.46	6.22
2100	1.44	6.53
2200	1.41	6.84
2300	1.39	7.16
2400	1.39	7.47
2500	1.34	7.78
2600	1.32	8.09
2700	1.32	8.40
2800	1.29	8.71
2900	1.27	9.02
3000	1.17	9.33
3500	1.22	10.89

[a]Total time (h) of application of each pulse with constant $R_t = t_+ /t_- = 2.5$ and constant $R_E = E_+/E_- = 2.0$.

tion by rare-cutting restriction enzymes, these genomes migrate as a smear where band inversion is difficult to detect without a good understanding of the effects of pulsed electric fields. As we showed earlier, band inversion and band broadening can be controlled to obtain a monotonic size–distance distribution for multimegabase molecules, as well as for smaller molecules.

Figure 18A shows a gel where pulses were chosen to give a long-range distribution from 100 kbp to 6000 kbp (the pulse sequences are similar to those used for separating chicken microchromosomes, *see* Chapter 10). DNA from a human cell line was completely digested by *Not* I, *Nru* I, *Not/ Nru*, *Sfi* I, and *BssH* II. The frequency of cutting varies greatly among the rare cutters as shown by the DNA distribution in each of the tracks. Note that no compression zone is present in the

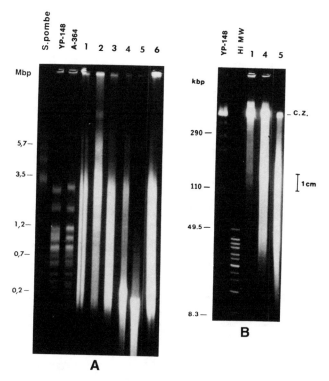

A

B

Fig. 18. Human genomic digest with *Not* I (1), *Nru* I (2), *Not* I/*Nru* I (3), Sfi I (4), *Not* I/*Sfi* I (5), and BssH II (6) was separated using a pulse train similar to Table 1, Chapter 10 for panel A, or by using ZIFE conditions (R_E = 3.2 and R_t = 0.714) of t_+ = 25 at E_t = 140 V/41 cm for 9 h; 3 s at 140 V for 29 h; 10 s at 140 V for 10 h; 15 s at 140 V for 10 h; 25 s at 140 V for 9 h; 40 s at 140V for 3 h; (panel **B**). In both cases, the electrophoresis was carried out using a BRL model H-1 gel box with 0.8% agarose gel in 1X TBE buffer. The compression zone (limiting mobility) is indicated by c.z.

100–6000-kbp range, allowing a reliable analysis of the complete or partial restriction digest on a single gel. When an enzyme shows a high frequency of cutting activity (*Sfi* I), a specific window of separation can be opened (using ZIFE conditions for 8–360 kbp in Fig. 18B) in order to decompress and spread the local DNA distribution. In this case the interpolated values are reliable, because the compressed zone caused by the locally high DNA concentration in Fig. 18A (lanes 4 and 5) has been stretched to over 12 cm in Fig. 18B. The ODPFGE method was successfully applied to the construction of the physical map of human chromosome band 13q14 *(13)*.

4. Notes

1. All examples of separations shown were obtained by using type NA Pharmacia agarose. The three other types of agarose, when used at the same concentration as that from NA Pharmacia, give the same results in terms of the window of resolution with a small difference (±5 to 10%) in the absolute mobility.
2. Yeast cells can be frozen at $-20°C$ in the presence of 15% glycerol for long-term storage.
3. The appropriate cell concentrations can be determined as follows: A 1-cm long plug (made using 3/32 in id tubing) has a volume of roughly 80 µL. For an initial concentration of 10^7 cells/mL in PBS, we have 5 $\times$ 10^6 cells/mL after dilution with agarose and 4 $\times$ 10^5 cells/1-cm plug. Half of this (0.5 cm) is loaded into a gel slot, i.e., 2 $\times$ 10^5 cell equivalents of DNA (~1 µg) would be contained in each slot of the agarose gel.
4. The use of lower agarose concentration or lower buffer concentration results in an increase of the absolute mobility of DNA molecules, as well as an increase in the size range of resolution. A dramatic band broadening also occurs. Using a higher agarose and buffer concentration will result in a slower mobility and a decrease of the size range of resolution. The bands obtained will be sharper, however.
5. The plugs can be sealed in the well by adding 0.5% agarose solution maintained at 42°C. This is to prevent the loss of the DNA still trapped in the well after the electrophoresis during the staining procedure.
6. Standard ODPFGE experiments are carried out at room temperature (=23–25°C) without cooling or recirculating the buffer. When fields higher than 4 V/cm are used, recirculation of the buffer is needed to maintain the temperature at 23°C.

References

1. Lalande, M., Kunkel, L. M., Flint, A., and Latt, S. A. (1984) Development and use of metaphase chromosome flow sorting methodology to obtain recombinant phage libraries enriched for parts of the human X chromosome. *Cytometry* **5**, 101–107.
2. Feinberg, A. P. and Vogelstein, B. (1983) A technique for radio-labelling DNA restriction endonuclease fragments to high specific activity. *Anal. Biochem.* **132**, 6–13.
3. Turmel, C., Brassard, E., Forsyth, R., Hood, K., Slater, G. W., and Noolandi, J. (1990) High resolution zero integrated field electrophoresis (ZIFE) of DNA, in *Current Communication in Molecular Biology. Electrophoresis of Large DNA Molecules* (Birren, B. and Lai, E., eds.), Cold Spring Harbor Laboratory, Cold Spring Harbor, NY, pp. 101–131.

4. Lalande, M., Noolandi, J., Turmel, C., Rousseau, J., and Slater, G. W. (1987) Pulsed-field electrophoresis: Application of a computer model to the separation of large DNA molecules. *Proc. Natl. Acad. Sci. USA* **84,** 8011–8015.

5. Slater, G. W., Rousseau, J., Noolandi, J., Turmel, C., and Lalande, J. (1988) Quantitative analysis of the three regimes of DNA electrophoresis in agarose gels. *Biopolymers* **27,** 509–524.

6. Turmel, C., Brassard, E., Slater, G. W., and Noolandi, J. (1989) Molecular detrapping and band narrowing with high frequency modulation of pulsed field electrophoresis. *Nucleic Acids Res.* **18,** 569–575.

7. Olson, M. V. (1989) Pulsed field gel electrophoresis, in *Genetic Engineering* (Setlow, J. K., ed.), Plenum, New York, pp. 183–227.

8. Heller, C. and Pohl, M. (1990) Field inversion electrophoresis with different pulse time ramps. *Nucleic Acids Res.* **18,** 6299–6304.

9. Birren, B. W., Lai, E., Clark, S. M. Hood, L., and Simon, M. I. (1988) Optimized conditions for pulsed field gel electrophoretic separations of DNA. *Nucleic Acids Res.* **16,** 7563–7582.

10. Fan, J. B., Chikashige, Y., Smith, C. L., Niwa, O., Yanagida, M., and Cantor, C. R. (1988) Construction of a Not I restriction map of the fission yeast *Schizosaccharomyces pombe* genome. *Nucleic Acids Res.* **17,** 2801–2818.

11. Lasker, B. A., Carle, G. F., Kobayashi, G. S., and Medoff, G. (1989) Comparison of the separation of *Candida albicans* chromosome-sized DNA by pulsed field gel electrophoresis techniques. *Nucleic Acids Res.* **17,** 3783–3793.

12. Chu, G. (1990) Pulsed field gel electrophoresis: Theory and practice. *Methods: A Companion to Methods Enzymol.* **1,** 129–142.

13. Higgins, M. J., Turmel, C., Noolandi, J., Neumann, P., and Lalande, M. (1990) Construction of the physical map for three loci in chromosome band 13q14: Comparison to the genetic map. *Proc. Natl. Acad. Sci. USA* **87,** 3415–3419.

PART II

AUXILIARY METHODS

CHAPTER 8

Preparation, Restriction, and Hybridization Analysis of Mammalian Genomic DNA for Pulsed-Field Gel Electrophoresis

Denise P. Barlow

1. Introduction

The aim of this chapter is to provide a very practical set of instructions that will enable workers with some experience in standard DNA preparation and electrophoresis techniques to prepare and analyze mammalian DNA by pulsed-field gel electrophoresis (PFGE; *see also* refs. *1* and *2* for additional PFGE protocols). Detailed methods are provided that describe: sources of mammalian DNA, preparation of DNA, restriction enzyme digests, standard PFGE running conditions, preparation of size markers, and preparation, hybridization, and analysis of blots. Notes about potential pitfalls accompany each section together with a trouble-shooting guide in Section 5.

2. Materials

All solutions should be sterilized before use and all chemicals should be of a grade suitable for molecular biology use.

From: *Methods in Molecular Biology, Vol. 12: Pulsed-Field Gel Electrophoresis*
Edited by: M. Burmeister and L. Ulanovsky
Copyright © 1992 The Humana Press Inc., Totowa, NJ

1. Agarose: Any standard molecular biology grade agarose can be used for pulsed-field gels. High quality low melting point (LMP) agarose is needed for embedding cells. Many companies now specialize in genome mapping products and they now sell agarose specifically for PFGE gels, and LMP agarose tested for compatibility with restriction enzymes.
2. Ammonium-Tris solution: 7.47 g NH_4Cl, 2.1 g Tris base/L.
3. Cell lysis buffer: 1% Sodium lauroyl sarcosine, $0.5M$ EDTA, pH 8.0, 2.0 mg/mL proteinase K. Allow EDTA to completely dissolve before addition of sarcosine. The solution is made up without proteinase K, filter sterilized, and stored at room temperature. Proteinase K is added just before use.
4. CPE: 120 mM Na_2HPO_4, 40 mM citric acid, 20 mM EDTA, pH 8.0.
5. CPES: CPE, 1.2M sorbitol, 5 mM DTT make up without DTT and add before use.
6. Denaturant buffer: 1.5M NaCl, 0.5M NaOH.
7. ETB buffer: 0.45M EDTA, pH 8.0, 10 mM Tris-HCl, pH 8.0, 7.5% β-mercaptoethanol.
8. Hemocytometer: Standard hemocytometer slide (e.g., improved Neubauer type) as used for counting cells for tissue culture work.
9. Homogenizer: 15-mL glass Dounce™ homogenizer plus a tight fitting "B" pestle.
10. LB (Luria Broth) medium: 10 g Bactotryptone, 5 g Bactoyeast extract, and 10 g NaCl/L at pH 7.5.
11. Mortar and pestle: Ceramic, 15 cm diameter, thick walled.
12. Novozym-234, available from Novo Biolabs, Denmark.
13. PFGE equipment and supplies: PFG boxes and plastic molds for embedding cells are available from many companies, e.g., Pharmacia-LKB Sweden, Clontech USA, BioRad USA.
14. Phosphate buffered saline: 8.0 g NaCl, 0.2 g KCl, 1.44 g $Na_2HPO_4 \cdot 7H_2O$ and 0.24 g KH_2PO_4/L at pH 7.2.
15. PMSF: phenylmethylsulfonylfluoride freshly prepared by dissolving 40 mg/mL PMSF in isopropanol at 37°C. PMSF is rapidly inactivated in aqueous solutions.
16. SCE: 0.1M Sodium citrate, 10 mM EDTA, pH 8.0.
17. Solution A (yeast prep 2): 1 vol of 50 mM EDTA and 1 vol of 2.5 mL SCE + 200 μL β-mercaptoethanol + 3.0 mg Zymolyase.
18. TE: 10 mM Tris-HCl, pH 8.0, 1 mM EDTA, pH 8.0.
19. TBE: Tris base (0.089M), Boric Acid (0.089M), EDTA (0.002M).
20. YPD: 1% yeast extract, 2% bactopeptone, 2% dextrose.
21. YPD plates: YPD + 2% agar.

3. Methods

3.1. Sources of Mammalian DNA

Human DNA can be prepared from tissue culture cells, white blood cells, sperm, or from solid tissues, e.g., tumor material. Mouse DNA can also be prepared from tissue culture cells or from organs, most commonly from spleen, thymus, testes, and bone marrow. To avoid transfer of communicable diseases from human samples, collect samples in a sterile manner, wear gloves to avoid direct skin contact with the sample and autoclave remnants before disposal. Parts of the following protocol were adapted from one described by Carle and Olson (2).

3.1.1. Tissue Culture Cells

1. Disperse tissue culture cells into single cells in a standard manner and count a diluted aliquot in a hemocytometer.
2. Pellet the cells by centrifugation at $200g$ for 10 min and resuspend at a concentration of $25-50 \times 10^6$/mL in phosphate buffered saline (PBS) at 4°C (*see* Note 1).

3.1.2. Human Blood

1. Place 20–50 mL whole blood into a sterile 50-mL graduated plastic tube containing 1 mL of sodium citrate (3.8%, w/v)/10 mL of blood. Human blood can be expected to contain $5-10 \times 10^6$ white blood cells/mL whole blood. Leave sample on ice for at least 1 h or red cell lysis is inhibited. The use of heparin as an anticoagulant is not advised as it can copurify with DNA.
2. Pellet the cells by centrifugation at $200g$ for 10 min at room temperature.
3. Decant off the supernatant (autoclave before disposal) and resuspend the cells in 30 mL of freshly prepared Ammonium-Tris solution/5 mL packed cell vol.
4. Mix by gentle inversion and place at 37°C until the red blood cells lyse. This will be seen as a clearing of the solution, which takes the appearance of dark red wine and can take up to 15 min.
5. Spin the cells as described earlier and decant off the supernatant. Resuspend the pellet in 20 mL of PBS at 4°C. Count a diluted aliquot in a hemocytometer using a vital stain or a microscope that is able to distinguish between red cell debris and white blood cells. Resuspend the cells at a concentration of $25-50 \times 10^6$ per mL in PBS at 4°C.

3.1.3. Sperm

1. Store sperm samples at 4°C for at least 1 h before processing to reduce motility, or alternatively use 10 mM EDTA to immobilize the sperm (A. Weith, personal communication).
2. Dilute the solution of sperm 10× by vol with PBS at 4°C and pellet by centrifugation at 500g for 10 min at 4°C, using a centrifuge that stops quickly and smoothly.
3. Quickly decant off the supernatant and resuspend the now immotile sperm in cold PBS. Count an aliquot using a hemocytometer and resuspend at 50–100 × 10^6/mL.

3.1.4. Solid Tissues

1. Fresh tissue is dissected into 2-mm cubes using sterile scalpels and broken into single cells by using approx 20 strokes from a tight fitting pestle and a glass homogenizer.
2. Count an aliquot using a hemocytometer, it is normally possible to distinguish single cells or cell clumps of 1–3 cells. If the cells appear to be clumped into larger aggregations, then the homogenization step should be repeated. Resuspend the cells at 25–50 × 10^6/mL in PBS.
3. Frozen samples must first be ground to a fine powder under liquid nitrogen. Precool a ceramic mortar and pestle on dry ice and grind the tissue to the consistency of fine coffee grounds. Scrape off the powder into PBS using a sterile metal spatula, immediately homogenize in a glass homogenizer and proceed as from Step 2, again it should be possible to distinguish single cells or clumps containing 1–3 cells. Samples can be processed in this way if they were initially frozen rapidly in liquid nitrogen and then maintained at –80°C. Tissues frozen otherwise may lyse when added to the PBS.

3.1.5. Mouse DNA

The procedure is the same for any organ of the mouse, however the spleen, thymus, testes, and bone marrow are more convenient because these are most easily dispersed into single cells.

1. The mouse is killed by cervical dislocation and the organs dissected and placed into a sterile Petri dish. Care should be taken to remove traces of the pancreas from the spleen as this is a rich source of nucleases. Approximately 100–150 million cells can be obtained from a normal adult spleen.
2. Dissect the organ into 2-mm cubes and homogenize in PBS using a tight fitting pestle and a glass homogenizer. Count a diluted aliqout of cells using a hemocytometer and resuspend the cells at a concen-

tration of 25–50 × 10^6/mL in PBS. Preparations from the spleen will contain some mature red blood cells but the numbers are insignificant compared to nucleated cells.

3.2. Preparation of Mouse and Human DNAs for Restriction Digests

DNA for PFGE must be of sufficient length so that it can be cut by enzymes into a size class of approx 0.05–3.0 Megabases (Mb). Clearly, the uncut DNA should be in a size class in excess of 20 Mb. Two facts, the susceptibility of long DNA molecules to mechanical shear forces, and the ubiquitous presence of nucleases that degrade DNA, act to reduce the size of uncut DNA. The action of mechanical shear can be reduced by preparing the DNA from cells embedded in agarose and the activities of nucleases can be considerably reduced by the use of inhibitors and by careful technique (*see* Note 2).

3.2.1. Embedding Cells in Agarose

1. This is done using plastic molds that have the same dimensions as the gel comb used for the pulsed-field gel apparatus. Commercial manufacturers supply molds with their PFGE equipment and because better results are obtained using agarose blocks with undamaged edges it is advisable to use these rather than homemade alternatives. The molds should be precooled to 4°C before use. An alternative method of embedding cells in which the agarose/cell mixture forms a suspension of agarose beads is described in Chapter 9.
2. Place 0.5 mL of the single cell suspension, normally at 25–50 × 10^6/ mL of diploid cells, in a 1.5-mL microfuge tube at room temperature. Add an equal vol of 1% LMP agarose dissolved in PBS and held at 50°C, mix the contents by gentle inversion but do not allow bubbles to form.
3. Immediately dispense 80 µL of the mixture (containing 1–2 × 10^6 cells) into the precooled plastic molds, avoid generating bubbles.
4. The agarose blocks of embedded cells are allowed to set for 20 min and then gently pushed out using a sterile bacterial 10-µL loop, or air from a 5-mL rubber pipet bulb into cell lysis buffer. Approximately 50 agarose blocks are incubated in 46 mL of lysis buffer in a 50-mL sterile plastic tube. Some 10 m*M* DTT is added to the cell lysis buffer for the digestion of sperm DNA *(3)*.
5. The agarose-embedded samples are incubated at 55°C for 48 h, with gentle agitation. The agarose blocks will be transparent at the end of this period, and can be seen more easily when viewed against a dark background.

6. Decant the agarose blocks into a sterile glass beaker and rinse briefly with four changes of 100 mL sterile TE. The blocks are then returned to a sterile plastic 50-mL tube and washed twice at 55°C in 50 mL of TE containing 40 µg/mL PMSF (phenylmethylsulfonylfluoride). This step is best done using a slow rocking platform inside an oven and serves to remove the cell lysis buffer and to inhibit proteinase K.

7. The DNA can be digested with restriction enzymes at this stage or transferred to 0.5*M* EDTA, pH 8.0, for long term storage at 4°C. Samples have been stored for up to 5 yr in this way with no sign of deterioration. Samples stored in EDTA need to be rinsed in two 50-mL vol of TE before enzyme digestion.

3.3. Restriction Enzyme Digests

PFGE electrophoresis is now able to separate fragments in the 0.05–6.0 Mb size range. Currently available restriction enzymes for PFGE are able to generate fragments in the 0.05–3.0 Mb range, sizes in excess of 3.0 Mb can normally be generated only through partial digests. Inasmuch as analysis of genomic DNA by pulsed-field gel electrophoresis depends very much on the enzymes used, it is worthwhile to try to understand how PFGE enzymes act to fragment DNA into large pieces. All the enzymes used for PFGE analysis produce large fragments because they cut at recognition sites that occur rarely in mammalian DNA or are cleaved rarely in vivo. Two classes of enzyme are available, those recognizing sites of eight or more nucleotides, and those recognizing sites that contain the dinucleotide CpG (*see* Table 1). The latter class is more useful at present because they generate larger fragments. PFGE enzymes recognizing sequences containing the dinucleotide CpG produce large fragments because CpG is rare in the mammalian genome *(4)* and because the enzymes are sensitive to cytosine methylation. The consequences of this are that restriction sites for this class of PFGE enzymes tend to fall inside what are known as *CpG islands (5)*. These are short regions of the genome approx 0.5–1.0 kb long, rich in the the dinucleotide CpG and lacking cytosine methylation, cytosines in a CpG dinucleotide outside an island tend to be methylated. Since CpG islands are frequently located in the 5' flanking sequences of a gene, one advantage of the clustering of PFGE enzyme sites within islands is that PFGE maps can be used to identify the position of a gene *(6)*. A disadvantage of this clustering is that frequently the island will contain more than one site for a particular enzyme, this tends to reduce the chances of bridging neighboring islands and producing continuous physical maps of chromosomes. This

Table 1
PFGE Enzymes

Enzyme	Site	Size	Partials
*Bss*HII	G/CGCGC	Medium	—
*Mlu*I	A/CGCGT	Large	p
*Sst*II	CCGC/GG	Medium	pp
*Nru*I	TCG/CGA	Large	p
*Not*I	GC/GGCCGC	Large	—
*Nar*I	GG/CGCC	Medium	p
*Nae*I	GCC/GGC	Medium	p
*Sfi*I	GGCCNNNN/NGGCC	Small	—
*Sma*I	CCC/GGG	Medium	pp
*Eag*I*	C/GGCCG	Medium	—
*Xho*I	C/TCGAC	Small	pp
*Pvu*I#	CG/ATCG	Large	—
*Cla*I	AT/CGAT	Small–Medium	pp
*Sal*I	G/TCGAC	Small–Medium	pp
*Sfu*I	TT/CGAA	Medium–Large	p
*Spl*I	C/GTACG	Large	p
*Rsr*II	CG/G(A/T)CCG	Large	p

p: partials often seen; pp: partials normally seen; large: fragments in the 0.8–3.0 Mb range; medium: fragments in the 0.5–1.0 Mb range; small: fragments in the 0.05–0.5 Mb range; * cuts within the *Not*I site; # frequently contaminated with *Pvu*II.

disadvantage can be slightly overcome by choosing enzymes that occur less often in islands because they contain A or T as well as the CpG dinucleotide in their recognition sequence *(7)*. An additional way to overcome this problem is by using tissue culture cells to complement analyses with animal material. One feature of CpG islands is that they always lack cytosine methylation in all tissues in the animal, however, recently, Antequera et al. *(8)* have shown that tissue culture cells can methylate the islands of nonessential genes. Thus PFGE mapping of DNA from tissue culture cells may allow islands to be bridged but may also fail to detect CpG islands.

3.3.1. Choice of Enzymes for Megabase Mapping

Table 1 lists some of the currently available enzymes for megabase mapping. The identification of additional PFGE enzymes and alternative methods to fragment DNA is under continuous development *(9)*. Many of the enzymes, especially those containing A or T in addition to CpG, naturally produce partial digests that can be useful in extending the size of map generated around any one marker. As a general approximation, enzymes containing only C and G in their recognition sequence tend to produce more complete digests. For the beginner, *Sal*I and *Cla*I are good enzymes because they are relatively cheap and produce small- to medium-sized fragments that usually give clear signals after hybridization. Table 1 also lists the size range of DNA fragments commonly generated with these enzymes, these values are an approximation and it should be noted that genomic regions have been identified that are deficient in sites for PFGE enzymes *(3)* or contain an excess of sites *(10)*. *See also* Chapters 12 and 18 for further information on PFGE enzymes.

3.3.2. Setting up the Enzyme Digest

1. Wear gloves and work with sterile solutions and equipment (*see* Note 3). To a 1.5-mL sterile microfuge tube add in the following order (final vol of 200 μL):
   a. Distilled H_2O: 80 μL (adjust for volume of enzyme)
   b. Bovine serum albumin (5 mg/mL): 20 μL (molecular biology grade)
   c. Restriction enzyme buffer (10X): 20 μL (as supplied with the enzyme)
   d. Enzyme: 20 U.
2. Mix well.
3. Add the intact agarose-embedded DNA block (80 μL) and ensure the agarose block slides down the tube and is covered by the solution.
4. Incubate for 4 h to overnight at the appropriate temperature, digestion is complete after 4 h but can be left overnight.
5. After digestion add EDTA to 50 m*M*.

Always include a control digest that contains everything except the enzyme when using untested materials, this control is valuable in order to assess the DNA quality and the type of hybridization signal obtained. Manipulate the agarose block using sterile plastic Petri dishes on a dark background and sterile bacterial loops. When using two enzymes with incompatible buffers, the first solution is removed after 4 h from the tube and the agarose block rinsed briefly in TE, then rinsed again for 10 min in 1.5 mL TE in the microfuge tube. A fresh buffer solution containing the second enzyme is then added. Difficulties are frequently experienced with double digests and it is advisable to monitor the efficiency of both enzymes using control probes.

3.3.3. Partial Enzyme Digests

Partial enzyme digests are a valuable tool that can be used to confirm linkage between two DNA markers, or when insufficient markers are available, it can be used to extend the size of restriction map generated around one DNA marker. The routine production of partial digests in agarose-embedded DNA by PFGE enzymes has been achieved by limiting the enzyme concentration, and by the use of intercalating dyes to inhibit enzyme activity *(11)* and by the use of specific CpG methylases (*12*, and *see also* Chapter 12).

3.4. Standard Pulsed-Field Gel Running Conditions

There exists a large and probably confusing variety of pulsed-field gel equipment at present. The basic device, OFAGE (orthogonal field alternate gradient electrophoresis, ref. *13, see also* Chapter 4), has been modified and the second generation of equipment includes FIGE (field inversion gradient electrophoresis, ref. *2, see also* Chapter 1), CHEF (contour-clamped homogeneous electric field, ref. *14, see also* Chapter 2), TAFE (transverse alternating field electrophoresis, ref. *15, see also* Chapter 5), and CFGE (crossed field gel electrophoresis, ref. *16*). All these devices subject DNA to two alternate field pulses and separation probably results from the differing abilities of DNA fragments of varying lengths to reorientate to a new field. For beginners, the CHEF and OFAGE systems are the simplest to set up and both are now available with support services from commercial suppliers. Accurate physical lengths can best be estimated by separating DNA over a relatively large distance in comparison with standard size markers (*see* later discussion). Two PFGE running conditions are routinely used that separate two size classes, 0.05–1.5 Mb and 1.0–6.0 Mb, over a 15-cm length. Both conditions can be fine tuned, simply by adjusting the pulse length to separate the intermediate-sized molecules within these size classes. Table 2 shows standard PFGE conditions for separating DNA fragments by OFAGE or CHEF in 20 × 20 cm × 0.9 cm, 0.8% agarose gels. Record sheets should be maintained that note the PFGE parameters and include a photograph of the gel. Adjusting any of the parameters listed in Table 2 will change the size class of molecules that are separated, however in practice adjusting the pulse length gives the best results. The digested, agarose-embedded DNA samples, are loaded intact into the sample wells

Table 2
Standard PFGE Electrophoresis Conditions

	OFAGE		CHEF	
Size class	0.05–1.5Mb	1.0–6.0Mb	0.05–1.5Mb	1.0-6.0Mb
Volts/cm	10	3	6	2
Field angle	90°–170°	90°–170°	120°	120°
Buffer	0.25X TBE	0.5X TBE	0.25X TBE	0.5X TBE
Buffer temp.	15°C	15°C	15°C	15°C
Pulse length	120 s	45 min	90 s	45 min
Run time	36 h	192 h	36 h	192 h

and can be manipulated by decanting the sample into a sterile plastic Petri dish and using either a sterile glass cover slip or sterile scalpel blades to transfer the block into the slot. The block should be pushed well into the slot using a sterile bacterial plastic loop to lie below the surface of the gel and air around the block removed using a fine pipet or syringe and needle. The sample should be sealed in place with LMP agarose dissolved in running buffer. An example of pulsed-field gels separating in these two size classes with appropriate markers, is shown in Fig. 1 (*see also* Note 4).

3.5. Size Markers for Pulsed-Field Gels

Size markers specific for different size classes are necessary and many are now commercially available (*see* Table 3). Attempts are currently being made to standardize pulsed-field gel mapping data and to include these data in current genome map data bases. Many of the size markers used are prepared from different yeast strains that vary slightly in chromosome length and sometimes chromosome number (*see* Fig. 2). To support data standardization it should be possible to use standard strains whose chromosome size has been well characterized (*see* Table 4) and then to calibrate other stains against these.

3.5.1. Preparation of Yeast Size Markers

METHOD 1 *(17)*

1. Streak out yeast on YPD plates and incubate at 30°C for 2 d. Yeast live about 3–4 wk on an agar plate and should be stored permanently as a glycerol stock at –80°C.
2. Grow a single large colony overnight at 30°C in 50 mL of YPD.

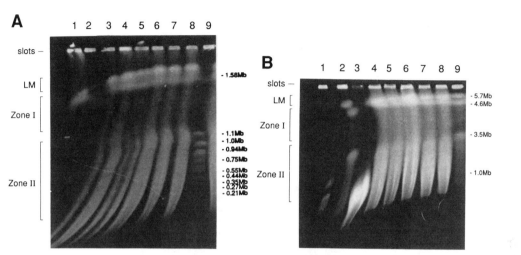

Fig. 1. Panel **A** shows ethidium bromide stained mouse genomic DNA separated by PFGE in the 0.05–1.5 Mb size range. Lane 1—uncut control, Lane 2—*S. cerevisiae* YP80, Lane 3—*Bss*HII (sample damaged on loading), Lane 4—*Bss*HII + *Not*I, Lane 5—*Bss*HII + *Mlu*I (sample damaged on loading), Lane 6—*Bss*HII + *Sac*II, Lane 7—*Bss*HII + *Sal*I, Lane 8—*Bss*HII + *Eag*I, Lane 9—*S. cerevisiae* YP80. Panel **B** shows ethidium bromide stained mouse genomic DNA separated by PFGE in the 0.5–4.0 Mb size range. Lane 1—*H. wingeii,* Lane 2—*S. pombe,* Lane 3—degraded DNA, Lane 4—*Mlu*I, Lane 5—*Nru*I, Lane 6—*Not*I, Lane 7—*Mlu*I + *Nru*I, Lane 8—*Mlu*I + *Not*I, Lane 9—*S. pombe.* Three features of both gels are indicated on the left hand side, these are the LM (limiting mobility) region, Zone 1, and Zone 2. All PFGE DNA separated in any type of apparatus will show these three features. Intact DNA will normally show fluorescence only in the LM region (Lane 1, panel A), degraded DNA will normally show no or reduced fluorescence in the LM region (Lane 3, panel B). DNA digested with PFGE enzymes will show fluorescence in all three regions as shown in Lanes 3–8 (panel A) and Lanes 4–8 (panel B). Material in the LM region has not been size separated, thus in panel A the 1.5 Mb yeast chromosome and in panel B the 4.6 and 5.7 Mb yeast chromosomes have not been resolved.

3. Pellet in benchtop centrifuge at 500*g* for 10 min. Wash once in 50 mL of 100 m*M* EDTA, pH 8, and then take up in 50 mL of CPES buffer.

4. Prepare the plastic molds, leave on ice to cool.

Table 3
Size markers for PFGE

Size class	Size marker
3.5–5.7 Mb	*Schizosaccharomyces pombe*
4.5 Mb	*E. coli genome (linearized)*
1.0–3.5Mb	*Hansenula wingeii*
0.02–1.6 Mb	*Saccharomyces cerevisiae*
0.04–0.8 Mb	*phage Lambda oligomers*

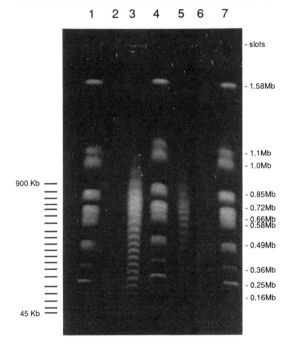

Fig. 2. Size markers for PFGE. Lanes 3 and 5 contain Lamda phage multimers spanning 45–900 kb. Lanes 1, 4, and 7 contain yeast *S. cerevisiae* GY22 chromosomes, only 12 bands out of 16 are resolved under these PFGE conditions.

5. Count cells using a hemocytometer and resuspend in CPES buffer at 2×10^9 cells/mL. Add Novozym to 4 mg/mL. Novozym is used to remove the yeast cell wall, producing what are called *spheroplasts,* and works well for the species listed in Table 3.
6. Immediately mix 0.5 mL of yeast cells with an equal vol of 1% LMP agarose made up with 1X CPE buffer and kept warm at 45°C. Dispense 80 µL into the molds and leave for 20 min to set.

Table 4
Chromosome Sizes of Standard Yeast Strains

	Size of chromosome	
Chromosome number	*S. cerevisiae* AB972[a]	*S. cerevisiae* YP148[b]
XII	does not normally band	(> 2.0 Mb)
IV (12)	1630 kb	1580 kb
XV (11a)	1170 kb	see band 11b
VII (11b)	1140 kb	1110 kb
		1025 kb*(XV)
XVI (10b)	1010 kb	1000 kb
XIII (10a)	955 kb	940 kb
II (9)	840 kb	829 kb
XIV (8)	830 kb	791 kb
X (7)	785 kb	752 kb
XI (6)	675 kb	681 kb
VIII (5a)	595 kb	598 kb
V (5b)	590 kb	550 kb
IX (4)	455 kb	441 kb
III (3)	355 kb	351 kb
VI (2)	280 kb	276 kb
I (1)	240 kb	213 kb
		92* (XV)
	H. wingeii	*S. Pombe* 972
	VII 3.3 Mb	5.7 Mb
	VI 2.9 Mb	4.6 Mb
	V 2.2 Mb	3.5 Mb
	IV 1.8 Mb	
	III 1.5 Mb	
	11 1.25Mb	
	1 1.03Mb	

Number in brackets is the band number for *cerevisiae* assigned from PFGE gels. *H. wingeii* (20), *S. pombe* (21),YP148 (22), AB972 (23). *Chromosome XV in YP148 is in two fragments of 92 kb and of 1025 kb, each containing pBR322 sequences.

7. Put the blocks into 50 mL of CPE buffer and agitate gently for 2 h at 37°C.
8. Transfer the blocks into 50 mL of cell lysis buffer solution for 2 d at 50°C.
9. Store yeast markers in 0.5M EDTA, pH 8.0 at 4°C; shelf life is from 1–3 y. Before use, equilibrate the markers in 1X running buffer.

METHOD 2 *(2)*

1. Proceed as in Method 1 for Steps 1–4.
5. Count the cells and resuspend in Solution A at a concentration of 2×10^9 cells/mL. This method works well for *S. cerevisiae* but less well for *S. pombe*.
6. Mix 0.5 mL cell suspension with an equal vol of 1% LMP agarose made up with 0.5X SCE buffer and dispense 80-µL aliquots into precooled plastic molds, leave 20 min to set.
7. Incubate agarose blocks for 24 h at 37°C in ETB buffer.
8. Rinse the blocks twice in TE and then incubate for 48 h in cell lysis buffer at 55°C.

3.5.2. Phage Lambda Multimers

Any Lambda phage of known size that lyses easily can be used.

1. Pick a phage plaque using the narrow end of a Pasteur pipet into 100 mL of LB Broth plus 10 mM MgCl$_2$. The ratio of the area of viable cells to the area of the plaque should be 5:95, otherwise phage-resistant cells will overgrow the culture.
2. Incubate overnight at 37°C with vigorous agitation. Cell lysis should be apparent after 12 h.
3. Add chloroform to 0.1% and agitate for 15 min at 37°C to complete the cell lysis.
4. Spin the culture for 10 min at 700g, in chloroform resistant tubes, to pellet the cell debris and transfer the supernatant to clean centrifuge tubes.
5. Pellet the phage by centrifugation at 4000g for 3 h. Decant the supernatant and resuspend the pellet in 800-µL TE using a bacterial plastic loop to resuspend the phage.
6. Add an equal vol of 1% LMP agarose prepared in TE and aliquot 80 µL into plastic molds as described earlier.
7. Incubate agarose blocks at 55°C for 48 h in cell lysis buffer, then rinse several times in TE and incubate in 0.1M EDTA, pH 8, for a further 48 h at 50°C. The multimers form during this second incubation period.
8. Store the lambda multimers in 0.5M EDTA, pH 8, at 4°C; shelf life is approx 6 mo.

3.6. Preparation and Hybridization of DNA Blots: Data Interpretation

Most standard DNA blotting and hybridization techniques work well for PFGE DNA. The construction of a PFGE map involves repeated hybridization of one DNA blot using several probes so it is necessary to prepare good quality blots that produce strong hybridization signals (*see* examples in ref. *18*). Special tricks are not needed but the steps involved should be optimized and for best results a few points should be noted. One key step for good PFGE blots has been found to be the DNA transfer by blotting, and full details are given further on. The DNA is blotted onto a nylon membrane, because these can be rescreened realistically up to 20 times, the membrane can be charged or uncharged but DNA should be crosslinked to an uncharged membrane by UV exposure. The slots of the gel should be included in the blot because these provide essential information to orientate the blot, align hybridization results from different probes, and to assess the extent of DNA digestion.

3.6.1. PFGE DNA Transfer

1. After electrophoresis, the gel is stained for 40 min in 0.5 µg/mL ethidium bromide and photographed using a long wavelength transilluminator with fluorescent rulers placed on either side of the gel. Gloves should be worn when handling solutions containing ethidium bromide and solutions should be disposed of carefully. It is important for the construction of restriction maps that both size markers and rulers are clearly visible in the photograph and it may be necessary to destain the gel in order to clearly visualize the size markers. Ethidium bromide added during the electrophoresis period will slow and probably alter the characteristics of the DNA migration.
2. Depurinate the DNA by placing the 20 × 20 cm gel for 20 min in 500 mL of 0.25*M* HCl at room temperature, use gentle agitation. Rinse the gel and container in distilled water to remove excess HCl.
3. Denature the DNA by shaking the gel for 40 min in 800 mL of denaturant buffer. Repeat once.
4. Prepare the nylon membrane by wetting in distilled water and then soaking in denaturant buffer.
5. A wet blot with a reservoir of denaturant buffer is used to transfer the DNA. Blot the gel from the back and ensure that the buffer only ascends through the gel. The gel should be blotted for at least 36 h using a weight of approx 400 g (light enough to prevent the gel from collapsing before transfer has occurred) but can be left longer since the DNA does not blot through the membrane. The towels and reservoir should be replenished as necessary.
6. After transfer the blot is carefully disassembled and the position of the slots marked on the membrane using a narrow gage needle; this

step is essential in order to analyze the data obtained from repeated hybridization of the blot. The blot is labeled using a waterproof pen on the DNA side.

7. The filter is neutralized by washing for 2 min in 500 mL of 50 mM sodium phosphate (pH 7.0), excess liquid is removed by blotting in Whatman 3M paper and the filter dried for 30 min, but not longer as this will reduce the hybridization signal, in a vacuum oven held at 80°C.

8. If uncharged nylon membranes are used the DNA should be crosslinked to the membrane by exposure to UV (260 nm). The time of UV exposure should be empirically determined for optimum results, this can be done using standard DNA blots dried as described in Step 7, and exposing the transferred DNA to 200–400 μWatts/cm^2 for intervals of 30–200 s. Standard UV$_{260}$ 15W germicidal lamps or a commercial UV crosslinking oven can be used. This step is not necessary for charged nylon membranes that can be hybridized directly after neutralization as described in Step 7. In general, however, uncharged membranes are preferable to charged membranes because of the ease with which the former can be stripped and reused. The old probe is removed by shaking the filter at room temprature in two changes of 100 mM NaOH (500 mL) each for 20 min, followed by neutralization in 500 mL of 50 mM sodium phosphate, pH 7.0, and rehybridization.

9. Filters can be stored sealed inside plastic bags at room temperature in the dark.

3.6.2. Data Interpretation

Construction of long-range maps from pulsed-field gel blots involves sequential hybridization of one PFGE blot with different DNA markers to assess if the markers hybridize to the same DNA fragments. To do this accurately, the autaradiographs have to be aligned with each other. *See* Note 5 and also Chapter 13 in this vol on computer imaging systems to analyze PFGE data. A simple way to do this is to orientate the autoradiograph with the filter using radioactive ink and then to mark the position of the slots on the autoradiograph from the holes made in the filter as described in Section 3.6.1. The information in the autoradiograph should be described in two ways. The size of the hybridizing fragment should be determined using appropriate size markers for the size class and any fragments corecognized by two probes should be identified. The fact that two different markers corecognize one DNA fragment is suggestive that the two markers are physically linked. However, probably because of the distribution of PFGE enzyme sites within CpG islands, it is a frequent finding that many DNA fragments produced by PFGE enzymes are the same size (*see* example in ref. *19*). The question that needs to be resolved is whether two DNA markers recognize the same DNA fragment or recognize two different DNA fragments of similar size that exactly comigrate. This question can be resolved by following some guidelines.

1. Corecognition of a single DNA fragment by two markers should be shown by using more than one enzyme.
2. If possible the position of the two DNA markers along the length of the large DNA fragment should be shown by using a second enzyme that cuts within the fragment generated by the first enzyme.
3. Corecognition of a single DNA fragment by two markers could be confirmed by generating partial digests or using natural partials that occur in different tissues, and showing that the larger partial digest product was also corecognized by the two markers.

Once it is established that two markers are physically linked on DNA fragments generated by some PFGE enzymes, then a restriction map should be constructed using all the data. It should then be possible to verify the initial interpretation by a few key double enzyme digests.

4. Notes

1. The ideal concentration of DNA for PFGE electrophoresis is between 10–20 µgs/lane, which corresponds to $1.0–2.0 \times 10^6$ cells. Concentrations in excess of this will migrate abnormally. The same sized fragment contained in a sample at one concentration will migrate more slowly compared to the same sized fragment contained in a sample of a lower concentration. PFGE systems that generate homogeneous fields (CHEF, FIGE, and TAFE; *see* Chapters 1, 2, and 5) are more subject to overloading effects compared to an inhomogeneous field system (OFAGE; *see* Chapter 4), and lower concentrations should be used. Concentrations of less than 10 µg/lane do not normally generate a strong enough hybridization signal to allow the repeated use of DNA blots that is necessary for megabase mapping. The cell number should be accurately determined at this point because it is not easy to adjust the concentration after the cells are embedded. If it is difficult to estimate cell number, for example from frozen tissues, then it is advisable to prepare two different sample concentrations. It is also important at this stage to ensure that the sample is well dispersed into single cells. Clumps of more than 2–3 cells will produce local

increased concentrations of DNA in the sample lane and distort the hybridization signal. Samples could be frozen after dissection into 2-mm cubes or when dispersed into single cells. Most experience has been with material rapidly frozen using liquid nitrogen, then maintained at −80°C and processed into PFGE DNA after grinding to a powder and quick thawing by dropping the frozen powder into PBS. PFGE DNA prepared from frozen tissue is, however, never as intact as that from fresh tissue. It is possible that the quality could be improved by freezing in the presence of tissue culture freezing solutions but this has not been reported on.

2. Damage to the agarose block will produce defects in the migration of the sample (*see* Fig. 1A, lanes 3 and 5). To prevent this, all manipulations should be done gently using a sterile plastic bacterial loop. Problems have been experienced using proteinase K from aliquots that have been repeatedly frozen and lost activity. It is advisable either to snap freeze the enzyme solution in small aliquots for storage, or to make the solution fresh each time. The agarose-embedded DNA is directly digested with enzymes but because not all LMP agarose sources are compatible, it is best to use one that has been tested. Many companies that sell restriction enzymes for PFGE now also sell LMP agarose for embedding cells. Once the DNA samples are free of the cell lysis buffer, care should also be taken to avoid contamination with nucleases, this can be done by using sterile buffers and glassware and by wearing gloves when handling the samples.

3. A frequent problem is contamination of enzymes by nucleases because of this it is advisable to set aside the enzymes used for PFGE work, aliquot them in smaller volumes, and handle them in a sterile manner. The results of this low level of contamination are not seen in other work involving the DNA sample, but it is sufficient to produce degraded DNA in the megabase size range that will result in weak hybridization signals. The activity of the enzymes should also be monitored using control DNA as described by the manufacturers.

4. Migratory artifacts can be seen if the block does not lie at the base of the slot and protrudes above the surface of the gel, if air was allowed to collect in front of, or behind the sample, or if the agarose block was damaged during manipulations (*see* example in Fig. 1, lanes 3 and 5). In addition, poor resolution and hybridization signals will result if the gel is not completely covered by the buffer or if the buffer is not adequately cooled. The PFGE electrophoretic chamber should be cleaned regularly with sterile distilled water, to inhibit nucleases and prevent the accumulation of buffer salts, and checked regularly to ensure that the expected voltage gradient is being applied across

the gel. If the voltage gradient is not equal from each set of electrodes then the integrity of the electrodes and the level of the gel tank should be checked.

5. Most of the pitfalls could be avoided with appropriate controls. Problems have been encountered because of carryover from the previous signal when PFGE blots are reused. Also with double digests when it might be necessary to confirm that both enzymes have cut, using control probes. Frequently a DNA marker will not hybridize to a clear band but instead hybridizes to the region of limiting mobility (*see* Fig. 1). This means that the DNA fragment is out of the size range that has been chosen for the PFGE electrophoresis. The band will be resolved if a larger size range is chosen. Frequently the probe recognizes a clear band in the size range of the gel but an additional signal in the limiting mobility. This means that an additional larger partially digested fragment could be resolved if a larger range was chosen.

5. Trouble Shooting

5.1. DNA Does Not Cut

Ethidium bromide staining shows fluorescence only in the slot or in the limiting mobility region (lane 1, Fig. 1A).

1. Check efficiency of the enzyme and buffers using control DNAs as recommended by the supplier.
2. Cell lysis buffer or EDTA may still be present in the agarose blocks. Do not rinse more than 10 blocks/50 mL of TE when removing the cell lysis buffer or the EDTA before digestion, in addition rinsing at 50°C is necessary to remove the sodium lauroyl sarcosine present in the cell lysis buffer.
3. Check quality of low melting point agarose using agarose embedded plasmid or phage DNA and standard enzymes such as *Eco* RI. Most LMP agarose is nowadays of good quality and several suppliers market LMP agarose specially for PFGE analysis.
4. Proteinase K will inactivate enzymes if insufficient or inactive PMSF is used.

5.2. DNA Is Degraded

Ethidium bromide staining does not show any fluorescence in the limiting mobility region (lane 3, Fig. 1B).

1. Enzyme or buffer solutions contaminated with nonspecific or specific nucleases. Identify the contaminated solution by repeating the digestion omitting either buffer or enzyme. Control digests should

normally be included with each experiment when using untested material.

2. Sample incompletely treated with proteinase K. This can happen with badly stored enzyme or when too many samples are processed at one time. Thus in the absence of the high concentrations of EDTA, the tissue nucleases act to degrade DNA. This can be seen if a control digest is performed without adding enzyme or buffer and can be rectified by treating the samples with fresh cell lysis buffer.

3. Treatment of DNA sample with proteinase K in the absence of EDTA. Some samples of proteinase K retain some nuclease activity, this is inhibited when the enzyme is used in the presence of EDTA as in the cell lysis buffer.

5.3. Multiple Partially Digested Bands Are Seen After Hybridizing Using a Single Copy Probe

Partial digests are commonly seen and generally are only troublesome if strong hybridizing signals cannot be obtained. The activity of the enzymes and buffers can be checked as described earlier to ensure that digestion is as complete as possible, or alternatively other tissues could be used for analysis, which would show different methylation patterns. If still in doubt then the filter could be hybridized using control probes known to recognize a single band. Some genomic regions are more prone to partials than others (*see also* Chapter 18), and some PFGE enzymes show site specificity in addition to methylation sensitivity (*see* product description by New England Biolabs, Beverly, MA, USA).

5.4. Pale or Indistinct Hybridizing Signals

1. DNA is partially degraded. The integrity of uncut DNA and also the presence of low levels of nucleases in enzymes and buffers should be checked by running appropriate controls.

2. Aspects of the DNA blot and hybridization should be checked and optimized, for example the gel could be stained after transfer to confirm that all DNA has been transferred. An alternative blotting procedure has sometimes given improved results. In this case the gel is stained in ethidium bromide as described earlier and photographed using a 260 nm transilluminator. The gel is exposed to the UV source for the minimum time necessary to obtain a photograph using Polaroid 667 film and no longer, during this period the DNA is UV-knicked sufficiently to allow good transfer. Prolonged exposure of the gel to UV_{260} reduces the hybridization signal. Care should be

taken to check that the filter and UV source for the transilluminator are correct. The gel is then placed in denaturant buffer for 60 min, this is repeated once and then the gel is blotted using a reservoir of denaturant buffer as described earlier.
3. Gel is overexposed to short wave UV from a 260 nm transilluminator. The DNA becomes crosslinked in the gel and less able to bind to the nylon membrane and to the hybridizing probe.
4. Buffer capacity is inadequate leading to poor focusing of fragments. This can happen as a result of poor cooling (the optimum temprature is from 12–15°C) or errors in buffer composition.
5. DNA concentration is too low.

Acknowledgments

The methods described here were developed with the help of Hans Lehrach (ICRF, London) during my time as a Post-Doctoral Fellow in his laboratory at the EMBL, Heidelberg. Special thanks go to Margit Burmeister (UCSF, San Francisco) for her contribution to many of the points described here, and to Andreas Weith (IMP, Vienna) for his comments on the manuscript.

References

1. Smith, C. L. and Cantor, C. R. (1987) Purification, specific fragmentation and separation of large DNA molecules. *Methods in Enzymology vol. 155* (Wu, R., ed.), Academic, London, p. 44.
2. Carle, G. and Olson, M. V. (1987) Orthogonal field alternation gel electrophoresis *Meth Enzymol vol. 155* (Wu, R., ed.), Academic, London, p. 468.
3. Burmeister, M., Monaco, A. P., Gillard, E. F., van Ommen, G. J. B., Affara, N. A., Ferguson-Smith, M. A., Kunkel, L. M., and Lehrach, H. (1988) A 10 megabasepair physical map of human Xp21, including the Duchenne Muscular Dystrophy gene. *Genomics* **2**, 189–202.
4. Sved, J. and Bird, A. P. (1990) The expected equilibrium of the CpG dinucleotide in vertebrate genomes under a mutation model. *Proc. Natl. Acad. Sci. USA* **87**, 4692–4696.
5. Bird, A. P. (1986) CpG-rich islands and the function of DNA methylation. *Nature* **321**, 209–213.
6. Lindsay, S. and Bird, A. P. (1987) Use of restriction enzymes to detect potential gene sequences in mammalian DNA. *Nature* **327**, 336–338.
7. Bird, A. P. (1989) Two classes of observed frequency for rare cutter sites in CpG islands. *Nucleic Acids Res.* **17**, 9485.
8. Antequera, F., Boyes, J., and Bird, A. P. (1990) High levels of de novo methylation and altered chromatin structure at CpG islands in cell lines. *Cell* **62**, 503–514.

9. Helene, C., Thuong, N. T., Behmoaras, T. S., and Francois, J. C. (1989) Sequence-specific artificial endonucleases. *Trends Biotechnol.* **7**, 310–315.

10. Bučan, M., Zimmer, M., Whaley, W. L., Poustka, A., Youngman, S., Allitto, B. A., Ormondroyd, E., Smith, B., Pohl, T. M., McDonald, M., et al. (1990) Physical maps of 4p16.3, the area expected to contain the Huntingtons disease mutation. *Genomics* **6**, 1–15.

11. Barlow, D. P. and Lehrach, H. (1990) Partial Not I digests generated by low enzyme concentration or the presence of ethidium bromide can be used to extend the range of pulsed-field gel mapping, *Technique* **2**, 79–87.

12. Hanish, J. and McClelland, M. (1989) Controlled partial restriction digestions of DNA by competition with modification methyltransferases. *Analytical Biochemistry* **179**, 357–360.

13. Schwartz, D. C. and Cantor, C. R. (1984) Separation of yeast chromosome-sized DNAs by pulsed-field gradient gel electrophoresis. *Cell* **37**, 67–75.

14. Chu, G., Vollrath, D., and Davies, R. W. (1986) Separation of large DNA molecules by Contour-Clamped Homogeneous electric fields. *Science* **234**, 1582–1585.

15. Gardiner, K., Laas, W., and Patterson, D. (1986) Fractionation of large mammalian DNA restriction fragments using vertical pulsed-field gradient electrophoresis. *Somatic Cell Mol. Gen.* **12**, 185–195.

16. Southern, E. M., Anand, R., Brown, W. R. A., and Fletcher, D. S. (1987) A model for dle separation of large DNA molecules by crossed-field gel electrophoresis. *Nucleic Acids Res.* **15**, 5925–5943.

17. De Jonge, P., De Jongh, F. C. M., Meijers, R., Steensma, H. Y., and Scheffers, W. A. (1986) Orthogonal field alternation gel electrophoresis banding patterns of DNA from yeast. *Yeast* **2**, 193–204.

18. Herrmann, B. G., Barlow, D. P., and Lehrach, H. (1987) A large inverted duplication allows homologous recombination between chromosomes heterozygous for the proximal t complex inversion. *Cell* **48**, 813–825.

19. Barlow, D. P., Bučan, M., Lehrach, H., Hogan, B. L. M., and Gough, N. (1987) Close genetic and physical linkage between the murine haematopoetic growth factor genes GM-CSF and Multi-CSF (IL3). *EMBO J.* **6**, 617–623.

20. Jones, C. P., Janson, M., and Nordenskjold, M. (1989) Separation of yeast chromosomes in the megabase range suitable as size markers for pulsed-field gel electrophoresis. *Technique* **1**, 90–95.

21. Smith, C. L., Matsumoto, T., Niwa, O., Klco, S., Fan, J. B., Yanagida, M., and Cantor, C. R. (1987) An electrophoretic karyotype for Schizosaccharomyces pombe by pulsed-field gel electrophoresis. *Nucleic Acids Res.* **15**, 4481–4489.

22. Gerring, S. L., Connelly, C., and Hieter, P. (1991) Positional mapping of genes by chromosome blotting and chromosomal fragmentation. *Methods in Enzymology vol. 194* (Guthrie, C. and Fink, G. E., eds.), Academic, London, pp. 57–77.

23. Link, A. J. and Olson, M. V. (1991) Physical map of the Saccharomyces cerevisiae genome at 110 kilobase resolution. *Genetics* **127**, 681–698.

CHAPTER 9

Encapsulation of Cells in Agarose Beads

Joan Overhauser

1. Introduction

For the analysis of large DNA fragments using pulsed-field gel electrophoresis (PFGE), it is necessary to first embed the cells in agarose
to prevent shearing of the DNA during protein extraction and restriction enzyme digestion. The speed and efficiency with which these steps
are performed is based on the diffusion rate of the buffers and proteins in each step.

Most store-bought PFGE apparatuses come with a slot former for
the production of agarose blocks containing the cells of choice. These
agarose blocks after treatment must usually be cut into smaller strips
for loading into the gel wells and only fit easily into the wells for which
the slot former was made.

An alternative method to the formation of agarose blocks has been
devised, which involves the embedding of cells in small agarose beads
(1,2). The advantage of these beads is twofold. First, because the beads
are much smaller than the agarose blocks, the diffusion of proteins
and buffer is much more rapid. Thus, the time needed to equilibrate
the beads in any solution is much faster than that for agarose blocks.
And second, the agarose beads can easily fill any well size.

This chapter presents the procedure for the encapsulation of cells
in agarose beads. Although the beads are a little more difficult to make
than agarose blocks, large volumes of beads can be made rapidly
and less restriction enzyme is normally necessary to digest the
embedded DNA.

From: *Methods in Molecular Biology, Vol. 12: Pulsed-Field Gel Electrophoresis*
Edited by: M. Burmeister and L. Ulanovsky
Copyright © 1992 The Humana Press Inc., Totowa, NJ

The protocols will be divided into three sections: (1) preparation of agarose beads containing yeast chromosomes; (2) preparation of agarose beads containing genomic DNA; and (3) restriction endonuclease digestion of DNA in agarose beads.

2. Materials

1. YEPD medium: 1 g/L yeast extract, 20 g/L bactopeptone, 2% (w/v) dextrose.
2. YEPD plates: YEPD medium, 15 g/L bactoagar.
3. SE buffer: 75 mM NaCl, 25 mM Na$_2$EDTA, pH 8.0.
4. 1% Low-melting point (LMP) agarose (Seaplaque, FMC, Rockland, ME) in SE buffer.
5. Mineral oil.
6. 5 mg/mL Lyticase (Sigma, St. Louis, MO) (freshly made in SE and centrifuged to remove undissolved protein).
7. β-mercaptoethanol.
8. ES buffer: 1% (w/v) sarcosyl, 25 mM Na$_2$EDTA, pH 8.0, 50 μg/mL proteinase K.
9. TE: 10 mM Tris-HCl, pH 8.0, 1 mM Na$_2$EDTA.
10. SB: 50% (w/v) sucrose, 25 mM Na$_2$EDTA, pH 8.0, 0.1% bromophenol blue, 0.1% xylene cyanole.
11. PBS: 3 mM KCl, 1.5 mM KH$_2$PO$_4$, 0.14M NaCl, 8 mM Na$_2$HPO$_4$.
12. 1% LMP agarose (Seaplaque, FMC, Rockland, ME) in PBS.
13. Mineral oil.
14. SDE: 1% SDS, 25 mM Na$_2$EDTA, pH 8.0.
15. 25 mM Na2EDTA, pH 8.0.
16. 0.1 mM PMSF (phenylmethylsulfonyl fluoride)in isopropanol.
17. Pasteur pipet with tip broken to increase bore size at the end.
18. 1X restriction enzyme buffer (according to manufacturer's recommendations).
19. Restriction enzymes.

3. Methods

3.1. Preparation of Agarose Beads Containing Yeast Chromosomes

1. Streak out yeast cells onto an YEPD plate and incubate 2 d at 30°C.
2. With a single colony inoculate a 20-mL YEPD culture in the morning and shake during the day at 30°C.
3. At the end of the day, take the OD$_{600}$ of the culture.
4. Inoculate a fresh 200-mL culture overnight such that the OD$_{600}$ is about 0.003 and shake overnight at 30°C.

5. Determine the OD_{600} of the culture in the morning and allow to grow until an OD_{600} of 1.5 is reached.
6. Pellet the cells in a centrifuge at $500g$ for 10 min.
7. Discard the supernatant and resuspend the yeast cells in 10 mL of SE and centrifuge as before.
8. Wash the cells twice more in SE.
9. Resuspend the cells in 4 mL of SE and transfer to a 125-mL Erlenmeyer flask and place in a 45°C water bath. (*See* Note 1.)
10. Melt a 1% LMP agarose solution in SE and place in the 45°C water bath. Place mineral oil in the 45°C water bath. (*See* Note 2.)
12. Place a 250-mL beaker containing 100 mL of cold SE in a bucket filled with ice, put on a stir plate set at medium speed, and allow to stir until the beaker is cold (10–15 min).
13. Add 4 mL of agarose solution to the yeast cells and swirl to mix.
14. Add 16 mL of the prewarmed mineral oil to the yeast cells. (*See* Note 3.)
15. Rapidly swirl the mixture for at least 30 s in the 45°C water bath until an emulsion is formed that does not rapidly disassociate. (*See* Note 4.)
16. Quickly pour the mixture into the cold SE buffer, which is still stirring. Large globules will immediately form and settle at the top of the stirring mixture. (*See* Note 5.) Continue stirring for several minutes. After the beaker is removed from the stir plate, most of the beads will float to the top and the SE solution will look clear.
17. Transfer the mixture to several 50-mL polystyrene conical tubes and centrifuge at $500g$ for 10 min.
18. Some of the agarose beads may be trapped near the top mineral oil layer. Remove the mineral oil layer at the top and disperse the beads that have not pelleted by repeated pipeting with a broken-off Pasteur pipet. Centrifuge the tubes as before.
19. Remove all of the supernatant including the beads that do not pellet the second time. (These contain encapsulated mineral oil.) Resuspend the pelleted agarose beads in 10 mL of SE, combine, and centrifuge as before. Remove the supernatant.
20. Add 0.5 mL of β-mecaptoethanol to the pellet, 1 mL of Lyticase, and SE to a final vol of 10 mL. Disperse the beads and incubate at 37°C for 2 h.
21. Pellet the beads as before and resuspend in 20 mL of ES and incubate overnight at 50°C. (*See* Note 7.)
22. Pellet the beads and resuspend in 25 mM EDTA, pH 8.0, containing 0.1 mM PMSF.
23. Pellet the beads and wash twice with 25 mM EDTA, pH 8.0, and store at 4°C. Rock the beads for 10 min before each centrifugation.
24. If the beads are to be used in the next several weeks, wash the beads several times with TE and store at 4°C. Rock the beads for 10 min

before each centrifugation. For long-term storage, keep the beads in 25 m*M* EDTA, pH 8.0, to minimize DNA degradation.

25. Directly before use, transfer the beads to an Eppendorf tube, and add an equal vol of SB. After 5 min, centrifuge the tubes for 1 min and remove the supernatant. The beads are ready to be loaded onto the gel as size markers. (*See* Notes 9 and 10.) Eliminate this step if the beads are to be digested with restriction enzymes.

3.2. Preparation of Agarose Beads Containing Genomic DNA

1. Trypsinize the cells such that a single cell suspension is achieved and centrifuge at 500*g* for 10 min.
2. Wash approx 1×10^8 cells with PBS, centrifuge, and resuspend in 5 mL of PBS.
3. Transfer the cells to a 125-mL Erlenmeyer flask and place in a 45°C water bath. (*See* Note 1.)
4. Melt a 1% LMP agarose solution in PBS and place in a 45°C water bath. Place mineral oil in the 45°C water bath. (*See* Note 2.)
5. Place a 250-mL beaker containing 100 mL of cold PBS in a bucket filled with ice, put on a stir plate set at medium speed and allow to stir until the beaker is cold (10–15 min).
6. Add 5 mL of agarose solution to the cells and swirl to mix.
7. Add 20 mL of the prewarmed mineral oil to the cells. (*See* Note 3.)
8. Rapidly swirl the mixture for at least 30 s in the 45°C water bath until an emulsion forms that does not rapidly disassociate. (*See* Note 4.)
9. Quickly pour the mixture into the cold PBS, which is still stirring. Large globules will immediately form and settle at the top of the stirring mixture. (*See* Note 5.) Continue stirring for several minutes. Remove the beaker from the stir plate. Most of the beads will float to the top and the PBS solution will clear.
10. Transfer the mixture to several 50-mL polystyrene conical tubes and centrifuge at 500*g* for 10 min. (*See* Note 6.)
11. Some of the agarose beads may be trapped near the top mineral oil layer. Remove the mineral oil layer at the top and disperse the beads that have not pelleted by repeated pipeting with a broken-off Pasteur pipet. Centrifuge the tubes as before.
12. Remove all of the supernatant including the beads that do not pellet the second time. (These contain encapsulated mineral oil.) Resuspend the pelleted agarose beads in 10 mL of PBS. Combine the pellets and centrifuge as before. Remove the supernatant and wipe the inside of the tube to remove any excess mineral oil.

13. Add 20 mL of SDE to the pellet. Resuspend the beads by repeated pipeting with a broken off Pasteur pipet. (*See* Note 7.)
14. Rock the suspension for 10 min. The beads will become more translucent during this time.
15. Pellet the beads as before and resuspend in 20 mL of ES and incubate overnight at 50°C.
16. Pellet the beads and resuspend in 25 m*M* EDTA, pH 8.0, containing 0.1 m*M* PMSF.
17. Pellet the beads and wash twice with 25 m*M* EDTA, pH 8.0, and store at 4°C for long-term storage (2–12 mo).
18. If the beads are to be used in the next month, eliminate the 25 m*M* EDTA, pH 8.0, washes and wash the beads 3X with 20 mL of TE. Rock the beads for 10 min before each centrifugation. (*See* Notes 9–11.)

3.3. Restriction Endonuclease Digestion of DNA in Agarose Beads

1. Transfer the agarose beads to an Eppendorf tube and centrifuge for 1 min to pellet the beads. The beads are very translucent and may be difficult to see. It is easiest to pour off the supernatant and tap the tube. The beads will stay at the bottom of the tube. Add additional bead slurry until the appropriate amount of beads (<800 µL) to be digested is obtained.
2. Fill the Eppendorf with 1X restriction buffer and vortex to resuspend the bead pellet. Allow the beads to sit for 5 min then centrifuge for 2 min.
3. Remove the buffer and wash the beads twice more with 1X restriction enzyme buffer. Remove the buffer after the last wash.
4. Add restriction endonuclease using similar concentrations used for digested genomic DNA (2–5 U/µg DNA). Incubate at the appropriate temperature for 4 h.
5. Remove 5 µL and transfer to an Eppendorf tube containing 5 µL of SB. Load the sample onto a standard 1% agarose minigel and electrophorese using normal conditions.
6. If the sample is digested, all of the DNA will migrate out of the well.
7. After digestion is confirmed, fill the tube with SB and let sit for 5 min before centrifugation.
8. Remove the supernatant. The beads are ready for loading. (*See* Notes 9–11.)

4. Notes

1. If the volumes are increased, make sure the size of the Erlenmeyer flask is also increased. Using too small of an Erlenmeyer flask prevents a good swirling motion that is necessary to produce the emulsion.

2. It is critical that all of the solutions be equilibrated at the correct temperature before bead formation. The temperature difference between the cell solution and the PBS (or SE) solution is very important.

3. Increasing the amount of mineral oil added to the solution can be performed if difficulty is experienced in forming the emulsion.

4. Vortexing the solution instead of swirling generates an emulsion that results in very small beads and is not recommended.

5. Formation of the oil emulsion is the most difficult step in the procedure. It is important to swirl the solution as rapidly as possible. It may be helpful to test out the procedure first with bacterial cells to ensure that the emulsion can be reproducibly produced and poured into a cold solution to generate beads.

6. The use of polystyrene tubes is recommended for washing the beads. As the cells are lysed, the beads become translucent and may be difficult to see in polypropylene tubes.

7. The addition of SDE or ES to the beads may result in a slightly viscous solution. This is owing to cells that have not been properly embedded. Vigorous pipeting will reduce the viscosity. Subsequent washing of the beads will remove most of the unembedded DNA from the solution. If the solution is extremely viscous, a good emulsion was not made, and many of the cells are not embedded in agarose.

8. The quality of the beads can be determined by electrophoresing a small amount of beads in a normal agarose gel. Only a small amount of DNA (2–5%) should migrate out of the well.

9. The most efficient and reproducible method for pipeting the digested beads is with a wiretrol (micropipet with plunger; Fisher, Pittsburgh, PA). The beads can be added to the well and evened out with buffer.

10. If the buffer is circulated during electrophoresis, it is important to seal the beads in the well with melted agar.

11. For each bead prep, a small amount is digested with *Eco*RI for 1 h and electrophoresed using a standard agarose gel apparatus. Analysis of the digested DNA can be performed to ensure that the final concentration of DNA in the bead sample is correct as well as ensuring that all of the EDTA has been removed from the bead sample before digestion with more expensive enzymes.

References

1. Jackson, P. A. and Cook, P. R. (1985) A general method for preparing chromatin containing intact DNA. *EMBO* **4,** 913–918.
2. Overhauser, J. and Radic, M. Z. (1987) Encapsulation of cells in agarose beads for use with pulsed-field gel electrophoresis. *FOCUS* **9,** 8,9.

CHAPTER 10

Preparation and Separation of Intact Chromosomes of Vertebrates by One-Dimensional Pulsed-Field Gel Electrophoresis (ODPFGE)

Jaan Noolandi and Chantal Turmel

1. Introduction

This novel way of preparing chromosomes for pulsed-field gel electrophoresis (PFGE) takes advantage of the fact that the whole chromosome population is synchronized in metaphase. This is a very important step toward their intact separation by PFGE; for instance, a standard preparation as used for digestion with rare-cutter enzymes shows a different pattern of resolution, characterized by diffuse bands and nonspecific migration (*see* Chapter 7). Here, vertebrate chromosomal DNA was prepared by a modified chromosome isolation procedure for flow cytometry *(1)*. This procedure involves lysis of cells (blocked in metaphase by colcemide) with digitonine in the presence of spermidine and spermine as described below. The structural integrity of metaphase-blocked chromosomes is given by the presence of spermine and spermidine, which act as chromosomal morphology stabilizers. The digestion with proteinase K is carried out as described before in order to eliminate chromosomal proteins. The pulse parameters for separating intact chicken microchromosomes by one-dimensional pulsed-field gel electrophoresis (ODPFGE) are also given.

From: Methods in Molecular Biology, Vol. 12: *Pulsed-Field Gel Electrophoresis*
Edited by: M. Burmeister and L. Ulanovsky

2. Materials

2.1. Materials for Preparation of Metaphase-Blocked Chromosomes for Separation by ODPFGE

1. Colcemide.
2. Spermine ($0.4M$). This stock solution can be aliquoted and frozen at –20°C.
3. Spermidine ($1M$). This stock solution can be aliquoted and frozen at –20°C.
4. PBS (Ca–Mg-free).
5. KCl ($75\ mM$).
6. Wash buffer: 15 mM Tris-Cl, pH 7.2, 2 mM EDTA, pH 8.0, 0.5 mM EGTA, 80 mM KCl, 20 mM NaCl, and 14 mM β-mercaptoethanol.
7. Isolation buffer: wash buffer containing 0.2 mM spermine, 0.5 mM spermidine, and 0.1% digitonine.
8. 1% Low-melting-point agarose dissolved in wash buffer.
9. $0.5M$ EDTA, pH 8.0.
10. Sarcosyl (*N*-lauroylsarcosine sodium salt, anhydrous mol wt 293.4).
11. Proteinase K.

2.2. Materials for Buffer, Gels, Gel Trays, Power Supplies Required for ODPFGE

As previously described. *See* Chapter 7.

3. Methods

3.1. Preparation for Intact Chromosome Separation by ODPFGE

1. Incubate growing cells confluent at 60% for ≅12 h in the presence of colcemide (0.04 mg/mL of media).
2. Pellet cells by centrifugation at $300g$ for 4 min.
3. Discard the supernate.
4. Suspend the pellet in 25 mL of PBS and centrifuge at $300g$ for 4 min.
5. Repeat Steps 3 and 4 twice.
6. Suspend the cell pellet in 5 mL of $0.075M$ KCl (hypotonic medium) to swell cells. Leave on ice for 10 min.
7. Centrifuge the suspension at $300g$ for 4 min.
8. Decant KCl solution and add 5 mL of cold wash buffer.
9. Centrifuge at $300g$ for 4 min.
10. Resuspend pellet in 1–2 mL of ice-cold isolation buffer solution.

11. Vortex twice for 10 s, then let the suspension stand on ice for 30 min. The structural integrity of metaphase-blocked chromosomes is given by the presence of spermine and spermidine, which act as chromosomal morphology stabilizers.

12. Take the clean supernatant leaving a loose pellet of debris and nuclei behind.

13. The chromosome suspension is then mixed with an equal vol of 1% LMP agarose (dissolved in wash solution and maintained at 42°C).

14. Pipet the suspension into autoclaved silicone tubing.

15 Place the tubing at 4°C for 5–10 min.

16. Allow agarose to slide out and cut into plugs 0.5-cm length. (Gentle pressure must be applied at the end of the tubing.)

17. Place the plugs in 5 vol of 0.5M EDTA, pH 8.0, 1% sarcosyl, 1 mg/mL of proteinase K.

18. Incubate at 50°C for 48 h.

19. Wash plugs three times in 0.5M EDTA before using or store at 4°C.

3.2. Separation of Intact Chicken Microchromosomes by ODPFGE

The cytological data obtained from different chicken cell lines show that they contain about 17 pairs of individually identifiable macrochromosomes, and 22 pairs of indistinguishable microchromosomes *(2)*; the macrochromosomes account for about 85% of the genome length and the microchromosomes *(3)* make up the remaining 15%.

The pulse durations t_+ and field intensities E_+ used for the separation shown in Fig. 1 are listed in Table 1; the ratios $R_t = t_+/t_- = 2.5$ and $R_E = E_+/E_- = 2.0$ were kept constant (*see* Chapter 7 for ODPFGE). The short-duration/high-intensity pulses separate the smaller molecules, whereas the long-duration/low-intensity ones serve to separate the larger ones. The total time for separation was 175.8 h. Note that the field intensity decreases as the pulse duration increases; a simple variation of either t_+ or E_+ alone does not work because at high fields, Mbp-size chromosomes have size-independent mobilities at high pulse frequencies, and frequently do not even form bands at lower pulse frequencies (data not shown). However, with a lower field and a 3000 s pulse duration, we were able to separate the three *S. pombe* chromosomes. The lower limit of this distribution of pulse durations was determined in earlier work, where $t_+ = 75$ s was shown to separate molecules in the

Noolandi and Turmel

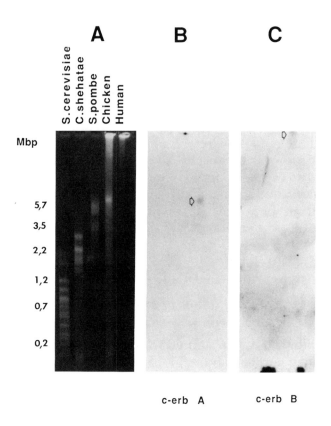

c-erb A   c-erb B

Fig. 1. Separation of yeast chromosomes and chicken microchromosomes by one-dimensional pulsed-field electrophoresis; the hybridization with chromosome-specific probes confirm the separation of the chicken microchromosomes from the macrochromosomes *(see text)*. The pulse conditions are given in Table 1. A BRL gel box model H-1, with 41 cm between the electrodes, was used with 0.8% agarose gel and 1X TBE buffer.

size range of 0.1–0.4 Mbp. Assuming that the critical pulse times scale roughly proportional to $(Mbp)^{+1}$ $(E_+)^{-1}$, the upper limit of 6000 s and 0.97 V/cm was selected to separate molecules of up to 10–40 Mbp, i.e., the range between the vertebrate macrochromosomes and the yeast chromosomes.

Figure 1 shows the ethidium bromide stained chromosome bands formed by the three yeast strains, as well as the band obtained from the preparation of intact chicken chromosomes from cell line DU-24 *(4)*. The yeast chromosomes are all resolved in a

Table 1
Pulse Conditions for Separation Shown in Fig. 1

Pulse duration, t_+, s	Electric field E_+ V/cm	Pulse duration t_+, s (continued)	Electric field E_+ V/cm (continued)
75	2	1800	1.51
100	2	2000	1.46
150	2	2100	1.44
180	2	2200	1.41
200	2	2300	1.39
250	2	2400	1.39
300	1.97	2500	1.36
350	1.95	2600	1.34
400	1.93	2700	1.32
450	1.90	2800	1.29
500	1.88	2900	1.27
650	1.85	3000	1.27
800	1.83	3500	1.22
950	1.83	4000	1.10
1100	1.76	4500	1.05
1250	1.71	5000	1.02
1400	1.66	5500	1
1600	1.59	6000	0.97

The 75–3000 s pulses were applied in total for 4.4 h each, and the 3500–6000 s pulses were applied in total for 7.3 h each, with $R_t = t_+/t_- = 2.5$ and $R_E = E_+/E_- = 2.0$ constant for each pulse of a given duration. Total time of separation was 175.8 h.

monotonic relationship between size and mobility (hybridization data not shown), i.e., there is no inversion effect, such as in FIGE (*see* Chapter 7), however the human chromosomes do not migrate. Figure 2 (circles) demonstrates that for the chosen pulse conditions, the mobility scales linearly with the log of mol size over the range of 0.5–6 Mbp.

To confirm that we had separated chicken microchromosomes from macrochromosomes, the gel of Fig. 1 was blotted and probed by hybridization with chicken-derived DNA probes assigned specifically to macro- and microchromosomes. Chicken-specific microchromosome DNA probes were provided by Björn Vennström (Department of Molecular Biology, Karolinska Institute, Stockholm). The pFID probe is specific for the c-*erb* A locus, which is located on a microchromosome *(5,6)* and contains the 1.3-kbp *Eco*RI/*Sma*I fragment. The

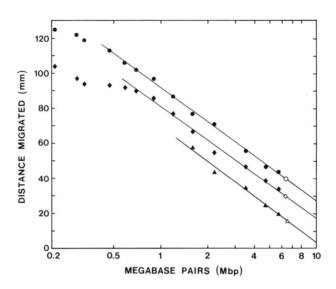

Fig. 2. Semi-log plot of distance migrated (in mm) vs log of mol size (in Mbp). The upper curve (circles) gives the position of the molecules on the gel of Fig. 1. The middle curve (diamonds) is for a similar separation that used the pulses of Fig. 1 for the following durations: 75–450 s, 4.4 h each; 500–2900 s, 4.0 h each; 3000 s for 10 h; 3500, 4000, 5000, 6000 s pulses for 11 h each. For the lower curve (triangles), the pulses and their durations were: 500–2900 s pulses, 4.4 h each; 3000 s pulses, 10 h; 3500–6000 s pulses, 73 h each. The open symbols give the position of the chicken microchromosome bands on the linear fits; the extrapolated sizes are given in the top corner.

CER probe specific to the c-*erb* B locus, which is located on a macrochromosome, contains the 4.2-kbp *Xho*I fragment *(6,7)*. DNA fragments used as probes were isolated from plasmid in low-melting-point agarose and radiolabeled by the random primer method *(8)*. Southern transfer to Gene Screen Plus (New England Nuclear Corp.) and hybridization were carried out as described before *(9)* with each probe separately.

Probe identity was confirmed by hybridization to a standard *Hind*III, *Bam*HI, and *Eco*RI chicken DNA blot (data not shown). Figure 1 demonstrates that the c-*erb* A probe, assigned to microchromosomes, hybridized only with the band resolved in the gel (panel B) and the c-*erb* B probe, assigned to macrochromosomes, gives a signal with the loading well (panel C). Crossed hybridization of those two filters showed the same results. This confirms the separation of the microchromosomes; the latter appear to be very close in size since the band formed is only 1–2 mm wide.

Figure 2 shows results for three different separations presented on a semilog plot of distance migrated vs log mol size. For the lower curve, only $t_+ \geq 500$ s pulses were used; as expected, the <1.5 Mbp molecules were then in a compression zone between 62 and 70 mm. For the middle curve, more weight was given to the long and short pulses of the sequence; the curve is linear for >1 Mbp molecules. The upper curve corresponds to Fig. 1, and provides a fairly uniform distribution of pulses, with slightly more weight given to the longer pulses. The open symbols for each separation give the position of the microchromosome band on the linear fit; from these points, the estimated size of the chicken microchromosomes appears to be equal to or larger than 6.4 Mbp. The successful separation of the intact microchromosomes may be related to the elimination of replication forks by blockage in metaphase. It has been suggested previously that the lack of resolution for these molecules could be related to nonlinear conformations, such as replication forks, but the cells were not blocked in metaphase in this case *(10,11)*. The metaphase-blocked chromosome preparation is clearly a very important step in the separation of the microchromosomes. Figure 3 shows the comparison between a metaphase-blocked preparation (lane marked chicken) and a normal preparation used for restriction enzyme digestion (lane marked chicken*). In the normal preparation, the band is so diffuse that it is barely visible.

The appropriate choice of pulse conditions as well as the sample preparation protocol made it possible to obtain a large range of mol sizes, from 0.1 Mbp to more than 6.0 Mbp, on the same gel without band inversion. This observation shows that one-dimensional pulsed-field electrophoresis can be properly applied to genomic mapping. The linearity of the plot of Fig. 3 for molecules up to ≈6–7 Mbp can be a powerful tool for the study of long-range restriction mapping, especially if coupled with a partial digestion strategy. Together with an electrophoretic process that permits a high resolution in a specific size range, this procedure helps to bridge the gap between conventional molecular mapping techniques and cytogenetic analysis.

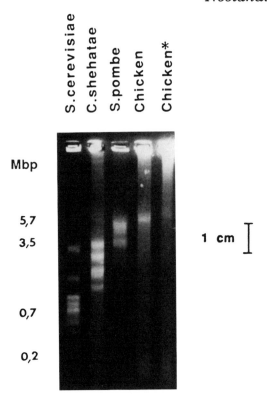

Fig. 3. The separation was carried out with pulse conditions similar to Fig. 1. The intact chicken chromosomes were prepared as described previously by blockage of the cells in metaphase (lane labeled chicken) and by the standard procedure used for genomic digest by restriction enzyme (lane labeled chicken*).

References

1. Lalande, M., Kunkel, L. M., Flint, A., and Latt, S. A. (1984) Development and use of metaphase chromosome flow sorting methodology to obtain recombinant phage libraries enriched for parts of the human X chromosome. *Cytometry* **5**, 101–107.
2. Owen, J. J. T. (1965) Karyotype studies of *Gallus domesticus*. *Chromosoma* **16**, 601–608.
3. Stubblefield, E. and Oro, J. (1982) The isolation of specific chicken macrochromosomes by zonal centrifugation and flow sorting. *Cytometry* **2**, 273–281.
4. Morais, R., Desjardins, P., Turmel, C., Zinkewich-Peoti, K. (1988) Development and characterization of continuous Avian cell line depleted of mitochondrial DNA. *In Vitro Cell Develop. Biol.* **24**, 649–658.

5. Sap, J., Muñoz, A., Damm, K., Goldberg, Y., Ghysdael, J., Leutz, A., Beug, H., and Vennström, B. (1984) The c-*erb*-A protein is a high-affinity receptor for thyroid hormone. *Nature* **324**, 635–640.

6. Symonds, G., Stubblefield, E., Guyaux, M., and Bishop, J. M. (1984) Cellular oncogenes (c-*erb*-A and c-*erb*-B) located on different chicken chromosomes can be transduced into the same retroviral genome. *Mol. Cell. Biol.* **4,**1627–1630.

7. Lax, I., Johnson, A., Howk, R., Sap, J., Bellot, F., Winkler, M., Ullrich, A., Vennström, B., Schlessinger, J., and Givol, D. (1988) Chicken epidermal growth factor (EGF) receptor: cDNA cloning, expression in mouse cells, and differential binding of EGF and transforming growth factor alpha. *Mol. Cell. Biol.* **8,** 1970–1978.

8. Feinberg, A. P. and Vogelstein, B. (1983) A technique for radio-labelling DNA restriction endonuclease fragments to high specific activity. *Anal. Biochem.* **132,** 6–13.

9. Higgins, M. J., Hansen, M. F., Cavenee, W. K., and Lalande, M. (1989) Molecular detection of chromosomal translocations that disrupt the putative retinoblastoma susceptibility locus. *Mol. Cell. Biol.* **9,** 1–5.

10. Smith, C. L., Warburton, P. E., Gaal, A., and Cantor, C. R. (1986) Analysis of genome organization and rearrangements by pulsed field gradient gel electrophoresis, in *Genetic Engineering* (Setlow, J. K. and Hollaender, A., eds.), Plenum, New York, vol. 8, pp. 44–71.

11. Barlow, D. P. and Lehrach, H. (1987) Genetics by gel electrophoresis: The impact of pulsed field gel electrophoresis on mammalian genetics. *Trends Genet.* **3,** 167.

CHAPTER 11

Isolation of Plant DNA for Pulsed-Field Gel Electrophoresis

Kun-Sheng Wu, Marion S. Röder, and Martin W. Ganal

1. Introduction

A prerequisite for physical mapping by pulsed-field gel electrophoresis (PFGE) is the saturation of a given genome or chromosomal region with single copy markers. One possible way to achieve this goal is by construction of a saturated restriction fragment length polymorphism (RFLP) map. Although RFLP maps are now available for many plant species, only a few systems provide the high density of markers (at least one marker every 1000 kb) required for long-range physical mapping using PFGE. At present, only four plant systems, *Arabidopsis* *(1,2)*, tomato *(3)*, potato *(4,5)*, and rice *(6)*, have a sufficient density of markers. These species are characterized by relatively small genomes when compared to other plants, with an average distance between individual RFLP markers of 400–800 kb. PFGE in combination with the digestion of DNA, using rare-cutting restriction enzymes, is able to bridge these gaps, and will allow the construction of long-range physical maps for regions of these plant genomes *(7)*. PFGE has already been used in a number of species to construct long-range restriction maps of a number of gene families and repeated DNA sequences *(8–12)*.

From: *Methods in Molecular Biology, Vol. 12: Pulsed-Field Gel Electrophoresis*
Edited by: M. Burmeister and L. Ulanovsky
Copyright © 1992 The Humana Press Inc., Totowa, NJ

The isolation and digestion of plant DNA for PFGE can be subdivided into three steps. First, since plant cells have a stable cell wall and a large vacuole, intact protoplasts are liberated from plant tissue or cell cultures via digestion with cell wall degrading enzymes, and purified. Protoplasts are then embedded into agarose blocks or beads, and the DNA is purified by extensive digestion with proteinase K. The resulting high-mol-wt DNA is then cut with rare-cutting restriction enzymes and separated on pulsed-field gels. It should be emphasized that, while step one is unique with respect to different plants, the remaining steps are identical regardless of the type of protoplasts or tissue.

The procedures reported here are a summary of the experience obtained in this laboratory over the last two years and, because of our focus on crop plants, do not consider *Arabidopsis*, for which the reader is referred to the published literature *(13,14)*. We provide here procedures for a number of important crop plants, including some for which no detailed procedure has been published to date. Furthermore, these procedures have been used for both physical mapping by means of PFGE, and the construction of yeast artificial chromosomes in tomato and potato (ref. *15, see also* Chapter 16 for methods for constructing such libraries).

This chapter provides three different protocols for the isolation of protoplasts from several dicotyledonous plants, rice, wheat, and barley, and one for suspension cell cultures of tomato. However, we recommend that the individual researcher becomes familiar with, not only the protocol for the plant of interest, but also with the procedures for the other plants described here. In addition, the Notes section contains many helpful hints detailed in the other procedures that might aid in trouble shooting. Finally, these protocols are not perfect, and by combining steps from different procedures it is certainly possible to further optimize the presented techniques.

2. Materials

2.1. Protoplast Isolation for Several Dicotyledonous Plants

1. Protoplast buffer: 0.5M mannitol, 20 mM 2[N-morpholino]-ethanesulfonic acid (MES). Adjust to pH 5.6 with KOH. Autoclave and store at room temperature.
2. Digestion buffer: Protoplast buffer with 1% cellulase Onozuka RS (Yakult Honsha Co., Ltd., Tokyo, Japan) and 0.05% Pectolyase Y-23

(Seishin Pharmaceutical Co., Ltd., Tokyo, Japan). Prepare protoplast buffer, then add enzyme powder and stir. Freeze at –20°C. Thaw just before use and mix well. It is possible to refreeze thawed digestion buffer.
3. 150-mm petri dishes.
4. 80-µm and 30–40-µm sieves made from polyester microfilament or equivalent.
5. Hemocytometer (0.2 mm depth).
6. 1% low-melting agarose (BRL ultrapure or FMC) in protoplast buffer. Boil and keep at 50°C.
7. Plug mold or slot-blot apparatus.
8. ESP: 0.5M EDTA, pH 9–9.5, 1% sarkosyl, 1 mg/mL proteinase K.

2.2. Protoplast Isolation from Rice

1. Protoplast buffer: 16% mannitol, 20 mM MES, 1% dextran sulfate, pH adjusted to 5.6 with KOH. Sixteen percent mannitol is difficult to dissolve at room temperature. This can be overcome by heating the solution to 80°C.
2. Enzyme solution: 1.5% cellulase Onozuka RS, 0.15% pectolyase Y-23 in protoplast buffer. Cool heated protoplast buffer to room temperature before adding the enzyme.
3. 150 mm petri dishes.
4. Miracloth (Calbiochem, La Jolla, CA).
5. Washing solution: 230 mM NaCl, 166 mM CaCl$_2$, 7 mM KCl, pH 5.6 with 1M NaOH.
6. Vacuum flasks.
7. Hemocytometer.
8. 80-µm and 30–40-µm sieves.
9. 1% low-melting agarose (BRL ultra pure or FMC) in washing solution. Boil and keep at 50°C.
10. ESP: 0.5M EDTA, pH 9–9.5, 1% sarkosyl, 1 mg/mL proteinase K.

2.3. Protoplast Isolation from Wheat and Barley

1. Protoplast buffer: 0.5M sorbitol. Adjust to pH 6.0 with KOH. Autoclave and store at room temperature.
2. Digestion buffer: Protoplast buffer with 0.05% Pectolyase Y-23 and either 1% Cellulase Onozuka RS or 1.5% Cellulysin (Calbiochem, La Jolla, CA). Freeze in aliquots at –20°C. Thaw just before use.
3. Follow steps 3–8 in Section 2.1., except that a 30–4-µm sieve is not needed.

2.4. Protoplast Isolation
from Suspension Cultures

1. Protoplast buffer: 0.5M mannitol. Adjust to pH 6.0 with KOH. Autoclave and store at room temperature.
2. Digestion buffer: 1% Macerase (Calbiochem, La Jolla, CA) and 2% Cellulysin (Calbiochem) in 0.5M mannitol, pH 5.5 with KOH.
3. Follow steps 3–8 in Section 2.1. except that a 30–40-μm sieve is not needed.

2.5. Restriction Enzyme Digests

1. TE 10/10: 10 mM Tris, 10 mM EDTA, pH adjusted to 8.0 with NaOH.
2. PMSF: Dissolve PMSF at a concentration of 100 mM in isopropanol and use it the same day. PMSF is very toxic. Wear gloves and a mask.
3. TE 10/1: 10 mM Tris, 1 mM EDTA, pH 8 adjusted with HCl.
4. 40 mM Spermidine (trihydrochloride).
5. 10 mM DTT.
6. Restriction enzyme buffer (10X); rare-cutting restriction enzymes.
7. ES: 0.5M EDTA, pH 9–9.5, 1% sarkosyl.

3. Methods
3.1. Protoplast Isolation
for Several Dicotyledonous Plants

1. Grow plants in the greenhouse or growth chamber (*see* Note 1). Usually they can be used until they reach the flowering stage, or as long as they have young, expanding leaves. Harvest young leaves and transfer them to protoplast buffer.
2. Cut leaves repeatedly from the midvein into strips of 1–2 mm so that the leaf is intact but feathered (*see* Note 2). Transfer into a petri dish with approx 75 mL of digestion buffer. Cover the petri dish with one dense layer of leaves (approx 2–4 g). Shake gently at room temperature.
3. Incubate in digestion buffer for the required time, i.e., approx 6 h for tomato leaves (*see* Note 3).
4. Remove residual leaf pieces and rinse them by shaking in a small amount of protoplast buffer to release all protoplasts. Filter the protoplasts sequentially through the 80-μm and 30–40-μm sieves. Rinse the petri plates thoroughly with protoplast buffer since protoplasts tend to settle to the bottom of the petri plates.
5. Transfer the protoplasts in digestion buffer into 50-mL centrifuge tubes and spin for 5 min at 200g at room temperature. Discard the supernatant and resuspend the protoplasts in 0.25–0.5 vol protoplast buffer.

6. Repeat Step 5 once (*see* Note 4).
7. Count an aliquot from the resuspended protoplasts using the hemocytometer and calculate the total number of protoplasts.
8. Spin protoplasts again. Remove the supernatant with a pipet. Resuspend the protoplasts in protoplast buffer so that a final concentration of $2–4 \times 10^7$ is achieved (*see* Notes 5–7).
9. Mix protoplasts with an equal vol of 1% low-melting agarose in protoplast buffer and transfer to a prechilled mold. Transfer the mold to 4°C for 10–15 min to solidify the agarose.
10. Transfer blocks into at least five times the combined block vol of ESP and incubate at 50°C for approx 36 h. Discard the ESP solution and add new ESP (same amount as before). Continue the incubation at 50°C for another 24 h.
11. Following Step 10, blocks can be stored in ESP for up to several months without significant degradation, or processed directly for digestion with restriction enzymes.

3.2. Protoplast Isolation from Rice

1. Grow rice plants in pots in a greenhouse until they show maximal tillering (*see* Note 8), which is approx 50 d after sowing.
2. Three days before protoplast isolation, cut off the leaf blades above the auricles (*see* Note 9). Keep the remainder of the plants in a dark room or cover them with black plastic bags.
3. Cut the plants at the base and wrap them in wet paper towels to prevent drying.
4. Remove all green leaf sheaths from each tiller. Take only the whitish portions (*see* Note 10) of the young leaf sheaths and transfer them into protoplast buffer.
5. Cut the sheaths with a razor blade into thin sections (2–3 mm) in protoplast buffer.
6. Collect the sheath pieces in a tea sieve and transfer them into a vacuum flask containing enzyme solution. For 10 g of tissue, use a minimum of 50 mL of enzyme solution. Apply a vacuum. Shake several times to mix and leave under vacuum for 20–30 min to infiltrate the enzyme (*see* Note 11).
7. Release vacuum and pour the tissue/enzyme mixture into a large plastic petri dish (150 mm), seal with parafilm and incubate on an orbital shaker (approx 40 rpm) for 3–4 h.
8. Check for release of protoplasts with a microscope after 3 h of incubation (*see* Note 12).
9. Pour the tissue/enzyme mixture sequentially through one layer of miracloth, 80-µm, and 30–40-µm sieves into a flask. Rinse the petri

dish with two or more vols of washing solution (*see* Note 13). Transfer to 50-mL centrifuge tubes.

10. Spin for 10 min at 200*g.*
11. Pour off the supernatant, replace with 10 mL of washing solution, and gently resuspend the pellets. Combine three tubes and fill up to 50 mL with washing solution. Centrifuge again using the above conditions.
12. Carefully pour off the supernatant. Resuspend pellets and combine them in 50 mL of washing solution.
13. Count protoplasts using a hemocytometer under a microscope.
14. Pellet protoplasts, pour off supernatant, and adjust the density with washing solution to 4×10^7/mL, or as desired (*see* Note 14).
15. Add an equal vol of 1% low-melting agarose, mix well, and transfer the mixture to the prechilled mold. Keep at 4°C for 10 min to solidify the agarose.
16. Transfer blocks into at least five times the vol of ESP and incubate at 50°C for 36 h. Discard the ESP and replace it with new ESP. Continue incubation at 50°C for another 24 h.

3.3. Protoplast Isolation from Wheat and Barley

1. Harvest the young basal part of fresh leaves (approx two-thirds of the leaf) from greenhouse plants (*see* Note 15), and cut the leaves into approx 5-cm long pieces using scissors. Do not take leaves that are infected with fungi or virus (*see* Note 16).
2. Slice leaf pieces along the veins with a razor blade and transfer to protoplast buffer. When all samples have been processed, exchange protoplast buffer with digestion buffer (*see* Note 17).
3. Incubate approx 4 h with gentle shaking at room temperature (*see* Note 18). Remove the remaining leaf pieces with forceps and filter the protoplast suspension through an 80-μm sieve.
4. Transfer the protoplast solution to centrifuge tubes and spin for 5 min at approx 200*g.* Remove the supernatant and resuspend protoplasts in protoplasting buffer.
5. Repeat Step 4, then count an aliquot of the resuspended protoplasts in a hemocytometer and calculate the total number of protoplasts.
6. Sediment protoplasts again, remove the supernatant with a pipet, and resuspend protoplasts to a final density of $1–2 \times 10^7$/mL in protoplasting buffer (*see* Note 19).
7. Proceed with Step 9 in Section 3.1.

3.4. Protoplast Isolation
from Suspension Cultures

1. Transfer suspension culture (*see* Note 20) to centrifuge tubes and sediment for 5 min at approx 200 *g*. Discard supernatant and resuspend cells in an equal vol of digestion buffer.
2. Incubate for approx 3 h at 25°C with gentle shaking (*see* Note 21). Filter protoplast suspension through an 80-μm sieve. Proceed with Step 5 of Section 3.1.

3.5. Restriction Enzyme Digests

1. Wash blocks three times for 60–90 min each, in at least 10–20 vols of TE 10/10 at 50°C. In the first two washing cycles, add PMSF to a final concentration of 1 m*M*.
2. Wash blocks at 50°C two additional times in TE 10/1 for 60 min (*see* Note 22).
3. For digestions, use 80–100-μL blocks. If your blocks are larger, cut them into pieces. The digestion scheme we use is as follows:
   a. Block (approx), 80 μL
   b. Water, 70 μL
   c. Restriction enzyme
   d. buffer (10X conc.), 20 μL
   e. DTT, 5 μL
   f. Spermidine, 20 μL
   g. Restriction enzyme, 30–40 U.
   Incubate several hours or overnight (*see* Notes 23 and 24).
4. Stop digests by the addition of 500 μL ES. Digested plugs can be stored in ES at 4°C for several months.

4. Notes

4.1 Protoplast Isolation
for Several Dicotyledonous Plants

1. We usually use young plant material from the greenhouse, however, it is also possible to use in vitro grown plants. The use of high intensity lights should be avoided, because often the plants are stressed and accumulate anthocyanins and starch in their tissue. If you find that your tissue contains excessive starch, place the plants in the dark for 1–2 d before protoplast isolation.
2. Feathering leaves along the veins reduces the amount of debris significantly and makes removal of undigested material and subsequent filtration much easier.

3. This procedure has been used successfully for several cultivars of the following plant species: Tomato (*Lycoperiscon esculentum*), potato (*Solanum tuberosum*), *Brassica oleracea,* cucumber (*Cucumis sativus*), mung bean (*Vigna radiata*), and zucchini (*Cucurbita pepo*). The only modification was the time for cell wall digestion. Initially, for a new cultivar of these species, a small amount of leaves should be cut, digested with the protoplasting enzymes, and monitored every 60 min under an inverted microscope. The ideal time point for use is when many protoplasts are released into the medium, but no lysis of the protoplasts has yet occurred. For tomato leaves, depending on the variety, this is after 5–7 h. Both potato and *Brassica* leaves need less time for protoplasting (approx 4–5 h). Primary leaves from cucumber, zucchini, and mung bean require approx 2–3 h for digestion. In vitro grown material, in general, needs much less time for digestion than greenhouse material. For a new plant species, it is advisable to try several different mannitol concentrations (0.2–0.7M) with different concentrations of enzymes (cellulase 0.5–2%, pectolyase 0.05–0.2%) on a small amount of tissue in 60-mm petri dishes. Digestion is easily monitored every hour under an inverted microscope to determine optimal concentration(s). According to these results the experiment can then be scaled up with 150-mm petri dishes.

4. If the digestion has proceeded too far, or if you have a lot of starch grains in your protoplast preparation, it sometimes helps to resuspend the protoplast pellet very gently and stop when you reach the starch pellet underneath the protoplast pellet. Transfer the resuspended protoplasts into a new centrifuge tube. We usually observe a reduced yield in such preparations.

5. At this point the pellet can also be resuspended in SCE (1M sorbitol, 0.1M sodium citrate, 0.06M EDTA, pH 7.0), and used for the preparation of microbeads (R. Wing, personal communication and Chapter 9).

6. Do not embed the protoplasts in blocks that are larger than 250 µL because it is very hard to lyse and wash them.

7. The amount of DNA is highly dependent on the genome size. For example, a final protoplast concentration of 1×10^7/mL in tomato (diploid genome size, 2 pg) is equivalent to approx 20 µg of DNA. For wheat, the same amount of protoplasts would result in approx 300 µg of DNA (diploid genome size, 30 pg). The important number is the number of protoplasts, since it reflects the number of copies of a given genome.

4.2. Protoplast Isolation from Rice

8. The optimal stage for protoplast isolation from rice is the time from maximum tillering to the prepanicle initiation stage. During this interval, the stems are vigorous and the structure of the sheath is "loose," making it easier for the enzymes to penetrate the tissue. Young plants usually have a hard texture that hinders the diffusion of the enzyme into the tissue, resulting in low protoplast yield. We have used 10-d-old seedlings with only limited success. To maintain the plants in the vegetative stage, one can keep rice plants under long day conditions.

9. Cutting the blades off and keeping them in the dark before harvesting the tissue proved to be an effective way to reduce starch granules in the cells. Accumulation of starch in the cells will cause bursting of the protoplasts during the isolation process, and in extreme cases, might interfere with restriction enzyme digestion.

10. The whitish young portions of the leaves and sheaths usually produce a high yield and good quality protoplasts. We routinely obtain about 2×10^7 protoplasts/g of tissue. Green sheaths can also be used, although the yield is not as high as with whitish tissue. In case one uses both green and whitish tissue, they should be kept separate, since green tissue requires a longer incubation time in the protoplasting enzymes.

11. A high osmoticum is necessary for rice. At such high osmotic pressure, the protoplasts lose water and become smaller in size than cells in the original tissue. The given osmotic conditions have been tested for several rice varieties and the results were comparable. Nevertheless, it is advisable to optimize the osmotic conditions when working with new varieties. An easy way to do this is to observe the behavior of a few protoplasts under the microscope for a few minutes. If the protoplasts are increasing in size or even bursting, the concentration of the osmoticum needs to be increased. On the other hand, if the protoplasts become wrinkled, the concentration of the osmoticum should be decreased. For many rice varieties the optimum mannitol concentration is between 13 and 16%.

12. The degree of digestion should be determined empirically. Short digestion times will result in a lower yield, whereas longer times create more debris. Optimal digestion is achieved when many individual protoplasts are suspended in the buffer and a few cell clumps are still visible. Floating epidermis on top of the enzyme solution is also an indicator of optimal digestion.

13. The protoplasts do not pellet in the enzyme solution. At least two vols of washing solution should be added to one vol of protoplast mixture to ensure pelleting. If the mannitol concentration is varied from the given protocol, the osmotic value of the washing solution must be adjusted, as described in Note 11.

14. Protoplasts isolated with this method are usually contaminated with some debris. However, this does not have any detectable effect on DNA quality or digestibility. Debris removal can be accomplished with a Ficoll gradient after the first washing as follows: Add 5 mL each of 8, 5, and 3% Ficoll solution in washing buffer as a step gradient to a 50 mL centrifuge tube, gently apply the protoplasts on top of this gradient, and spin 5 min at 100g. High quality protoplasts will float at the top, and debris will settle to the bottom. Carefully transfer the protoplasts from the top layer to a new tube and continue the washing steps as described. This step could cause the loss of up to 30% of the protoplasts. Therefore, we do not recommend this step unless it is absolutely necessary.

4.3. Protoplast Isolation from Wheat and Barley

15. We have had the best experiences with leaves from plants that were 4–6 wk old. Leaf material from fresh tillers of older plants can also be used. Protoplasts from primary leaves turned out to be very fragile. Here, the incubation time in digestion buffer has to be reduced to approx 2 h, with the resulting protoplasts being very sensitive to over-incubation. Timing problems can result when many different samples have to be handled.

16. It is possible to harvest leaves a day before the experiment and store the leaf pieces floating on water at room temperature over night.

17. We use 150 mm petri dishes for this step and float the leaf pieces on the surface of the buffer. We use 40–50 mL of digestion buffer per petri dish, and cover the surface completely with leaf material.

18. Monitor the release of protoplasts under an inverted microscope at regular intervals. The best time to stop the digestion process is after many protoplasts have been released into the medium, but before protoplast lysis occurs.

19. We observed that the quality of the protoplasts correlates strongly with the amount and quality of the high-mol-wt DNA that can be obtained.

4.4. Protoplast Isolation from Suspension Cultures

20. This procedure was used for a densely growing suspension culture of the tomato cultivar, VFNT cherry. For best results, use the culture when it is in the logarithmic growth stage.
21. Monitor release of protoplasts under an inverted microscope. Stop digestion before lysis of protoplasts occurs, even if cell clumps are still visible.

4.5. Restriction Enzyme Digests

22. After these five washing steps, the blocks should be either clear or, if your sample contains a considerable amount of starch, white. Usually, in our experience, a small amount of starch does not inhibit the digestion with restriction enzymes.
23. The amount of enzyme and time of digestion should be determined empirically. The quality of rare-cutting enzymes sometimes varies between manufacturers, and we have observed degradation after overnight digestion with some enzymes.
24. The usefulness of rare-cutting restriction enzymes has to be evaluated individually for any given plant species. The frequency of cutting is, however, mostly dependent on several major factors. First, the GC-content has a major influence on the cutting frequency. For example, tomato and potato have a GC-content of 37–38% (16), whereas the GC-content of rice is 42%. This results in a twofold difference in average fragment size between tomato/potato and rice. A second factor is the amount of cytosine methylation in plants, since most rare-cutting restriction enzymes are sensitive to this type of methylation. Again, as an example, the extent of C-methylation in tomato/potato is approx 25% (16), whereas the rice genome is less methylated (18.5%). Taking these factors and others (for example, the existence of GC-islands) together, it is obvious that the average fragment size with rare-cutting enzymes is highly dependent on the species one is working with. Though rare-cutting restriction enzymes that are useful in mammalian systems are generally also suitable with tomato (Table 1), the situation is less transferable to rice. In rice, the average fragment size, even with eight bp recognition site enzymes (*Not*I, *Sfi*I), is very small (Table 1), and therefore difficult to use for the construction of long-range physical maps. To date, we have not found an enzyme that results in an average fragment size of more than 500 kb for rice (*see also* Chapters 12, 14, and 18 for more information on enzymes and additional techniques).

Table 1
Cutting Frequencies for Rare-cutting
Restriction Enzymes in Tomato and Rice

Enzyme	Average fragment size in kb*	
	Tomato	Rice
*Not*I	1000 (10)	150 (24)
*Sfi*I	650 (8)	190 (23)
*Rsr*II	800 (4)	305 (10)
*Csp*I	730 (4)	200 (7)
*Bss*HII	950 (9)	130 (24)
*Sal*I	640 (20)	150 (22)
*Sma*I	550 (12)	110 (19)
*Sac*II	760 (11)	130 (10)
*Mlu*I	670 (20)	110 (22)
*Nae*I	570 (8)	140 (9)
*Nar*I	590 (6)	150 (9)
*Nru*I	690 (4)	170 (15)

*Number in brackets is the number of RFLP markers hybridized onto blotted pulsed-field gels, to date. In the case of multiple bands for a given clone (e.g., due to methylation), only the largest informative band is reported.

Acknowledgments

We thank Steven D. Tanksley, in whose lab most of this work was carried out, for his continuous support. We also thank Gregory Martin, Rod Wing, and James Giovannoni for critically reading this manuscript and their comments. Additional support was provided by the USDA-DOE-NSF Plant Science Center at Cornell, which is gratefully acknowledged.

References

1. Chang, C., Bowman, J. L., DeJohn, A. W., Lander, E. S., and Meyerowitz, E. M. (1988) Restriction fragment length polymorphism linkage map for *Arabidopsis thaliana. Proc. Natl. Acad. Sci. USA* **85**, 6856–6860.
2. Nam, H. -G., Giraudat, J., den Boer, B., Moonan, F., Loos, W. D. B., Hauge, B. M., and Goodman, H. M. (1990) Restriction fragment length polymorphism linkage map of *Arabidopsis thaliana. Plant Cell* **1**, 699–705.
3. Tanksley, S. D. and Mutschler, M. A. (1990) Linkage map of the tomato (*Lycopersicon esculentum*) (2N=24), in *Genetic Maps* (O'Brien, S. J., ed.), Cold Spring Harbor Laboratory Press, Cold Spring Harbor NY, pp. 6.3–6.15.

4. Bonierbale, M. W., Plaisted, R. L., and Tanksley, S. D. (1988) RFLP maps based on a common set of clones reveal modes of chromosomal evolution in potato and tomato. *Genetics* **120**, 1095–1103.

5. Gebhardt, C., Ritter, E., Debener, T., Schachtschnabel, U., Walkemeier B., Uhrig, H., and Salamini, F. (1989) RFLP analysis and linkage mapping in *Solanum tuberosum. Theor. Appl. Genet.* **78**, 65–75.

6. McCouch, S. R., Kochert, G., Yu, Z. H., Wang, Z. Y., Khush, G. S., Coffman, W. R., and Tanksley, S. D. (1988) Molecular mapping of rice chromosomes. *Theor. Appl. Genet.* **76**, 815–829.

7. Ganal, M. W., Young, N. D., and Tanksley, S. D. (1989) Pulsed field gel electrophoresis and physical mapping in the *Tm-2a* region of chromosome 9 in tomato. *Mol. Gen. Genet.* **215**, 395–400.

8. Sørensen, M. B. (1989) Mapping the *Hor2* locus in barley by pulsed field gel electrophoresis. *Carlsberg Res. Comm.* **54**, 109–120.

9. Jung, C., Kleine, M., Fischer, F., and Herrmann, R. G. (1990) Analysis of DNA from a Beta procumbens chromosome fragment in sugar beet carrying a gene for nematode resistance. *Theor. Appl. Genet.* **79**, 663–672.

10. Van Daelen, R. A. J., Jonkers, J. J., and Zabel, P. (1989) Preparation of megabase-sized tomato DNA and separation of large restriction fragments by field inversion gel electrophoresis (FIGE). *Plant Mol. Biol.* **12**, 341–352.

11. Ganal, M. W., Lapitan, N. L. V., and Tanksley, S. D. (1991) Macrostructure of the tomato telomeres. *Plant Cell* **3**, 87–94.

12. Ganal, M. W., Bonierbale, M. W., Röder, M. S., Park, W. D., and Tanksley, S. D. (1991) Genetic and physical mapping of the patatin genes in potato and tomato. *Mol. Gen. Genet.* **225**, 501–509.

13. Guzman, P. and Ecker, J. R. (1988) Development of large DNA methods for plants: Molecular cloning of large segments of *Arabidopsis* and carrot DNA into yeast. *Nucleic Acids Res.* **16**, 11091–11105.

14. Ecker, J. R. (1990) PFGE and YAC analysis of the *Arabidopsis* genome. *Methods* **1**, 186–194.

15. Martin G. B., Ganal, M. W., and Tanksley, S. D. (1991) Construction of a yeast artifical chromosome library of tomato and identification of cloned segments linked to two disease resistance loci. Submitted.

16. Messeguer, R., Ganal, M. W., Steffens, J. C., and Tanksley, S. D. (1991) Characterization of the level, target sites and inheritance of cytosine methylation in tomato nuclear DNA. *Plant. Mol. Biol.* **16**, 753–770.

CHAPTER 12

Methyltransferases as Tools to Alter the Specificity of Restriction Endonucleases

Bruno W. S. Sobral and Michael McClelland

1. Introduction

Pulsed-field gel electrophoresis (PFGE) has allowed the resolution of very large DNA fragments from any organism. To apply PFGE to practical problems, such as genetic mapping and map-based gene cloning, it is necessary to specifically *generate* large DNA fragments that can then be separated by PFGE. Ideally, restriction enzymes would exist that could generate DNA fragments of desired sizes. Other factors being equal, and supposing that DNA sequences were random, then enzymes with larger target sequences should produce larger DNA fragments. In practice, no restriction enzymes are known to have larger than 8-bp-long target sites and, of course, DNA sequences are not random. These realities severely limit the observed sizes of DNA fragments produced by restriction enzymes acting on genomic DNA, particularly in the case of eukaryotic genomes, which are large and complex. This chapter describes enzymatic strategies to generate large DNA fragments and statistical tools that can aid researchers in choosing the restriction enzymes that are most likely to generate large fragments in the genome in question, if a sequence data base can be investigated. A recent review by McClelland and Nelson *(1)* focuses on the same technology as described here.

From: *Methods in Molecular Biology, Vol. 12: Pulsed-Field Gel Electrophoresis*
Edited by: M. Burmeister and L. Ulanovsky
Copyright © 1992 The Humana Press Inc., Totowa, NJ

2. Materials

2.1. Chemicals, Buffers, and Solutions

Unless otherwise noted, we purchased FastLane (for PFGE gels) and SeaPlaque (for making plugs) agarose from FMC (Rockland, ME), chemicals from Sigma Chemical Co. (St. Louis, MO) or Fisher Scientific Co. (Pittsburgh, PA), and restriction enzymes from Stratagene Cloning Systems, Inc. (La Jolla, CA), Boehringer Mannheim Biochemicals (Indianapolis, IN), or New England Biolabs (Beverly, MA).

1. 10X Potassium glutamate buffer (KGB): $1M$ Potassium glutamate, 250 mM Tris-acetate, pH 7.6, 100 mM magnesium acetate, 1 mg/mL acetylated, nuclease-free bovine serum albumin. Filter-sterilize and store in aliquots at –20°C. To use, add 1 mM DTT to 10X stock, then dilute according to McClelland et al. *(2)*, as shown in Tables 1 and 2. Potassium glutamate $(1M)$ can be substituted by $1M$ potassium acetate, if desired.
2. 10X Potassium glutamate buffer for methylases (KGBM) *(3)*: KGB without magnesium acetate, supplemented with 100 mM Na$_2$EDTA. After preparing the 1X solution for a reaction, add 100 µM *S*-adenosylmethionine (SAM) as the methyl donor. It is important that the SAM solution be prepared and immediately stored in small aliquots (10–50 µL) at –20°C; SAM is unstable, particularly at or above room temperature or at alkaline pH, so unused portions should not be repeatedly frozen and thawed.
3. Modified TBE PFGE running buffer (5X) *(4)*: 5M Tris-borate, pH 8.3, 1 mM Na$_2$EDTA.
4. THE plug wash/storage buffer (1X) *(4)*: 10 mM Tris-HCl, pH 8.0, 10 mM Na$_2$EDTA. Sterilize by autoclaving.
5. ESP buffer: 0.5 M Na$_2$EDTA, pH 9.0, 1% *N*-lauryolsarcosine, sodium salt, and 1 mg/mL proteinase K. Mix the components, except the proteinase K, and sterilize by filtration. Add the proteinase K and predigest for at least 1 h at 60°C before using on plugs.

2.2. Pulsed-Field Electrophoresis Equipment

Chapters 1 through 7 explain the general types of PFGE equipment and how they work. A variety of these are available from different companies. Most are expensive and some use proprietary connectors, making it difficult to mix and match equipment made from different suppliers. Laboratories on a low budget can use field inversion gel electrophoresis (FIGE *[5]; see* Chapter 1). FIGE can be done in conventional electrophoresis chambers, provided that buffer recirculation and cooling can be provided. We typically maintain the running temperature of our gels at approx 10–12°C. For many purposes, especially when the DNA fragments are smaller than 1 Mbp in size, FIGE is sufficient. To resolve larger DNA fragments without large investments, the electrophoresis device (ED) described by Schwartz et al. *(6)* is a viable option. ED units can be built cheaply by skilled shop personnel that are usually available at most universities. We have been particularly pleased with transverse alternating-field electrophoresis (TAFE), which is commercially available through Beckman Scientific Co. (Palo Alto, CA), or can be built as described by Gardiner et al. *(7)*. The newer TAFE II allows a larger number of samples to be run in one experiment, thereby solving one of the greatest limitations of the previous TAFE unit. Pulsing parameters and other PFGE parameters also are described in Chapters 1 through 7.

3. Method

3.1. Numerical Methods for Estimating Average Fragment Size for Restriction Enzymes

It has been known for 30 years that the distribution of bases in a naturally occurring DNA sequence is not random *(8,9)*. Since then, various studies have substantiated and further investigated nonrandom nucleotide distribution in natural DNAs and attempted to explain its biological significance (for examples and further study, *see* refs. *10–22*). For our purposes, suffice it to say that nonrandomness is a reality that makes choosing restriction enzymes that cut infrequently more difficult and less reliable. If DNA sequence data are available, then it is possible to make predictions about which restriction enzymes are expected to cut the DNA least frequently.

3.1.1. Markov Chain Analysis

To do Markov chain analyses for prediction of frequency of target sites for a given genome, it is necessary to first generate a data base containing the sequenced regions of the genome(s) in question. The data base will then be used to extract mono-, di-, or trinucleotide frequencies, depending on whether zero-, first-, or second-order Markov chains will be generated. The larger the data base available, the closer the mono-, di-, and trinucleotide frequencies are to their respective probabilities of occurrence in the genome in question *(10,23,24)*.

Table 1
The Effect of Potassium Glutamate Buffers on the Activity of Restriction Endonucleases[a]

Enzyme	0.5X	1.0X	1.5X	2.0X KGB	Enzyme	0.5X	1.0X	1.5X	2.0XKGB
AatII	3	3	3	2	MboII	2	3	3	2
AccI	3	3	1	0	MluI	3	3	4	4
AhaII	0	2	1	1	MspI	2	1	1	0
AluI	2	3	3	1	MstII	3	3	2	1
ApaLI	3	1	0	0	NarI	2	1	0	0
AAvaI	2	2	2	1	NciI	3	3	2	1
AvaII	1	3	2	0	NcoI	2	3	3	3
AvrII	3	3	3	2	NdeI	2	3	2	1
BaiI	2	2	3	3	NheI	3	3	3	2
BamIII	3	3	2	2	NlaIV	4	4	3	2
BanI	3	3	3	2	NotI	2	2	2	2
BanII	2	3	3	2	NruI	2	3	3	3
BbvI	3	3	3	2	NsiI	1	1	2	3
BdiI	2	3	1	0	PflMI	2	3	4	3
BglI	4	4	4	3	PstI	3	3	3	3
BglII	2	3	2	1	PvuI	2	3	3	3
BspI286	1	1	1	1	PvuII	3	3	2	1
BspMII	1	2	3	3	RsaI	3	3	2	1
MssHIII	3	3	2	1	RsrII	3	1	0	0
BstEII	3	3	3	3	SacI	3	3	1	0
BstNI	2	2	2	1	SacII	3	2	2	1
BstXI	3	3	2	1	SalI	0	1	3	3
ClaI	2	3	2	2	Sau3A	2	3	3	4
DdeI	2	3	3	3	Sau961	3	3	2	1
DpnI	2	2	2	2	ScaI	1	3	3	2
DraI	3	3	1	1	ScrFI	1	2	2	2
EaeI	3	3	2	1	SfaNI	0	1	2	3
EagI	0	1	2	3	SfiL	3	3	2	1
EcoO109	2	3	1	1	SmaI	3	3	3	2
EcoRI	3	3	3	3	SnaBI	3	2	1	0
EcoRV	2	3	3	3	SspI	2	2	3	2
FnuDII	2	2	1	1	StuI	1	2	3	3
Fnu4H	2	2	2	1	StyI	2	2	1	1
FokI	3	3	1	0	TaqI	2	3	3	1
FspI	3	3	3	1	XbaII	3	3	3	1
HaeII	3	3	2	1	XhoI	2	3	3	1
HaeIII	3	3	2	1	XmnI	3	3	0	0
HgaI	3	3	2	1	T4 DNA				
HgiAI	0	1	2	3	Polymerase	2	2	3	2
HhaI	2	3	2	1	E. coli DNA				
HindII	4	4	4	3	Polymerase I	2	2	3	3
HindIII	3	3	3	3	Polymerase I				
HinfI	2	3	2	1	Klenow	2	3	3	3
HpaI	2	3	2	1	T4 DNA ligase				
HpaII	3	1	1	1	(1 mm ATP)	3	2	1	1
HphI	2	3	1	0	Reverse				
KpnI	3	2	0	0	transcriptase	2	2	2	2
MboI	3	2	0	0					

Table 2
The Effect of Potassium Glutamate Buffers
and Mg^{2+} on the Activity of DNA Methylases[a]

Enzyme	STD (Mg^{2+})	0.5X KGB	1X KGB	1X KGB (no Mg^{2+})
M.AluI	2	3	3	3
M.BamII	2	3	3	3
M.ClaI	3	3	3	3
M.dam	2	4	4	4
M.EcoRI	0	4	4	3
M.HaeIII	3	2	2	2
M.HhaI	3	3	3	4
M. HpaII	3	3	3	4
M.HphI	3	3	3	3
M.PstI	3	3	4	3
M.TaqI	2	3	3	2

[a]Activities were compared to results found in the New England Biolabs standard recommended buffer (STD). The STD (Mg^{2+}) buffer has 10 mM MgCl$_2$ instead of the the recommended EDTA. 1X KGB is 100 mM potassium glutamate, 25 mM Tris-acetate (pH 7.6), 10 mM magnesium acetate, 50 µg/mL bovine serum albumin, 1 mM 2-mercaptoethanol, and 100 µM S-adenosylmethionine(2). The 1X (no Mg^{2+}) buffer is 1X KGB without the 10 mM magnesium acetate. Activities are relative to STD: 4 = >100% activity; 2 = 50–80% activity; 1 = 20–50% activity; 0 = < 2-% activity.

3.1.1.1. GENERATION OF THE SEQUENCE DATA BASE

This can usually be done by accessing the EMBL or NIH data bases and searching for all the sequences that have been entered for your favorite organism (or group of organisms). For prokaryotes, it is only necessary to take the precaution of deleting duplicate entries, splicing the sequences head-to-tail, and adding a space at the end of every sequence so that new sequences are not created by the union of two sequenced portions of the genome that are not truly contiguous. In the case of eukaryotes, not only duplicate entries must be spliced, but

Table 1 *(see opposite)*

[a]2X KGB denotes a buffer (K$^+$ glutamate ‑ buffer)(2) that contains 200 mM potassium glutamate, 50 mM Tris-acetate (pH 7.6), 20 mM magnesium acetate, 100 µg/mL bovine serum albumin, 1 mM 2-mercaptoethanol (filter sterilized and stored at 4°C). 1.5X, 1.0X, and 0.5X KGB are dilutions of 2X KGB with distilled water. Symbols represent activity relative to recommended buffer: 4 = more active in KGB; 3 = 80–100% activity in KGB; 2 = 50–80% activity in KGB; 1 = 20–50% activity in KGB; 0 = 20% activity in KGB.

also sequences originating from cytoplasmatic genomes, such as mito-chondria and chloroplasts, must also be excluded since they do not represent the nuclear genome. Various DNA sequence analysis pro-grams exist that allow for this task; these are reviewed in Bishop and Rawlings *(25)*.

3.1.1.2. ANALYSIS OF THE SEQUENCES

Mono-, di-, and trinucleotide frequencies can be either calculated by using sequence analysis programs *(25)* or, for some organisms, mono-nucleotide frequencies (base compositions) can be found in refer-ence books, such as Fasman *(26)*. These frequencies are used to calculate the estimated frequency of a restriction enzyme target site as follows. For a 4-bp recognition site, the frequency of the sequence $N_1N_2N_3N_4$ (N_1 to N_4 are specific bases in the recognition sequence) can be calculated from dinucleotide frequencies in the data base by the following formula:

$$p(N_1N_2).p(N_2N_3).p(N_3N_4)/p(N_2).p(N_3).$$

If we used trinucleotide frequencies, then the formula would be:

$$p(N_1N_2N_3).p(N_2N_3N_4)/p(N_2N_3).$$

We can use the same general formula for restriction enzymes that recognize hexameric or octameric target sites, such as *Not*I (5'-GCGGCCGC-3'):

$$p(GC).p(CG).p(GG).p(GC).p(CC).p(CG).p(GC)/$$
$$p(C).p(G).p(G).p(C).p(C).p(G),$$

is the formula for calculating the expected frequency from dinucle-otide frequencies. A sample output from such calculations done on a data base containing nuclear sequences from the genome of the flow-ering plant *Arabidopsis thaliana* is shown in Table 3. We have also searched the data base for sites of the various restriction enzymes for which we made predictions. Data from PFGE experiments using these enzymes show that most of the predictions actually underestimate the sizes of the observed fragments (*4,27*, and data not shown).

An alternative strategy for calculating the expected frequency of target sites that requires only dinucleotide frequencies and compares well with Markov chain analysis is presented by Hong *(15)*.

Table 3
Predicted Average Fragment Size from Second-Order Markov
Chains of Potential Rarely Cutting Restriction Enzymes
in the *Arabidopsis thaliana* Nuclear Genome

Enzyme	Target site	Predicted average fragment size (in kb)			
		sample[a]	mono[b]	dinuc[c]	trinuc[d]
*Bss*HII	(GCGCGC)	18	14	37	56
*Csp*I	(CGGWCCG)	18	14	35	26
*Eag*I (EclXI)	(CGGCCG)	15	14	25	21
*Kpn*I (Asp718)	(GGTACC)	?[e]	7	10	8
*Mlu*I	(ACGCGT)	45	7	16	21
*Nae*I	(GCCGGC)	13	14	22	18
*Nar*I	(GGCGCC)	15	14	22	29
*Nhe*I	(GCTAGC)	11	7	12	12
*Not*I	(GCGGCCGC)	>91	349	789	814
*Nru*I	(TCGCGA)	10	7	10	13
*Pac*I	(TTAATTAA)	15	17	22	33
*Pvu*I (BspCI)	(CGATCG)	15	7	10	10
*Pvu*II	(CAGCTG)	5	7	6	7
*Sac*II (KspI)	(CCGCGG)	?e	14	25	24
*Sal*I	(GTCGAC)	2	7	8	10
*Sfi*I	(GGCCN₅GGCC)	30	349	470	414
*Sma*I	(CCCGGG)	13	14	15	17
*Xho*I	(CTCGAG)	6	7	7	5

[a]Number of occurrences in a data base of 90,571 bp of *Arabidopsis thaliana* nuclear
genomic DNA with ribosomal and repeat DNA removed,
[b]Calculated from mononucleotide frequency of the data base.
[c]Calculated from dinucleotide frequency of the data base.
[d]Calculated from trinucleotide frequency of the data base
? Signifies not searched in data base.

One of the main sources of inaccuracy in these methods is that
regions of the eukaryotic genome vary markedly in their sequence
organization. By adding the frequency of di- or trinucleotides from
disparate regions of the genome, one can mask significant differences.
The other main source of error is the necessarily finite nature of the
sequence in question. In finite-length sequences, there is actually a
distribution of di- and trinucleotide (or higher order) sequences, each
having frequencies that are not necessarily equal to the actual probabilities of these di- and trinucleotides in the genome. For example,
let us assume a totally random DNA sequence; there is no sequence of

length 5 nucleotides that has even one of the nucleotide frequencies equal to its statistical probability of 0.25. Likewise, there is no sequence of length 17 in which all doublets occur in equal frequency. Even much longer sequences do not accurately reflect the distributions of higher-order nucleotide sequences in the genome *(4,10–16,23,24)*. Therefore, there is a length-dependent bias between frequencies and probabilities *(10)*, which must be taken into consideration whenever numerical methods are used to estimate the frequency of restriction enzyme target sites in a given genome.

3.2. Enzymatic Methods to Alter Restriction Enzyme Specificity

Some methyltransferases are available from commercial sources, such as New England BioLabs. When the methylase is commercially available, we have included the source in parenthesis. It is important that the enzymes used for the experiments described herein be as pure as possible (*see* Notes 1 and 5). In some cases, it is necessary to purify methylases, either because they are not commercially available, or because the purity of commercial products is insufficient. Purification of methylases is described in detail by Nelson and McClelland *(28)*.

3.2.1. DpnI-Based Strategies

This technique relies on the restriction enzyme *Dpn*I (5'-G ^{m}ATC-3'), which only cuts double-stranded DNA molecules efficiently if both strands are N^6-methylated at adenine *(28;* but *see* Note 3). Effective recognition sequences *(29)* of 8.0 *(30)*, 9.0 *(31)*, 10.0 *(31,32)*, and 12.0 bp *(33,34)* have been reduced to practice by using *Dpn*I in combination with various methyltransferases. *Dpn*I is commercially available from various sources. It is critical to know whether *Dpn*I digests genomic DNA from the organism in question before using *Dpn*I-based strategies to produce large DNA fragments; this is easily done by testing conventionally prepared DNA for digestability with *Dpn*I and running the products on a conventional agarose gel. Fortunately, the majority of species do not methylate their genomes at 5'-G ^{m}ATC-3'.

Preparation of DNA in agarose beads or plugs is described in Chapters 9 and 10. Preparation of genomic DNA from plants is described in Chapter 11 and in Sobral et al. *(4)*. For the basic reaction, take plugs directly from ESP *(4)* and:

1. Wash three times with at least 5 vol of THE buffer, changing buffer at 1-h intervals.
2. Cut a slice of the plug using a glass cover slide and equilibrate twice for 20 min each in 1 mL of 1X KGBM.
3. Add 300 µL of freshly prepared KGBM and add approx 5 U of methylase/µg of genomic DNA in the plug (or use a plasmid containing the methylase gene in vivo *[34]; see also* Note 6).
4. Incubate for 2 h overnight at room temperature or at 30°C (M. *Taq*I requires 50°C). To digest the methylated DNA with *Dpn*I, carry out the following steps.
5. Remove the KGBM and replace it with 1 mL of ESP.
6. Incubate for 1–2 h at 50°C.
7. Remove the ESP and replace it with 1 mL of 1X KGB.
8. Incubate for 20 min at room temperature and repeat Steps 7 and 8 at least thrice.
9. Add 300 µL of 1X KGB and at least 5 U of *Dpn*I/µg of DNA.
10. Incubate at 37°C for 3 h, then replace the buffer and enzyme and repeat Step 9 once more.
11. Load the sample onto a PFGE gel (*see* Notes 1 and 2).

3.2.2. Crossprotections

Restriction enzymes are usually sensitive to methylation of bases within their target site *(35)*. This fact can be used to generate large DNA fragments by a method called "cross-protection" *(36)*, in which a defined subset of restriction target sites is blocked from cleavage at partly overlapping methylase/restriction endonuclease sites by prior methylation.

Crossprotections can be exemplified using the restriction enzyme *Not*I (5'-GCGGCCGC-3'), which is blocked by methylation at 5'-GCGGC ᵐCGC-3' *(35)*. Modification by M.*Fnu*DII (NEB) or M.*Bep*I (5'-ᵐCGCG-3') blocks *Not*I cleavage at overlapping sites (5'-CGCGGCCGC-3', which is equivalent to 5'-GCGGCCGCG-3') and increases the apparent specificity of *Not*I digestion about twofold *(37)*. For crossprotections:

1. Wash the plugs at least three times in at least 5 vol of THE buffer.
2. Cut a slice of plug and add 1 mL of 1X KGBM; equilibrate for 20 min at room temperature. Repeat at least once.
3. Add 100 µL of 1X KGBM and 20–160 U of M.*Fnu*DII or M.*Bep*I for each µg of genomic DNA (or use plasmid in vivo *[34]; see* Note 6).
4. Incubate overnight at room temperature or at 30°C.

5. Inactivate the methylase by incubating at 50°C for 15 min. To digest methylated DNA with *Not*I carry out the following steps.
6. Change buffer by adding 1 mL of 2X KGB supplemented with 0.1% Triton X-100 (*see* Note 4) and equilibrating for 20 min at room temperature. Repeat at least once more.
7. Add 100 µL 2X KGB + 0.1% Triton X-100 and 12 U/µg DNA *Not*I.
8. Incubate for 5 h at 37°C.
9. Stop the reaction by adding 1 mL of ESP and incubating at 50°C for 1 h.
10. Equilibrate plug in 1 mL PFGE running buffer at least twice for 20 min at room temperature.
11. Load the sample onto a PFGE gel (*see* Notes 1 and 2).

Other possible combinations that have been demonstrated (35) for crossprotections are: *Csp*I (*Rsr*II) (5'-CGGWCCG-3') and M.*Hpa*II (5'-C ᵐCGG-3') (NEB); *Ecl*XI (*Eag*I) (5'-CGGCCG-3') and M.*Hpa*II; and *Nae*I (5'-GCCGGC-3') and M.*Bsp*RI (5'-GG ᵐCC-3').

3.2.3. Competitions

Competitions were developed as alternatives for restriction-enzyme-limited or Mg^{2+}-limited partial digestions (38), as the latter strategies do not always work well if the DNA substrate is embedded in agarose blocks (3). In competitions, partial cleavage of DNA with a restriction endonuclease is achieved by using a methylase that has the same target site specificity as the restriction endonucleases (3,39). The competition is controlled because the restriction enzyme and methylase are used in a specific ratio, both in excess of the amount required to completely methylate or cleave the DNA sample (39). This results in a uniform partial digest (39).

Competition reactions can be exemplified using *Not*I and M.*Bsp*RI (5'-GG ᵐCC-3').

1. Wash the plug at least three times for 20 min each using 1 mL THE buffer at room temperature, with gentle agitation.
2. Equilibrate the plug at least twice at room temperature in 1 mL of 1.5X KGB without DTT or ß-mercaptoethanol (39).
3. Add 100 µL of 1.5X KGB that has been supplemented with 100 µ*M* SAM (freshly added) and 1–10 U/µg DNA of M.*Bsp*RI, along with 8 U/µg DNA of *Not*I. 3–4 U of M.*Bsp*RI/8 U *Not*I has produced good results on *Escherichia coli* and *Bacillus subtilis* DNA (39).
4. Incubate at room temperature for 8 h.
5. Load the sample onto a PFGE gel (*see* Notes 1 and 2).

Other examples of competitions are: *Nru*I (5'-TCGCGA-3') or *Mlu*I (5'-ACGCGT-3') and M.*Fnu*DII (5'-CG ᵐCG-3') (3,39).

3.2.4. Other Strategies

Besides strategies that rely on the joint use of modification methyltransferases and restriction endonucleases, there are other methods for generating large DNA fragments. One method is to create synthetic DNA-cleaving reagents by combining a sequence-specific DNA-binding domain with a DNA-cleaving moiety *(40)*. In this strategy, sequence-specific cleavage of double-helical DNA is accomplished by binding of modified homopyrimidine EDTA oligonucleotides that form a triple helix structure and, in the presence of Fe(II) and dithiothreitol (DTT), cleave one (or both) strands of the "double-stranded" DNA at a specific site. So far, the practical limitation of this strategy is that cleavage is a very low-efficiency reaction and well-defined ends are not generated. The *trp* repressor gene from *E. coli* has also been chemically converted into a site-specific endonuclease by covalently attaching it to the 1,10-phenanthroline–copper complex *(41)*; once again, however, cleavage is not very efficient.

A second general method for producing large DNA molecules makes use of the large specificities of repressor proteins (they usually recognize operators of approx 20 bp in length) *(42,43)*. By combining the activities of a site-specific repressor, such as the *lac* repressor, which has an effective recognition sequence of 20 bp in length, a modification methylase, and the cognate restriction endonuclease, it is possible to produce a double-stranded cut only at the selectively protected site. This has been demonstrated for *E. coli* and yeast by cutting their respective genomes at a single, introduced target site *(43)*.

4. Notes

1. It is important to include genomic DNA from a well-characterized prokaryotic organism as a control for the reactions described. This is because the specific bands that are observed with prokaryotic DNA, because of the smaller genome size, serve as an indicator for the purity of the enzymes, as well as for the degree to which the reactions are performing as expected. Most of these reactions will be unsuccessful if enzymes of low purity are used.
2. When doing experiments with eukaryotic DNA of high complexity, final confirmation of the results must come by using Southern hybridization with a single copy genomic sequence, since it is not possible to accurately assess the degree to which the reactions are working simply by visual inspection of the resulting PFGE gels.
3. *Dpn*I has recently been shown to cleave hemimethylated GATC sites at approx 1/10–1/100 the rate of cleavage for fully methylated

substrates *(34)*. This implies that *Dpn*I will also cleave at more frequently occurring sequences than desired *(34)*; however, we have not found this to be a problem under our reaction conditions.

4. The use of 0.1% Triton X-100 in the restriction buffer is limited to *Not*I.
5. Failure to produce good results using these strategies generally is attributable to use of low-purity methylases. If the methylases are contaminated with other nucleases, then methylation steps can cause degradation of high-complexity genomic DNA. If you purify your own methylases, we suggest that extensive testing be done using well-characterized DNA dissolved in liquid medium first, before trying with plugs. Once you are satisfied that the titer and purity of the methylase is sufficient, then proceed to prokaryotic DNA in agarose plugs to optimize *in situ* reactions.
6. It is technically easier to methylate using the methylase gene cloned into the organism in question and expressed in vivo *(34)*. However, this has only be done so far with prokaryotic genomes. Also, problems can arise if the host genome has a strong restriction system directed at the methylated sequence, such as the *mrr* system *(34)*.

References

1. McClelland, M. and Nelson, M. (1987) Enhancement of the apparent cleavage specificities of restriction endonucleases: Applications to megabase mapping of chromosomes, in *Gene Amplification and Analysis, vol. 5 Restriction Endonucleases and Methylases* (Chirikjian, J. G., ed.), Elsevier, Amsterdam.
2. McClelland, M., Hanish, J., Nelson, M., and Patel, Y. (1988) A single type of buffer for all restriction endonucleases. *Nucleic Acids Res.* **15,** 364.
3. Hanish, J. and McClelland, M. (1989) Controlled partial restriction digestions of DNA by competition with modification methyltransferases. *Anal. Biochem.* **179,** 357–360.
4. Sobral, B. W. S., Honeycutt, R. J., Atherly, A. G., and McClelland, M. (1990) Analysis of the rice (*Oryza sativa* L.) genome using pulsed-field gel electrophoresis and rare-cutting restriction endonucleases. *Plant Mol. Biol. Rep.* **8,** 252–274.
5. Carle, G. F., Frank, M., and Olson, M. V. (1986) Electrophoretic separation of large DNA molecules by periodic inversion of the electric field. *Science* **232,** 65–68.
6. Schwartz, D. C., Smith, L. C., Baker, M., and Hsu, M. (1989) ED: Pulsed electrophoresis instrument. *Nature* **342,** 575,576.
7. Gardiner, E., Lass, W., and Patterson, D. (1986) Fractionation of large mammalian DNA restriction fragments using vertical pulsed-field gradient gel electrophoresis. *Somatic Cell Mol. Genet.* **12,** 185–195.
8. Josse, J., Kaiser, A. D., and Kornberg, A. (1961) Enzymatic synthesis of deoxyribonucleic acid. VII. Frequencies of nearest neighbor base sequences in deoxyribonucleic acid. *J. Biol. Chem.* **236,** 864–875.

9. Swartz, M. N., Trautner, T. A., and Kornberg, A. (1962) Enzymatic synthesis of deoxyribonucleic acid. XI. Further studies on nearest neighbor base sequences in deoxyribonucleic acids. *J. Biol. Chem.* **237**, 1961–1967.
10. Almagor, H. (1983) A Markov chain analysis of DNA sequences. *J. Theor. Biol.* **104**, 633–645.
11. Blaisdell, B. E. (1983) A prevalent persistent nonrandomness that distinguishes coding and noncoding eucaryotic nuclear DNA sequences. *J. Mol. Evol.* **19**, 122–133.
12. Blaisdell, B. E. (1985) Markov chain analysis finds a significant influence of neighboring bases on the occurrence of a base in eucaryotic nuclear DNA sequences both protein-coding and noncoding. *J. Mol. Evol.* **21**, 278–288.
13. Elton, R. A. (1975) Doublet frequencies in sequenced nucleic acids. *J. Mol. Evol.* **4**, 323–346.
14. McClelland, M. (1983) The frequency and distribution of methylatable DNA sequences in leguminous plant protein coding sequences. *J. Mol. Evol.* **19**, 346–354.
15. Hong, J. (1990) Prediction of oligonucleotide frequencies based upon dinucleotide frequencies obtained from the nearest neighbor analysis. *Nucleic Acids Res.* **18**, 1625–1628.
16. Churchill, G. A., Daniels, D. L., and Waterman, M. S. (1990) The distribution of restriction enzyme sites in *Escherichia coli. Nucleic Acids Res.* **18**, 589–597.
17. Salser, W. (1977) Globin messenger-RNA sequences-analysis of base-pairing and evolutionary implications. *Cold Spring Harbor Symp. Quant. Biol.* **42**, 985–1103.
18. Jukes. T. H. (1978) Codons and nearest neighbor nucleotide pairs in mammalian messenger RNA. *J. Mol. Evol.* **11**, 121–127.
19. Bird, A. P. (1980) DNA methylation and the frequency of CpG in animal DNA. *Nucleic Acids Res.* **8**, 1499–1504.
20. Nussinov, R. (1980) Some rules in the ordering of nucleotides in the DNA. *Nucleic Acids Res.* **8**, 4545–4562.
21. Nussinov, R. (1981) The universal dinucleotide asymmetry rules in DNA and amino acid codon choice. *J. Mol. Evol.* **17**, 237–244.
22. Erickson, J. W. and Altman, G. (1979) A search for patterns in the nucleotide sequence of the MS2 genome. *J. Math. Biol.* **7**, 219–230.
23. Phillips, G. J., Arnold, J., and Ivarie, R. (1987) Mono- through hexanucleotide composition of the *Escherichia coli* genome: A Markov chain analysis. *Nucleic Acids Res.* **15**, 2611–2626.
24. Phillips, G. J., Arnold, J., and Ivarie, R. (1987) The effect of codon usage on the oligonucleotide composition of the *E. coli* genome and identification of over- and underrepresented sequences by Markov chain analysis. *Nucleic Acids Res.* **16**, 9185–9198.
25. Bishop, M. J. and Rawlings, C. J. (eds.) (1987) *Nucleic Acid and Protein Sequence Analysis: A Practical Approach.* IRL, Oxford, UK.
26. Fasman, G. D. (ed.) (1976) *CRC Handbook of Biochemistry and Molecular Biology.* CRC, Boca Raton, FL.

27. Sobral, B. W. S., Honeycutt, R. J., and Atherly, A. G. (1991) The genomes of the *Rhizobiaceae* family: Size, stability, and rarely cutting restriction endonucleases. *J. Bacteriol.* **173,** 704–709.
28. Nelson, M. and McClelland, M. (1987) Purification and assay of type II DNA methylases. *Methods Enzymol.* **155,** 32–41.
29. McClelland, M. (1987) Site-specific cleavage of DNA at 8-, 9-, and 10-bp sequences. *Methods Enzymol.* **155,** 22–32.
30. McClelland, M., Kessler, L., and Bittner, M. (1984) Site-specific cleavage of DNA at 8- and 10-base-pair-sequences. *Proc. Natl. Acad. Sci. USA* **81,** 983–987.
31. McClelland, M., Nelson, M., and Cantor, C. R. (1985) Purification of *Mbo*II methylase (GAAG^{m}A) from *Moraxella bovis:* Site specific cleavage of DNA at nine and ten base pair sequences. *Nucleic Acids Res.* **13,** 7171–7182.
32. Weil, M. and McClelland, M. (1989) Enzymatic cleavage of a bacterial genome at a 10-base-pair recognition site. *Proc. Natl. Acad. Sci. USA* **86,** 51–55.
33. Patel, Y., Van Cott, E., Wilson, G. G., and McClelland, M. (1990) Cleavage at the twelve-base-pair sequence 5'-TCTAGATCTAGA-3' using M.*Xba*I (TCTAG ^{m6}A) methylation and *Dpn*I (G ^{m6}A/TC) cleavage. *Nucleic Acids Res.* **18,** 1603–1607.
34. Hanish, J. and McClelland, M. (1991) Enzymatic cleavage of a bacterial chromosome at a transposon-inserted rare site. *Nucleic Acids Res.* **19,** 829–832.
35. Nelson, M. and McClelland, M. (1991) Site-specific methylation: Effect on DNA methyltransferases and restriction endonucleases. *Nucleic Acids Res.* **19,** 2045–2071.
36. Nelson, M., Christ, C., and Schildkraut, I. (1984) Alteration of apparent restriction endonuclease recognition specificities by DNA methylases. *Nucleic Acids Res.* **12,** 5165–5173.
37. Qiang, B.-Q., McClelland, M., Podar, S., Spokauskas, A., and Nelson, M. (1990) The apparent specificity of *Not*I (5'-GCGGCCGC-3') is enhanced by M.*Fnu*DII or M.*Bep*I methyltransferases (5'-^{m}CGCG-3'): Cutting bacterial chromosomes into a few large pieces. *Gene* **88,** 101–105.
38. Albertsen, H. M., Le Paslier, D., Aberrahim, H., Dausset, J., Cann, H., and Cohen, D. (1989) Improved control of partial DNA restriction enzyme digest in agarose using limiting concentrations of Mg^{++}. *Nucleic Acids Res.* **17,** 808.
39. Hanish, J. and McClelland, M. (1990) Methylase-limited partial *Not*I cleavage for physical mapping of genomic DNA. *Nucleic Acids Res.* **18,** 3287–3291.
40. Moser, H. E. and Dervan, P. B. (1987) Sequence-specific cleavage of double helical DNA by triple helix formation. *Science* **238,** 645–650.
41. Chen, C.-H. and Sigman, D. S. (1987) Chemical conversion of a DNA-binding protein into a site-specific nuclease. *Science* **237,** 1197–1201.
42. Koob, M., Grimes, E., and Szybalski, W. (1988) Conferring operator specificity on restriction endonucleases. *Science* **241,** 1084–1086.
43. Koob, M. and Szybalski, W. (1990) Cleaving yeast and *Escherichia coli* genomes at a single site. *Science* **250,** 271–273.

CHAPTER 13

Desktop Digital Imaging

*Application to Detection of Length Heterogeneity
After Hyperresonant Pulsed-Field Gel Electrophoresis
of Mature Bacteriophage P22 DNA*

*Gary A. Griess, Elena T. Moreno,
and Philip Serwer*

1. Introduction

The multisample slab gels used for pulsed-field gel electrophoresis (PFGE) can produce gel patterns of hundreds of samples per day
(*see*, for example, ref. *1*). Because of this efficiency in the production
of data, the rate-limiting steps in obtaining information at times are
analysis, organization, reproduction, storage, and retrieval of data.
Digital image recording and processing has the capacity for increasing, by at least an order of magnitude, the efficiency (in time and cost)
of these processes. For the quantitative analysis of gel patterns, both
spatial and densitometric measurements are simplified by use of digital image processing (reviewed in refs. *2* and *3*). Microcomputers have
now developed to the point that digital image processing can be a
desktop procedure *(4)* that requires equipment (including computer,
video camera, storage device, and laser printer) that costs less than
$15,000. In the present communication, we describe procedures for
assembling and using such a microcomputer-based system for digital
image processing. The example that we use to demonstrate the image

From: *Methods in Molecular Biology, Vol 12: Pulsed-Field Gel Electrophoresis*
Edited by: M. Burmeister and L. Ulanovsky
Copyright © 1992 The Humana Press Inc., Totowa, NJ

processing is analysis of band widths after production of unusually high DNA length resolution, by use of a recently demonstrated (5) resonance in DNA separation by PFGE (hyperresonance). The DNAs that were chosen for the demonstration of digital image processing are the mature DNAs of four linear, double-stranded bacteriophages: λ, P22, T7, and a T7 deletion mutant (indicated by T7Δ) that has 8.4% less DNA than wild-type T7. During morphogenesis in a bacteriophage-infected cell, all of these DNAs are cut from a longer, concatemeric DNA. However, λ and T7 DNAs are cut at unique sites and have a unique length (reviewed in ref. 6). Complete nucleotide sequences are known for both λ (48.502 kbp; 7) and T7 (39.936 kbp; 8). In contrast, P22 DNA is sized by filling a bacteriophage capsid with DNA and cleaving DNA that is outside of the capsid (headful mechanism). The headful sizing mechanism of P22 yields mature DNA molecules that, by use of restriction endonuclease analysis, are found to vary in length by ±800 bp (the mean length is 43.4 kbp; 9). In the present communication, for the first time, PFGE has been used to observe this variability for intact P22 DNA. In addition, digital image processing has been used to quantify the P22 DNA length variability, observed by use of PFGE.

2. Materials

The materials that we have used for digital image processing were selected by a process in which attempts were made to minimize laboratory space occupied, maximize performance, and minimize cost. The exact equipment that was used is indicated in the following, together with the most important specifications.

1. Microcomputer with at least two expansion slots and at least two megabytes of random accessible memory (RAM): Macintosh IIci, Apple Computer, Inc., Cupertino, CA. All other members of the Macintosh II series can also be used.
2. High resolution color monitor: AppleColor 13-inch, high resolution RGB monitor (N0401).
3. Video digitizer board (frame grabber) that produces a linear response to both light intensity (8-bit resolution) and spatial dimensions when combined with the video camera described in Section 2.4: DT2255, Data Translations, Marlboro, MA.
4. Monochrome charged coupled device (CCD) video camera with manual gain control, no gamma correction, and 12–25 mm focal length lens. The CCD camera must have:

a. a linear response to both light intensity and spatial dimensions, and

b. an array of pixels not less than 500 × 450: Philips 56470 Series, Philips Components, Slatersville, RI.

This camera has a 610 × 480 square-pixel array. Before using the camera, place the camera in a case (Radio Shack #270-239, Fort Worth, TX). Disable the automatic gain control and acquire manual gain control by applying voltage from a voltage divider (0–11 V) across pin 8 (a circuit diagram is shown in Fig. 1). On the camera, place a c-mount lens that has a manual aperture control. For photographing ethidium-stained gels, place a Tiffen 23A filter on the lens.

5. Source of illumination for both ultraviolet and visible light:

a. Ultraviolet: Spectroline XX-15F, Spectronics Corp., Westbury, NY.

b. Visible: Porta-Trace #PT-2860, Gagne Associates, Inc., Binghamton, NY.

Enclose both the ultraviolet and visible light sources in a cabinet that has a dull black interior; use a cabinet that has a door, so that room light can be excluded during the digitization (capturing) of images. The design of the cabinet currently used is sketched in Fig. 2.

6. Printer that produces a linear response to spatial dimensions, when combined with the CCD camera and video digitizer board used. Because the printer provides rapid access to (nonpublication quality) hard copy used for either storage or review, use of a laser printer is advantageous for the purpose of bypassing photography: Personal LaserWriter NT, Apple Computer, Inc., Cupertino, CA.

7. System for storage: 600-megabyte removable cartridge, magneto-optical drive, Relax Technology, Union City, CA. Although we initially stored images (0.3 megabytes each) on floppy disks, the limits on the storage capacity of floppy disks (only 2–3 images each) caused problems in the storage and retrieval of the floppy disks. These problems prompted the use of the increased storage capacity. All 600 megabytes are simultaneously accessible to the microcomputer.

8. Software sufficient to capture images, manipulate images, perform length measurements, and quantify intensity of light: NIH Image (Rasband, W.; *see* ref. *10*). This program is available, without cost, on Internet. Enter ftp and either alw.nih.gov or 128.231.128.251. Enter anonymous for user name; use anything as the password. Enter: get /pub/image. Alternatively, Image can be copied from another current user. Alternative software has more recently been described *(11)*.

Costs for equipment described here are estimated in Note 1.

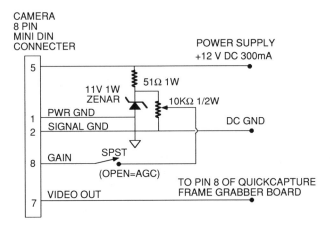

Fig. 1. Circuit diagram for converting automatic gain control to manual gain control in the CCD camera.

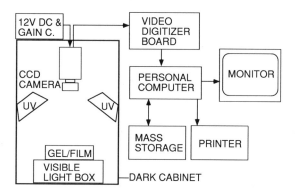

Fig. 2. Sketch of the placement of components needed for capturing of images. The components are labeled on the figure.

3. Methods

Image processing can be performed by use of gels stained by any method that yields a visible signal. In addition, images obtained through either a light microscope or an electron microscope can be processed. However, because the majority of applications for PFGE will be analyses of ethidium-stained gels in which DNA has been fractionated, the methods described here were designed for use in the analysis of ethidium-stained gels (a specific example is described in Note 2).

1. Insert the video digitizer board in its expansion slot and mount the CCD camera, lens, and filter in the cabinet. Connect the output of the CCD camera to the video digitizer board (15-pin connector) after placing the output lead through a hole in the cabinet. Both the CCD camera in the cabinet and the equipment outside of the cabinet are sketched in Fig. 2.

2. Open the aperture of the CCD camera to its widest. Launch Image. To activate the frame grabber, hold the command (apple) key while pressing key G; the live image is then visible on the monitor. Adjust the height of the CCD camera to encompass the needed field of view.

3. Place an ethidium-stained gel on a nonreflecting black surface, for example, black vinyl from a notebook cover. Place the gel under the camera and focus the camera by use of visible ambient light. Close the door to the cabinet and illuminate the gel obliquely from above with ultraviolet light (*see* Notes 3–5).

4. To control the background intensity of the image on the monitor, adjust the manual gain control that, for convenience, is on the outside of the cabinet (Fig. 2).

5. To capture an image that has been averaged over n frames, open "special" from the menu bar, select "average frames" and release the mouse. To determine n, open "options" from the menu bar, select "preferences," and enter n in the box next to "frames to average." To capture an image that has not been averaged, press the G key, while holding the command key.

6. Process the image. Follow instructions that come with both "Image" and other image-processing programs (*see* Note 6).

4. Notes

1. The cost of the equipment described here will vary with both the time and source of purchase. However, an approximate cost (in 1990) for frame grabber, modified video camera, and light box is $3,800 (US dollars). The cost for the computer and laser printer, often present in laboratories for other purposes, is $6,000. The optical disk drive costs an additional $3,000, but should be considered optional. Between 1988 and 1991, the average cost of components has decreased.

2. To illustrate the processing of a captured image, fluorescence as a function of distance from the electrophoretic origin has been determined for bands formed by mature bacteriophage DNAs, after PFGE. The form of PFGE used here was field inversion (asymmetric time, symmetric magnitude of the field; reviewed in ref. *12*) for which

tuning of the reverse pulse was performed to produce an unusually large (hyperresonant) separation of DNA approx 40 kbp in length. As shown in Fig. 3A, a scan of fluorescence intensity for the bands formed by the mature DNAs of bacteriophages T7 and λ (unique length) revealed a peak 0.6–0.7× as wide as the peak formed by the mature DNA of bacteriophage P22 (the width of a peak in pixels is indicated above the peak). The scans of Fig. 3A were obtained by using "Image" first to outline the region to be scanned and then to scan in the direction of migration, while integrating across the band. In Fig. 3B, a captured photograph of the same gel shows the original bands; the box for the lane of T7 DNA indicates the zone that was scanned. The captured image of Fig. 3B and the plots of Fig. 3A were arranged, cut, and marked as a digitized image. A photographic reproduction was produced from a floppy disk by use of a Matrix Slidewriter (Agfa-Matrix, Dallas, TX).

The data of Fig. 3 are the first to resolve, by use of intact mature P22 DNA, the heterogeneity in its length. The variability of length found in Fig. 3 agrees with the ±800 bp previously found by use of restriction endonuclease analysis.

3. Although the procedure described here specifies illumination from above, illumination from below the gel has also been used *(3)*. However, to obtain uniform illumination from below, an ultraviolet light source designed for this purpose must be constructed. To simplify both the achievement of uniform illumination and the avoidance of patterns of infrared light (the CCD camera is sensitive to infrared light), illumination from above is preferred.

4. Images of either autoradiograms or gels stained by use of nonfluorescent stains can be captured and processed by use of the equipment described here. For autoradiograms, the only differences in procedure are:
   a. elimination of the Tiffen filter on the CCD camera and
   b. illumination from below, by use of the visible light source indicated in Fig. 2.

   When multiple images of the same gel have been obtained (for example, images of both ethidium fluorescence and autoradiographic intensity, the latter possibly obtained by use of more than one radioactive probe), the program, Image, can be used to juxtapose these images at the same magnification. In our opinion, published collages of this type should always include a solid line that separates the images obtained before juxtaposition.

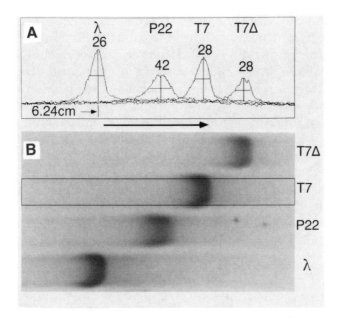

Fig. 3. Application of digital image processing. **(A)** Fluorometric scans
of the intensity of ethidium staining across bands of bacteriophage DNAs
separated by use of PFGE; **(B)** Image of the gel from which the scans in
(A) were obtained. Electrophoresis of 30 ng DNA per lane was performed
by use of 2 V/cm and 15 ± 2°C, through a 1.5% Seakem LE agarose gel
(FMC Bioproducts, Rockland, ME), cast in 0.01M sodium phosphate,
pH 7.4, 0.001M EDTA. During electrophoresis, the following pulses were
produced in two stages, by use of field inversion performed at constant
field magnitude *(12)* (a four-port device from DNAStar, Madison, WI
was used): Stage 1, 1.5-s reverse pulses interspersed with forward pulses
linearly ramped from 4.5 s to 1.8 s for 100 h; Stage 2, 1.5-s reverse pulses
interspersed with 1.8-s forward pulses for 100 h. This sequence of pulses
is derived from the finding of a hyperresonance in response of linear
DNA separation, when the DNA is about 40 kbp long and the reverse
pulse is 1.5 s (other conditions are the same as those used here) *(5)*. In
terms of the ratio of distance migrated, resolution would be improved if
the 1.8-s forward, 1.5-s reverse pulsing (Stage 2) were used throughout
the electrophoresis. However, use of the ramp in Stage 1 yields sharper
bands that separate from each other more rapidly. The direction of electrophoresis is indicated by an arrow whose length represents 1 cm on
the gel. The distance to the origin (not shown) is indicated. The lengths
of the DNAs used are (kbp): λ, 48.502; P22, 43; T7, 39.936; T7D, 36.6. At
the top of each peak in **A** is indicated the width at half height, in pixels.

5. For quantification of band intensities, direct capture of fluorescence from ethidium-stained gels is more accurate than capture of a photographic image of the gel *(3)*. However, by use of the procedure described here, the sensitivity of direct capture is, thus far, 0.3–0.5× the sensitivity of photography. This problem has been overcome by use of equipment more expensive than the equipment used here *(3)*.
6. The use of digital imaging simplifies the performance of image manipulations other than those that could easily have been performed by use of photography. If such advanced manipulations are performed, the authors of the manipulations are, in our opinion, obliged to indicate what manipulations are performed. If digital images are used for purposes of either art or commerce, this principle should apply whenever the suggestion is made that the images have scientific merit.

Acknowledgments

The authors thank Sherwood Casjens for providing P22 DNA, Shirley J. Hayes for providing λ DNA, Alvin J. Julian and Katharine R. Myers, Division of Educational Communications, for photographic reproduction of digitized images, and Linda C. Winchester for typing this manuscript. Support was received from the National Science Foundation (DMB 9003695) and the National Institutes of Health (GM 24365).

References

1. Serwer, P., Watson, R. H., and Son, M. (1990) Role of gene 6 exonuclease in the replication and packaging of bacteriophage T7 DNA. *J. Mol. Biol.* **215,** 287–299.
2. Boniszewski, Z. A. M., Comley, J. S., Hughes, B., and Read, C. A. (1990) The use of charged-coupled devices in the quantitative evaluation of images, on photographic film or membranes, obtained following electrophoretic separation of DNA fragments. *Electrophoresis* **11,** 432–440.
3. Sutherland, J. C., Sutherland, B. M., Emrick, A., Monteleone, D. C., Ribeiro, E. A., Trunk, J., Son, M., Serwer, P., Poddar, S. K., and Maniloff, J. (1991) Quantitative electronic imaging of gell flourescence with CCD cameras: Applications in molecular biology. *BioTechniques* **10,** 492–497.
4. Griess, G. A. and Serwer, P. (1991) A desktop video image recording and processing system for quantitative gel electrophoresis, in *First International Conference on Electrophoresis, Supercomputing and the Human Genome* (Cantor, C. R. and Lim, H. A., eds.), World Scientific, Singapore, p. 86.
5. Serwer, P. (1990) Sieving by agarose gels and its use during pulsed-field electrophoresis. *Biotechnol. Genet. Eng. Rev.* **8,** 319–343.

6. Casjens, S. (1985) Nucleic acid packaging by viruses, in *Virus Structure and Assembly* (Casjens, S., ed.), Jones and Bartlett, Boston, p. 75.

7. Sanger, F., Coulson, A. R., Hong, G. F., Hill, D. F., and Peterson, G. B. (1982) Nucleotide sequence of bacteriophage λ DNA. *J. Mol. Biol.* **162**, 729–773.

8. Dunn, J. J. and Studier, F. W. (1983) Complete nucleotide sequence of bacteriophage T7 DNA and the locations of T7 genetic elements. *J. Mol. Biol.* **166**, 477–535.

9. Casjens, S. and Hayden, M. (1988) Analysis in vivo of the bacteriophage P22 headful nuclease. *J. Mol. Biol.* **199**, 467–474.

10. O'Neill, R. R., Mitchell, L. G., Merrill, C. R., and Rasband, W. S. (1989) Use of image analysis to quantitate changes in form of mitochondrial DNA after x-irradiation. *Appl. Theor. Electrophor.* **1**, 163–167.

11. Redman, T. and Jacobs, T. (1991) Electrophoretic gel image analysis software for the molecular biology laboratory. *BioTechniques* **10**, 790–794.

12. Olson, M. V. (1989) Pulsed-field gel electrophoresis, in *Genetic Engineering, Principles and Methods*, vol. 11 (Setlow, J. K., ed.), Plenum, New York, p. 183.

PART III

BIOLOGICAL APPLICATIONS

CHAPTER 14

Bacterial Genome Mapping by Two-Dimensional Pulsed-Field Gel Electrophoresis (2D-PFGE)

Wilfried Bautsch

1. Introduction

Two-dimensional pulsed-field gel electrophoresis (2D-PFGE) is a powerful PFGE technique for the restriction mapping of bacterial genomic DNA. The method consists of two sequential steps of restriction endonuclease digestion and separation of the fragments by PFGE:

1. Step A, in which partially or completely digested bacterial DNA is separated by PFGE in the first dimension; and
2. Step B, in which a gel slice containing the separated DNA of the first dimension is cut out from the first gel, redigested with the same or a different restriction enzyme, and separated by PFGE in the second dimension, i.e., perpendicular to the first dimension.

The final gel displays a two-dimensional pattern of DNA spots whose distribution is determined by the localization of the restriction sites in the genomic DNA. Hence, it should be possible to deduce from this pattern the position of most, if not all, restriction sites in the original DNA.

To illustrate this rationale, imagine bacterial genomic DNA partially digested with a restriction enzyme and separated by one-dimensional PFGE. If all individual fragment sizes of the completely digested

From: *Methods in Molecular Biology, Vol 12: Pulsed-Field Gel Electrophoresis*
Edited by: M. Burmeister and L. Ulanovsky
Copyright © 1992 The Humana Press Inc., Totowa, NJ

DNA were exactly known, the constituent fragments of a given partial-digestion fragment could then, theoretically at least, be deduced from its size. Obviously, all fragments linked in a partial digestion fragment must also be linked in the chromosome. Given the limited accuracy of molecular weight determination by PFGE, however, this kind of analysis is only applicable to cases with very few restriction sites.

One way to facilitate the interpretation of more complex partial digestion patterns is to use different hybridization methods (1). Alternatively, one may simply cut out a partial-digestion fragment from the gel, redigest it to completion with the same enzyme, and separate the reaction products by PFGE. The constituent fragments of the partial-digestion fragment can now be identified by comparision with the fragments of a complete restriction digest of the bacterial DNA with the same enzyme. Essentially, this is already a two-dimensional approach, but it is cumbersome to perform for all fragments of interest and impossible for fragment clusters. However, using a gel slice from the first-dimension lane as described above (i.e., performing a partial/complete 2D-PFGE) is much less labor-intensive and ensures a complete analysis.

Similar arguments as for the partial-digestion analysis hold for a double-digestion analysis, in which single and double digestions with two enzymes A and B are analyzed in a single PFGE run. If all fragment sizes were exactly known, one could deduce unequivocally which fragments generated by enzyme A have been cut by B and vice versa. This knowledge allows the construction of a restriction map (though ambiguities are often encountered). However, with increasing numbers of fragments this analysis becomes impossible, too, for the same reason as for partial-digestion fragments. Cutting out a gel slice containing the separated fragments generated by enzyme A, digesting it with enzyme B, and separating it by PFGE again (i.e., performing a complete/complete 2D-PFGE) allows for the analysis of each fragment of a digestion with enzyme A separately. But a second 2D-PFGE gel (this time with an enzyme B digestion in the first dimension) is required for a complete analysis.

These two basic methods will be presented in this chapter. Sometimes, they will already suffice to construct a complete genomic macrorestriction map (2). The 2D approach is technically easy, rapid, and does not require any hybridization probes. Therefore, it should be applicable to all bacteria irrespective of their prior genetic characterization. However, if larger numbers of fragments need to be mapped, even 2D-PFGE gels will become difficult to interpret. Even though hybridization techniques may then become necessary, the 2D-PFGE will still considerably reduce the complexity of the problem. In addition, many useful variations of the basic 2D-methodology may be envisaged, like end-labeling of fragments and fragment fingerprinting *(3)*, combinations of PFGE with subsequent conventional gel electrophoresis *(4)*, and so on. After becoming familiar with the methods described in this chapter, the readers are encouraged to develop such variations themselves. The following methods section is divided into three parts:

1. One-dimensional pulsed-field gel electrophoresis: Before starting a 2D-PFGE mapping project you need many data on your bacterial DNA, suitable restriction enzymes, and running conditions, which will most easily be obtained from one-dimensional PFGE gels. This part will therefore provide a short protocol about the preparation of agarose blocks of bacterial DNA and subsequent restriction endonuclease digestion. Emphasis is placed on those steps in which the protocols for bacteria differ from those for eukaryotes. For a thorough understanding of the steps involved, the reader is advised to consult those chapters first.
2. Two-dimensional pulsed-field gel electrophoresis: (1) partial/complete 2D-PFGE. (2) complete/complete 2D-PFGE.
3. Interpretation of 2D-PFGE gels: This part provides some additional information on the interpretation of 2D gels.

2. Materials

High-mol-wt DNA is very sensitive to digestion by nucleases. Scrupulous care should therefore be taken to minimize contaminations. All solutions and equipment should be sterilized wherever possible. Always wear gloves when handling the DNA inserts.

2.1. Reagents

1. Agarose (low electroendosmosis grade): For a 1% (w/v) agarose gel, dissolve 1 g agarose in 100 mL 0.5X TBE by boiling the mixture for at least 5 min. Readjust the volume by adding distilled water up to the original weight. Cool to 65°C in a water bath before pouring the gel.
2. Agarose (low-gelling): For a 2% (w/v) solution, dissolve 2 g low-gelling agarose in 100 mL SE by boiling the mixture for at least 5 min and readjust the volume by adding distilled water up to the original weight. Keep the solution liquid by placing it in a water bath at 42°C.
3. Restriction enzymes.

4. Proteinase K: Dissolve proteinase K at 20 mg/mL in distilled water. Dispense into aliquots and store at –20°C.
5. Dithiothreitol (DTT): Dissolve DTT in 20 mM sodium acetate, pH 5.2, at 1M and sterilize by filtration. Dispense into aliquots and store at –20°C.
6. Ethidium bromide (EtBr): Dissolve EtBr at 0.5 mg/mL in distilled water. **Wear face mask and gloves when weighing EtBr!** Store in a dark bottle at 4°C. Always wear gloves when handling EtBr solutions as it is a carcinogen.
7. Lambda ladder (as a size standard): For preparation *see* Chapter 8.

2.2. Buffers

1. SE: 75 mM Sodium chloride, 25 mM EDTA, pH 7.5.
2. ES: 0.5M EDTA, 1% (w/v) N-lauroylsarcosine, pH 7.5.
3. 1X Restriction buffer: Prepare according to the manufacturer's instructions. Use 1M stock solutions of Tris, MgCl$_2$, KCl, and NaCl, respectively. Adjust the pH by addition of dilute HCl (remember that a shift from 20 to 37°C decreases the pH by about 0.3) and autoclave. Do not add BSA, gelatine, DTT, or β-mercaptoethanol.
4. 10X TBE: Dissolve 108 g Tris base and 55 g boric acid in 800 mL distilled water, add 4 mL 0.5M EDTA, pH 8.0, adjust the volume to 1L and autoclave. Dilute 1:20 with distilled water before use to give 0.5X TBE.
5. 1X TE: 10 mM Tris-HCl, 10 mM EDTA, pH 8.0.

2.3. Equipment

1. Standard laboratory equipment: Incubator, Eppendorf centrifuge (or equivalent), adjustable pipets, and pipet tips (1 μL–1 mL), 1.5 mL Eppendorf tubes (or equivalent), Pasteur pipets, scalpels/scalpel blades, stirring rods, glass plates 20 × 20 cm and 15 × 15 cm, aluminium foil, adhesive tape.
2. Two-bladed scalpel: Two scalpel blades screwed together on a 2-mm plastic spacer (*see* Fig. 1 later in chapter).
3. Incubation chamber: A plexiglass block with a 15 × 1 × 1 × = cm trough. (These are the dimensions for 15 × 15 cm PFGE gels; *see* Fig. 2 later in chapter.)
4. Pulsed-field gel electrophoresis equipment: Straight lanes are essential prerequisites for 2D-PFGE, which are generated by most commercially available PFGE configurations (like FIGE, CHEF, TAFE, and so on). The original OFAGE, however, is not suitable.

3. Methods

3.1. One-Dimensional Pulsed-Field Gel Electrophoresis

1. Inoculate 20 mL of a suitable growth medium with a fresh colony of your bacterial strain and incubate overnight at 37°C with agitation (*see* Note 1).
2. Pellet the cells by centrifugation (8000*g*, 15 min, 4°C), discard the supernatant, and wash the cells once with SE. Carefully resuspend the bacteria in 1/5–1/10 of the original volume in SE and determine the optical density at 600 nm.
3. Dilute the bacteria suspension to OD_{600} = 1.0, 2.0, and 4.0 in SE, respectively (*see* Note 2). Mix equal volumes of these with liquid 2% (w/v) low-gelling agarose in SE at 42°C and immediately pour the mixture into the molds. Let the agarose blocks (called inserts) solidify for at least 15 min at 4°C.
4. Transfer the inserts into Eppendorf tubes. To do this, you may have to dissect them with a scalpel into smaller ones of 1 × 10 × 10 mm at most (depending on the mold you use). Add 0.78 mL ES and 20 µL proteinase K, mix by inverting the tubes several times and incubate overnight at 52–56°C (*see* Note 3).
5. Wash the inserts at least thrice in TE for 30 min at room temperature (*see* Note 4). They may be stored at 4°C for at least 6 mo.
6. Cut the inserts with a scalpel into their final size (3-mm height). Transfer one or two of them into a 1.5-mL Eppendorf tube and equilibrate them three times for 30 min at room temperature in about 1 mL of the respective restriction buffer. For the restriction digestion, add 1 µL DTT (1*M*), 1 µL BSA (20 mg/mL), and 10 U of the restriction enzyme (*see* Note 5) to 150 µL restriction buffer. Mix the contents thoroughly by pipetting them carefully up and down and incubate the tube for 12–16 h at the recommended temperature. Stop the reaction by replacing the restriction buffer with 1 mL TE.
7. Pour a 1% (w/v) agarose gel (15 × 15 cm) in 0.5X TBE; keep a few mL agarose solution liquid at 65°C. Fill the slots of the solidified gel with 0.5X TBE and load the digested inserts into the slots. One lane should contain a lambda ladder as a size marker. Carefully remove the buffer with a sterile pipet and fix the inserts by adding liquid 1% (w/v) agarose to the slots. Run the gel in 0.5X TBE. If you have no prior information about the range of fragment sizes use pulse times

that will resolve fragments of 50–1000 kbp. At least once in your screening experiments you should also run an undigested DNA sample of your bacterium. This provides a convenient control for possible DNA damage during your preparation and this helps to become aware of the presence of single restriction sites, large plasmids, genomes with more than one chromosome *(5)*, or a linear bacterial chromosome *(6)*.

8. After the run, stain the gel in distilled water containing 0.5 µg/mL EtBr for 30min at room temperature with gentle agitation. Destain it twice with distilled water and take a standard polaroid. Check
   a. The quality of the PFGE run (resolution of the lambda ladder).
   b. The quality of your DNA preparation (any successful digestion?) and determine the right DNA concentration (*see* Note 2).
   c. The enzyme quality (*see* Note 6).

   Hopefully, you have found suitable enzyme(s) for restriction mapping. You should then roughly determine the total genome size of your bacterial strain by simply determining the individual fragment sizes in comparision to the lambda ladder (linear plot!) and adding them up. Estimate more appropriate pulse times for further PFGE experiments with this/these enzyme(s) (*see* Note 7).

9. For partial/complete 2D-PFGE of an appropriate enzyme, determine suitable conditions of enzyme concentration and incubation time for a partial digestion. Use inserts with at least twice the minimum DNA content (Note 2) and digest them with 1/2–1/5 of the minimum enzyme concentration necessary for a complete digestion for 30 min to 4 h. A good starting point for most enzymes is 1 U for 30 min, 1 h, 2 h, and 4 h. Stop the reaction by replacing the restriction buffer with TE. Run a total restriction digest with the same enzyme as a size standard and use pulse times that will resolve a partial-digestion fragment of a size equal to the sum of the two largest fragments of the complete digest. Otherwise, proceed as described in Steps 6–8.

   Look for a partial-digestion pattern on the final polaroid in which you can just discriminate the smallest fragments of a total restriction digestion. Use these partial-digestion conditions for partial/complete 2D-PFGE.

10. For complete/complete 2D-PFGE of two appropriate enzymes, determine suitable pulse times for resolution of the double-digestion products. Basically, proceed as described in Steps 6–8 with the following modifications: Perform the double digestion sequentially with three TE washes between the first and second digestion. Run complete single digestions of both enzymes and a lambda ladder together

with the double-digestion products on the gel and use pulse times that will resolve the largest fragment of the respective single digestions.

From the final polaroid, estimate more appropriate pulse times for resolution of the double-digestion products (these are needed for complete/complete 2D-PFGE).

3.2. Two-Dimensional Pulsed-Field Gel Electrophoresis

3.2.1. Partial/Complete 2D-PFGE

1. Prepare at least two partially digested inserts (*see* Note 8) using the conditions determined in Section 3.1., Step 9.
2. Pour a 1% (w/v) agarose gel in 0.5X TBE. Use a comb with wide teeth (preferably 10 mm). Load the partially digested insert for the 2D-PFGE into one of the central pockets of the gel as any curvature of the lanes after the run is least prominent there. Apply another partially digested insert to an outer lane to provide a check for the successful digestion of your DNA sample and a proper PFGE run (*see* Note 9). Run the gel under the predetermined conditions for a partial digestion with this enzyme (*see* Section 3.1., Step 9).
3. After the run, transfer the gel to a clean glass plate. Cut out a gel slice right from the middle of your insert lane using the technique illustrated in Fig. 1 (*see* Note 10). Transfer the gel slice to the incubation chamber as illustrated in Fig. 2. Wash it once with TE to remove minor contaminations that might have been introduced while handling the gel slice.
4. Stain the remaining two gel pieces in 0.5 µg/mL EtBr for half an hour at room temperature. Destain twice for 30 min at room temperature in distilled water and take a standard polaroid. Check that:
   a. The control lane displays the expected digestion pattern. Otherwise, repeat the whole experiment.
   b. On one edge of each gel piece DNA remnants of the cut lane are visible. Make sure that you see these remnants along the total length of the gel (*see* Note 9).
5. Trim the gel slice by cutting off gel pieces from both ends of the slice that do not contain any DNA fragments (use the polaroid from Step 4). Equilibrate the remaining gel slice three times in about 5–10 mL restriction buffer for 30 min each at room temperature. Remove the buffer with a sterile Pasteur pipet. Add just enough restriction buffer to cover the gel slice (about 3 mL); carefully note the volume. Add

Fig. 1. Technique of cutting a gel slice. With the help of a ruler, a second glass plate on top of the gel is oriented parallel to the gel edge covering about 2 mm of the lane to be cut. A 2-mm gel slice of the lane is cut out along the edge of the glass plate using a double-bladed scalpel.

Fig. 2. Technique of transporting the gel slice. With the help of a clean stirring rod (or equivalent) the gel slice is pushed carefully onto a glass plate. It will adhere to it and can subsequently be transported without rupturing.

BSA (final concentration 100 µg/mL), DTT final concentration 1 mM), and enough restriction enzyme for a complete digestion. Mix by pipetting the buffer several times up and down with a sterile Pasteur pipet.

Cover the incubation chamber with aluminum foil and seal it with adhesive tape to prevent evaporation during the incubation period. Incubate for 16–24 h at the recommended temperature (*see* Note 11). Stop the reaction by replacing the restriction buffer with TE.

6. Pour a 1% (w/v) agarose gel in 0.5X TBE. Keep about 2–3 mL agarose solution liquid in a water bath at 65°C. Use a one-well preparative comb (3-mm wide), which should touch the surface of the casting gel stand. After the gel has solidified, remove the comb, fill the slot with 0.5X TBE, and let the gel slice slide into the pocket (for transport *see* Fig. 2). Apply a complete restriction digest of your bacterial DNA (of the same enzyme to be mapped) to both outmost lanes. Remove the TBE buffer with a Pasteur pipet and fix the slice and the two inserts with liquid 1% (w/v) agarose. Run the gel under the predetermined pulse conditions for a complete restriction digest.

7. After the run, stain the 2D-PFGE gel in distilled water with 0.5 µg/mL EtBr for 30 min at room temperature with gentle agitation. Destain twice in distilled water for 30 min each and take a standard polaroid. (Subsequently, you may take a polaroid negative, too.)

3.2.2. Complete/Complete 2D-PFGE

To perform complete/complete 2D-PFGE, you need two different restriction enzymes A and B, which ideally should generate about the same number of fragments. Basically, the protocol is identical to the protocol for partial/complete 2D-PFGE described above, with modifications in the following steps:

1. Prepare at least two inserts completely digested with enzyme A.
2. Apply these completely digested inserts together with a lambda ladder as a size standard to the gel. Run the gel under the predetermined conditions for a complete digestion with enzyme A.
5. Use enzyme B for the digestion of the gel slice.
6. Apply at least two lambda ladders as size standards to the outmost lanes of the 2D gel. In addition, you may run a complete single digest with enzyme B and a complete double digest with enzymes A and B as further size markers in the outmost lanes. Use pulse times that are optimal for the resolution of a double digestion of your DNA with enzymes A and B (*see* Section 3.1., Step 10).

Remember that you need two complete/complete 2D-PFGE gels: One gel with an enzyme A digestion in the first and an enzyme B digestion in the second dimension, and a second gel with the opposite order of the restriction enzyme digestions.

3.3. Interpretation of Two-Dimensional Gels

For the detailed interpretation of 2D-PFGE gels, prints of the format 13 × 18 cm or larger should be prepared.

3.3.1. Partial/Complete 2D-PFGE Gels

1. Every 2D-PFGE gel contains a complete restriction digest of the bacterial DNA in both dimensions as a convenient size marker. In the first dimension these completely digested fragments are located on the upper contour line, in the second dimension in the two outmost lanes. Any constituent fragments of a partial-digestion fragment are located on a straight line in the second dimension. Extrapolating this line onto the contour line determines the position of the original partial-digestion fragment with respect to the completely digested fragments; its size can then be conveniently determined. Fragments are identified by comparision with the completely digested fragments in the two outmost lanes whose sizes are known.

2. Any difference between the size of the partial-digestion fragment and the sum of the sizes of its constituent fragments is indicative of the presence of additional fragments (partial-digestion fragments, double spots, and/or additional tiny fragments with fluorescence intensities too low to be detectable).

3. In cases of partial-fragment clustering, use mass law in combination with already confirmed links to sort out the correct order. Sometimes, grossly differing fluorescence intensities of the DNA spots provide additional clues regarding which fragments were originally linked in a partial-digestion fragment.

3.3.2. Complete/Complete 2D-PFGE Gels (Two Gels)

1. The general evaluation scheme for complete/complete 2D-PFGE gels is illustrated in Fig. 3. Identify all identical DNA spots on both 2D gels (horizontal lines). Their order is given by their linkage in single-digestion fragments (vertical lines). By switching to and fro between the two 2D gels most double-digestion fragments can be unambiguously ordered (fragments 1–6).

2. Always perform the following checks for partial digestion, double spots, and/or additional tiny fragments:

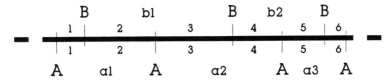

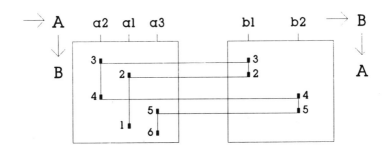

Fig. 3. Schematic representation of complete/complete 2D-PFGE. A short segment of a hypothetical DNA sequence is shown with the corresponding 2D gels below. Horizontal lines connect identical double-digestion products on both 2D gels, vertical lines connect the constituent fragments of each single-digestion fragment. Only a few relevant spots without any size standard are represented. A, B: restriction sites of enzymes A and B. a_n, b_n: fragments generated by enzymes A and B, respectively. Double-digestion fragments are numbered from 1 to 6.

    a. Determine the number of DNA spots on both 2D gels. It should be the same irrespective of the digestion order, because each 2D gel contains all fragments of a complete double digestion with the same two enzymes.

    b. Determine the sum of the sizes of all fragments of a second-dimension lane. It should be equal to the known size of the first-dimension fragment.

3. Each single-digestion fragment in Fig. 3 is cut once and only once by the second enzyme. However, in addition, the following two other arrangements are frequently encountered:

    a. Fragments that have not been cut by enzyme B (on a 2D gel with an enzyme A digestion in the first dimension) are located on the upper contour line of the second dimension. Obviously, these fragments are generated entirely by enzyme A. In the reciprocal 2D gel (with an enzyme B digestion in the first dimension), these fragments must be part of a fragment with at least two restriction sites for enzyme A, i. e., located in a lane with at least three DNA spots.

b. Lanes with more than three digestion products in the second dimension contain two (or more) fragments generated entirely by one enzyme only. The order of such fragments is inherently indeterminate since any permutation of their order would yield the same 2D pattern *(2)*. Again, in the reciprocal 2D gel these fragments are located on the upper contour line.

4. Notes

1. This protocol may be used for the culture of most aerobic bacterial species. At the end of the incubation period the bacteria are in the stationary phase. For slowly growing bacteria the incubation time must be prolonged accordingly. A few species do not grow to stationary phase in liquid culture (e.g., activation of an autolytic system in meningococci). Such bacteria must already be harvested in the early to middle logarithmic phase. Alternatively, you may resuspend freshly grown bacterial colonies from Petri plates in SE *(7)*. DNA near the replication terminus may be underrepresented on your PFGE gels under these conditions. This rarely presents a relevant practical problem, but you may nevertheless prefer to add chloramphenicol to the incubation medium before harvesting the cells *(1)*.

2. This procedure serves to determine the optimal DNA concentration. The most reliable PFGE data are obtained with the lowest DNA concentration that allows all DNA bands to be visualized. For a partial-digestion analysis at least twice that concentration should be used. You may have to pour another round of inserts if you see that some other DNA concentration would yield a better resolution.

   If you want to know the exact DNA concentration per band, you must either determine the OD_{600}: c.f.u./mL ratio (which varies from strain to strain) and convert these data into DNA content per insert with your determined genome-size data, or use densitometric evaluation against a DNA standard of known concentration

3. This protocol will probably work for all gram-negative bacteria. For gram-positive bacteria, pretreatment with cell wall-lysing enzymes may become necessary, like lysozyme *(7)* or stapholysin *(8)*.

4. You may prefer to add phenylmethylsulfonyl fluoride (PMSF) to the first TE wash to inactivate the proteinase K irreversibly *(7, 8)*. In most cases, however, this will not be necessary.

5. At this stage, you will probably have to screen a whole battery of different restriction enzymes to find suitable ones for 2D mapping. The larger the number of fragments the more difficult it will turn out to

obtain a complete macrorestriction map. Any number between 10 and 20 (–30) fragments per enzyme, therefore, is probably the best compromise between mapping efforts and sufficient informativeness of a bacterial macrorestriction map. Less than five restriction sites per genome do not require 2D-PFGE. They can be mapped conveniently by one-dimensional techniques. Enzymes with very few restriction sites per genome, however, are very helpful for checking the correctness of a macrorestriction map and providing reliable genome size data.

Bacterial DNA has a remarkably variable GC-content, which you may look up for your (or a closely related) strain in Bergey's manual *(9)*. Use the following rules of thumb to select candidate rare-cutting enzymes for a first-round screening:

Until now, there are nine restriction enzymes commercially available with an 8-bp recognition sequence. Of these, *Not*I (GC/GGCCGC), *Sfi*I (GC CC NN NN/NGGCC), *Srf*I (GCCC/GGGC) (Stratagene), and *Asc*I (GG/CGCGCC) (Biolabs) are rare-cutters in genomes with a GC-content of about 35–55%. Below and above these margins the number of fragments becomes too small and too large, respectively. *Pac*I (TTAAT/TAA) and *Swa*I (ATTT/AAAT) (Boehringer) *(10)*, on the other hand, should be useful especially for genomes with a GC-contents in the range of 45–65%. The other enzymes are *Sse*8387I (CCTGCA/GG) (Amersham) *(11)*, *Pme*I (GTTT/AAAC) (Biolabs) and *Sgr*AI (C (A/G)/CCGG/(T/C)G) (Boehringer) with a (probably) less stringent dependence on the GC-content. Further enzymes, e. g., *Fse*I (GGCCGG/CC) (Boehringer) *(13)* will soon become available.

The 4-bp pair sequence 5'-CTAG-3' seems to be selected against in most bacterial genomes *(13)*. This tetranucleotide is part of the recognition sequence of *Spe*I (A/CTAGT), *Xba*I (T/CTAGA), *Nhe*I (G/CTAGC) and *Avr*II (C/CTAGG).

Bacterial DNA with a high GC-content should be screened with enzymes like *Dra*I (TTT/AAA), *Ssp*I (AAT/ATT), and *Ase*I (ATT/AAT), with a high AT-content in their recognition sequence, whereas enzymes like *Rsr*II (CG/G(A/T)CCG), *Sac*II (CCGC/GG), *Sma*I (CCC/GGG), or *Eag*I (C/GGCCG) should be tried on AT-rich genomes.

If you still cannot find a suitable enzyme, you may screen *Hpa*I (GTT/AAC) and *Afl*II (C/TTAAG), each containing a stop codon in their recognition sequence, *Sph*I (GCATG/C) and *Nsi*I (ATGCA/T), each containing an ATG initiation codon, or *Bsp*HI (T/CATGA), containing both.

Finally, you may try methylases to introduce rare-cutting sites *(14)* or to alter the apparent recognition specificity of some restriction enzymes *(15)*. Knowledge of the codon usage and the oligonucleotide frequencies (*see* Chapter 12) of the bacterium will facilitate the selection of appropriate rare-cutters *(16)*. In any case, you will have to search yourself to find the enzyme(s) of choice. Sometimes even *Bam*HI (G/GATCC), *Kpn*I (GGTA/CC), and *Xho*I (C/TCGAG) are rare-cutters with less than ten restriction sites/Mbp *(17)*. However, many enzymes known as rare-cutters in eukaryotic genomes, like *Mlu*I (A/CGCGT), *Nru*I (TCG/CGA), and *Pvu*I (CGAT/CG), have no special significance for bacterial genome mapping because there does not exist any specific selection against the dinucleotide 5'-CG-3' in bacterial DNA.

6. Many enzyme charges are too contaminated with nonspecific nucleases to be of any use for PFGE experiments. If this is the case, try batches from other manufacturers. If you have found a good lot of a suitable enzyme, it is wise to buy a large amount of this lot for your further experiments because there is sometimes a considerable batch to batch variation of the same manufacturer's enzyme preparations.

With some enzymes, especially GC-rich ones like *Sfi*I, it is difficult to obtain complete digestions. Usually, this is easily recognized by gross variations in the fluorescence intensity of consecutive DNA bands (which should normally increase linearly with the fragment size) and huge discrepancies between the genome sizes determined with other enzymes. Gradually increase the enzyme concentration in your following experiments up to 100 U. If you still do not get a complete digestion of your bacterial DNA, it is probably wiser to skip this enzyme rather than bothering with the difficult interpretation of partial digestion patterns.

Methylation, though very common as, e.g., part of restriction/modification systems, mismatch repair, and so on *(18)*, is an all-or-none process in bacteria. Therefore, partial-digestion patterns attributable to differential methylation cannot normally be observed.

7. In further one-dimensional PFGE experiments, determine:
   a. The total genome size as precisely as possible. Run several gels with different pulse times that allow optimal resolution in every size range of your fragments. Always run a lambda standard in a lane immediately adjacent to the one to be sized.

   Apart from theoretical problems associated with genome size data obtained by PFGE *(19,20)*, there are three major practical problems to consider: partial-digestion fragments (*see* Note 6), clustering of fragments, and low-mol-wt fragments (i.e., fragments with a size below 20 kbp). To determine the number of fragments within a cluster is sometimes difficult. There are conflicting data in the literature regarding whether densitometric evaluation of your PFGE gels will help in solving that question *(19,21)*. 2D-PFGE itself is one means to determine this figure accurately and sometimes, you can infer it from precise genome size data obtained with a different enzyme (or better, two different enzymes).

Fragments below 10–20 kbp are difficult to detect in PFGE, because they do not contribute considerably to the total genome size and their fluorescence intensity is low. In addition, they may have diffused out of your agarose blocks during the incubation. To prove the completeness of a map, however, is an inherent problem to all macrorestriction maps published so far. You may become aware of the presence of additional tiny fragments by careful evaluation of your 2D-PFGE gels (e.g., by observing a difference between the size of a partial-digestion fragment and the sum of the sizes of its constituent fragments). To visualize such fragments, you may try to digest inserts with a high DNA content and run the PFGE gel immediately after the digestion, or use other techniques (conventional agarose gel electrophoresis, cloning techniques). If you have any hints of an uneven distribution of the restriction sites (e.g., clustering of fragments below 30 kbp), you should use a different enzyme.

b. The total genome size with a second enzyme, if you can find one. The two genome sizes thus determined usually agree within 1–2%. Differences above that are indicative of a systematic error, so you should search for further enzymes until you can trust your genome size data.

c. The minimum enzyme concentration necessary for complete digestion of your enzyme–substrate pair, especially when dealing with an expensive restriction enzyme. For complete digestion in the second dimension you need about 20 times as much enzyme!

8. It is critically important to use exactly the same partial-digestion conditions as determined in your screening experiments, especially the same enzyme lot and insert preparation. Scale-up the volumes of restriction buffer and enzyme proportional to the number of inserts to be digested, or digest several inserts in different tubes in parallel under the same conditions. Even then, it may be difficult to reproduce exactly the original pattern, though minor variations are tolerable.

Alternatively, perform a second round of a partial-digestion screening with several inserts per tube this time. Partially digested inserts may be stored in TE at 4°C for at least 6 mo.

9. It is wise to run an additional third insert to have a second try of cutting a proper slice if the first one fails.

10. You should practice this technique with some agarose gel before starting a real 2D experiment. Gel slices may even be cut with a conventional single-bladed scalpel. though it is much more difficult to obtain even slices.

11. The stability of the different restriction enzymes at their recommended incubation temperatures is quite variable. For a few it has been determined *(22,23)*. If your enzyme has a short half-life, it is wise to shorten the incubation time proportionately. Longer incubation times only increase the risk that contaminating nucleases damage your DNA. Incubation times longer than 24 h are rarely necessary.

References

1. Smith, C. L. and Condomine, G. (1990) New approaches for physical mapping of small genomes. *J. Bacteriol.* **172**, 1167–1172.

2. Bautsch, W. (1988) Rapid physical mapping of the *Mycoplasma mobile* genome by two-dimensional field inversion gel electrophoresis techniques. *Nucleic Acids Res.* **16**, 11461–11467.

3. Römling, U., Grothues, D., Bautsch, W., and Tümmler, B. (1989) A physical genome map of *Pseudomonas aeruginosa* PAO. *EMBO J.* **8**, 4081–4089.

4. Woolf, T., Lai, E., Kronenberg, M., and Hood, L. (1988) Mapping genomic organization by field inversion and two-dimensional gel electrophoresis: Application to the murine T-cell receptor γ gene family. *Nucleic Acids Res.* **16**, 3863–3875.

5. Suwanto, A. and Kaplan, S. (1989) Physical and genetic mapping of the *Rhodobacter sphaeroides* 2.4.1. genome: Presence of two unique circular chromosomes. *J. Bacteriol.* **171**, 5850–5859.

6. Ferdows, M. S. and Barbour, A. G. (1989) Megabase-sized linear DNA in the bacterium *Borrelia burgdorferi*, the Lyme disease agent. *Proc. Natl. Acad. Sci. USA* **86**, 5969–5973.

7. Nuijten, P. J. M., Bartels, C., Bleumink-Pluym, N. M. C., Gaastra, W., and van der Zeijst, B. A. M. (1990) Size and physical map of the *Campylobacter jejuni* chromosome. *Nucleic Acids Res.* **18**, 6211–6214.

8. Goering, R. V. and Duensing, T. D. (1990) Rapid field inversion gel electrophoresis in combination with an rRNA gene probe in the epidemiological evaluation of staphylococci. *J. Clin. Microbiol.* **28**, 426–429.

9. Bergey's Manual of Systematic Bacteriology, vol. 1 (1984), vol. 2 (1986), vols. 3 and 4 (1989), Williams and Wilkens, Baltimore, MD.

10. Sobral, B. W. S., Honeycutt, R. J., Atherby, A. G., and Clelland, M. (1991) Electrophoretic separation of the three Rizobium meliloti replicons. *J. Bacteriol.* **173,** 5173–5180.
11. Kotani, H., Nomura, Y., Kawashima, Y., Sagawa, H., Takagi, M., Kita, A., Ito, H., and Kato, I. (1990) Sse8387I, a new type-II restriction enzyme that recognizes the octanucleotide sequence 5'-CCTGCAGG-3'. *Nucleic Acids Res.* **18,** 5637–5640.
12. Meyertons Nelson, D., Miceli, S. M., Lechevalier, M. P., and Roberts, R. J. (1990) *Fse*I, a new type II restriction endonuclease that recognizes the octanucleotide sequence 5'-GGCCGGCC-3'. *Nucleic Acids Res.* **18,** 2061–2064.
13. McClelland, M., Jones, R., Patel, Y., and Nelson, M. (1987) Restriction endonucleases for pulsed field mapping of bacterial genomes. *Nucleic Acids Res.* **15,** 5985–6005.
14. Weil, M. D. and McClelland, M. (1990) Enzyme cleavage of a bacterial genome at a 10-base-pair recognition site. *Proc. Natl. Acad. Sci. USA* **86,** 51–55.
15. Nelson, M. and Schildkraut, I. (1987) The use of DNA methylases to alter the apparent recognition specificities of restriction endonucleases. *Methods Enzymol.* **155,** 41–48.
16. Rogerson, A. C. (1989) The sequence asymmetry of the *Escherichia coli* chromosome appears to be independent of strand or function and may be evolutionarily conserved. *Nucleic Acids Res.* **17,** 5547–5563.
17. Pyle, L. E. and Finch, L. R. (1988) A physical map of the genome of *Mycoplasma mycoides* subspecies *mycoides* Y with some functional loci. *Nucleic Acids Res.* **16,** 6027–6039.
18. Marinus, M. G. (1987) DNA methylation in *Escherichia coli. Ann. Rev. Genet.* **21,** 113–131.
19. Cantor, C. R. (1988) Pulsed-field gel electrophoresis of very large DNA molecules. *Ann. Rev. Biophys. Biophys. Chem.* **17,** 287–304.
20. Neimark, H. C. and Lange, C. S. (1990) Pulse-field electrophoresis indicates full-length mycoplasma chromosomes range widely in size. *Nucleic Acids Res.* **18,** 5443–5448.
21. Bautsch, W., Grothues, D., and Tümmler, B. (1988) Genome fingerprinting of *Pseudomonas aeruginosa* by two-dimensional field inversion gel electrophoresis. *FEMS Microbiol. Lett.* **52,** 255–258.
22. Crouse, J. and Amorese, D. (1986) Stability of restriction endonucleases during extended digestions. *FOCUS* **8,** 1–3.
23. New England Biolabs, Inc., Beverly, MA, USA (1990) 1990–1991 Catalog, p. 131.

CHAPTER 15

Protozoan Genomes

*Karyotype Analysis, Chromosome Structure,
and Chromosome Specific Libraries*

Lex H. T. Van der Ploeg, Keith M. Gottesdiener, Stanley H. Korman, Michael Weiden, and Sylvie Le Blancq

1. Introduction

Protozoa represent a diverse group of single-celled eukaryotes, many of which have parasitic life styles, infecting hundreds of millions of people. Unique aspects of their biology relate to their distinct evolutionary position and their complex life cycles, frequently involving different hosts.

Overall, single-celled eukaryotes have relatively small genomes, ranging in size from several Mbp to several hundreds of Mbp. These small genomes should be excellent model systems for studying eukaryotic genome function and organization. However, such studies have been hampered by the absence of chromosome condensation during metaphase, and by our inability to perform genetic crosses. The application of pulsed-field gel electrophoresis (PFGE) to the study of protozoan genomes has bypassed some of these problems. This has led to a preliminary understanding of protozoan genome organization, and has enhanced our knowledge of numerous aspects of cell function (*1–8, see* ref. *9* for a general review of PFGE).

From: *Methods in Molecular Biology, Vol. 12: Pulsed-Field Gel Electrophoresis*
Edited by: M. Burmeister and L. Ulanovsky
Copyright © 1992 The Humana Press Inc., Totowa, NJ

The genomic organization of several protozoa has now been studied in detail, and some characteristics of the protozoan genome have been defined. However, our knowledge is still rudimentary and even such basic information as the ploidy, number, and size of the chromosomes, has not been accurately determined for most species. Although it is not yet clear to what extent the protozoa studied thus far are representative of other species, some common features of their genome organization are worth noting (only a few, directly pertinent references have been listed).

1.1. The Genomes of Protozoa Are Relatively Small

The total genome size of most protozoa does not exceed 100 Mbp, and many protozoa have less complex genomes. The DNA is organized into multiple, linear chromosomes, ranging in size from 25 kbp to several Mbp *(see next sections)*. Circular, episomal DNA has also been identified in some protozoa as a consequence of drug selection and gene amplification *(10–12)*.

1.2. The Structure of Protozoan Chromosomes Appears Similar to the Chromosomes of Higher Eukaryotes

The analysis of a subset of chromosomes, the minichromosomes (MC) of *Trypanosoma brucei* (50–150 kbp in size), was particularly revealing. Physical mapping of these chromosomes and analysis of the majority of their DNA nucleotide sequence elements showed that these MC mainly consist of simple sequence DNA and they are linear molecules with telomere repeats at their ends. MC appear to be mitotically stable molecules. An electronmicroscopic analysis of MC revealed that they lacked abnormal structures (loops, knots, cruciforms) and that they have most likely replicated from a single bidirectional origin of replication located in the center of the MC (*13; see* Note 1).

1.3. Protozoan Karyotypes

1.3.1. Variation Among Species

The genomic complexity, and the number and size of the chromosomes (karyotype, which includes the mitochondrial DNA), differ widely when various species of protozoa are compared *(see below)*; even a comparison of closely related species reveals significant differences in genome content and organization (for examples of kinetoplastidae *[5–8,14–19]; T. cruzi [20]*; for examples of *Eimeria [21]*; *Pneumocystis carinii [22]*; *Theileria [23]*; *Plasmodium [24–26]*; *Giardia, [27–29]*).

1.3.2. Variation Within Species

A karyotype that is "characteristic" for each species can often be identified; and in some cases, PFGE could differentiate subspecies associated with specific protozoan diseases *(30)*. Chromosome alterations could also be detected when isolates from the same protozoan species were compared *(5,31–41)*. However, the variability in chromosome organization is limited and homologous chromosomes can be defined in the "basic karyotype" of *T. brucei* stock 427-60 (*39,40,42,43; see* specific applications below). Interestingly, however, even within a single clone, these homologous chromosomes differed by about 20% in size, and could occasionally be shown to vary by as much as 50%.

Karyotype analysis within species is thus complicated by the presence of a significant number of chromosomal alterations that were detected when isolates from the same species were compared *(5, 31–41,43–45)*.

1.3.3. Stability During the Life-Cycle

Many protozoa have complex life cycles, with different morphological stages associated with intermediate and definitive hosts. An analysis of the karyotypes of different morphological stages of *Plasmodium (7,46)* and of several of the kinetoplastidae *(4–6)* revealed that these were relatively stable during differentiation, and did not undergo abrupt organizational changes associated with an altered pattern of gene expression. In *G. lamblia,* for instance, our preliminary data indicate that the karyotype heterogeneity observed among isolates from a single geographic area was not associated with differentiation from trophozoite to cyst to trophozoite *(27)*.

1.4. Repetitive Elements

Protozoan genomes encode numerous repetitive elements, and even protein-coding genes are frequently repeated and arranged in tandem arrays in several of the trypanosomatids. This organization hinders the analysis of such regions, hence chromosome-specific libraries derived from chromosomes isolated by PFGE (*see* specific applications below) are commonly used to investigate the genomic structure of specific protozoan genes.

Below, we present methods for the fractionation of protozoan genomes and we describe several specific examples that led to more detailed understanding of protozoan genome dynamics. We outline some of the methods used to generate chromosome-specific libraries and a PCR method to clone regions that were initially selected against in chromosome-specific libraries (only a few, directly pertinent references are listed).

2. Protozoan Karyotype Analysis

2.1. Determination of a "Characteristic" Karyotype

To standardize the karyotype for a specific isolate, we numbered the chromosome bands and mapped genetic markers to specific bands. We initially focused on the genome of the protozoan *T. brucei (2,4,6)*, causative agent of sleeping sickness in humans. The life-cycle stages of this parasite identified thus far in its mammalian host and its insect vector, the tsetse fly, are all diploid with the total amount of DNA per nucleus estimated at about 70,000 kb *(47)*. *T. brucei* has an unusually large number of chromosomes owing to the presence of about 100 minichromosomes, which comigrate and are collectively numbered band 1 of the PFGE gels (Fig. 1; the MC mainly consist of simple sequence DNA, and are structurally similar *[13]*). In addition to the minichromosomes, *T. brucei* contains about 18 larger chromosomes, which encode housekeeping genes (Fig. 1, bands 2–19). Figure 1 shows five individual PFGE panels (of *T. brucei* stock 427-60) run at varying conditions designed to optimize chromosomal separation in the different size classes. Fourteen of the 18 larger chromosome bands have thus far been grouped into seven pairs of chromosome homologs *(43)*.

2.2. Comparison of Karyotypes from Different Protozoan Species

The complexity of protozoan genome organization can be observed in a side-by-side comparison of karyotypes from different protozoan species. Figure 2 illustrates (from left to right) the karyotypes of the kinetoplastid species *T. cruzi* (Y strain), *Leishmania braziliensis guyanensis*, *T. brucei* (stock 427-60), and the unrelated protozoan *Giardia lamblia* (WB clone F9-H7; *29*). The overall size distribution of chromosomes ranges from approx 200 kbp to over 6 Mbp (excluding the MC of *T. brucei*). The number of chromosomes ranges from at least 8 chromo-

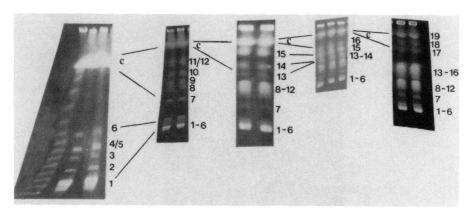

Fig. 1. Karyotype of *T. brucei* stock 427-60 (variant 118 clone 1). From left to right, ethidium bromide-stained panels are shown with chromosome-sized DNA separated under different conditions: Panel 1, 10 V/cm, 50-s pulse frequency, 24 h; panel 2, 5 V/cm, 900-s pulse frequency, 120 h; panel 3, 5 V/cm, 1800-s pulse frequency, 120 h; panel 4, 3 V/cm, 3600-s pulse frequency, 168 h; and panel 5, 3 V/cm, 6000-s pulse frequency, 240 h. The different ethidium-stained bands are numbered (1 being the smallest at 50–150 kbp and 19 the largest at about 5.7 Mbp) to facilitate their comparison. C indicates the compression zone in the different gels; dark lines connect regions of the gels that are separated with different efficiencies in the different panels. Lane 1 in panel 1 is a lambda ladder size standard.

somes for *G. lamblia* to over 120 chromosomes in *T. brucei* (though this cannot be ascertained from a single gel).

2.3. Karyotype Analysis: Intraspecies Variability

A rapid accumulation of changes in chromosome organization seems to be a hallmark of the genomes of some protozoan species *(6, 27,29,31–41,43–45)*. However, the extent to which the karyotype can differ varies from species to species. For instance, in most of the kinetoplastidae, subspecies or even isolates of a single species can differ by several chromosomal changes; however, an isolate or species-specific pattern remains.

An extreme example of chromosome variability is seen in the protozoan *G. lamblia*, where 60% of the subclones from a single isolate had heterogeneous karyotypes *(29)*. These karyotype alterations specifically involve ribosomal RNA encoding chromosomes (Fig. 3A). *G. lamblia* may, however, represent an exceptional case with an unusually high rate of mutation.

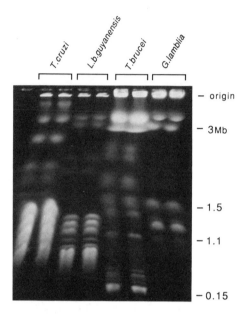

Fig. 2. Ethidium bromide-stained gel with two lanes each containing DNA from *T. cruzi* (Y strain), *L. braziliensis gyanensis*, *T. brucei* (stock 427-60), and *G. lamblia* (WB-F9-H7). DNA was separated at 5 V/cm for 120 h with a 800-s pulse frequency.

Different clinical isolates of *G. lamblia*, obtained over a short period of time, from a single geographic area have drastically different banding patterns affecting almost all chromosomes (Fig. 3B; *27*). The biological significance of this variability in the *Giardia* karyotype is unclear, but we can speculate that it may play a role in antigenic variability of *G. lamblia*.

As stated earlier, karyotype variability has also been documented in other protozoa, although changes appear to accumulate at a slower rate than in *G. lamblia*. The function of many of these karyotype modifications is unclear. In *T. brucei*, several of the changes in the chromosome banding pattern could be associated with alterations in the expression of the telomerically located variant cell surface glycoprotein (VSG) genes *(6,48,49)*. In *Plasmodium*, chromosomal rearrangements *(7,35,37,38,41,44,45)* have been identified, which were shown to affect the expression of a histidine-rich protein coding sequence *(44)*, whereas others occurred in response to drug selection *(38)*. In the few examples that have been studied, these chromosome rearrangements thus appear to be associated with changes in the control of gene expression.

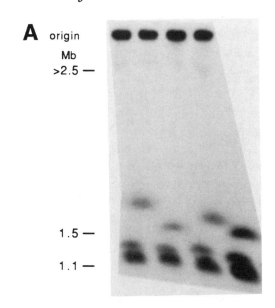

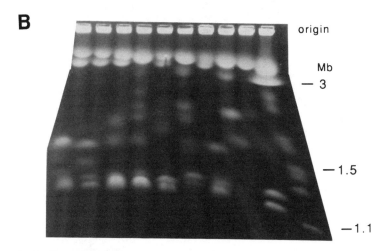

Fig. 3. **A:** Size heterogeneity of rRNA containing chromosomes among subclones of a *Giardia lamblia* cloned line. PFGE gel (pulse frequency, 950 s, 5 V/cm, for 130 h) separating the chromosomes of a series of subclones of *G. lamblia* derived from the cloned line WB-F9 *(29)*. Lane 1, WB-F9-D11; lane 2, WB-F9-A6; lane 3, WB-F9-E12; and lane 4 WB-F9-G3. The DNA was blotted to nylon membranes and hybridized with a rRNA probe *(29,63)*. **B:** Karyotype heterogeneity among different clinical isolates of *G. lamblia (22)*. Ethidium bromide-stained PFGE gel (pulse frequency, 950 s, 5 V/cm, for 130 h) separating the chromosomes of *G. lamblia* isolates cultured from nine different individuals in Jerusalem, Israel. Lane 10 is *T. brucei* DNA comigrated as a size-standard.

2.4. Reverse Genetics and the Study of Protozoan Genome Organization

Recently, transfection techniques *(50–53)* and expression vectors *(54,55)* have been developed for the transient and stable transfection *(54,56–59)* of several different kinetoplastid species. In addition, gene disruption and gene transfer by homologous recombination are highly efficient events in different protozoan species. Since linearized plasmids efficiently find their homologous targets in the trypanosome genome and integrate either by a double crossover or by gene conversion, an accurate dissection of the functional elements of the trypanosome genome is now possible. In addition, chromosomes can now be tagged with selectable markers that will allow transfer and selection in yeast. This will avoid the problems associated with chromosomes that share numerous repetitive elements.

2.5. Cells and PFGE Conditions

Most protozoa can be readily lysed and diffusion of detergent, proteinase K, and EDTA can efficiently release chromosomal DNA. Removal of cell walls or preparation of nuclei prior to embedding the cells in agarose is therefore usually unnecessary (*see* Note 2).

The effects of DNA concentration, pulse time, field strength, ionic strength of buffers, temperature, gelling material, choice of size-standard, and type of electrophoresis unit on separation efficiency and quality have been extensively discussed in other chapters and will not be addressed here. We will only discuss considerations that may aid in setting up PFGE separations for different protozoa.

In general, a cell density that is consistent with a DNA concentration in the agarose blocks of 50–500 μg/mL is recommended (*see* Notes 3 and 4). The cells are normally embedded in low-melting-point (LMP) agarose by diluting the cell suspension (in PBS) with an equal vol of 1% low-melting-point agarose (INCERT Agarose, FMC Corporation) at 42°C. For example, we use trypanosomes (total genome size, 70,000 kbp) at a final concentration of 2×10^9 cells/mL, whereas *Giardia* (total genome size estimated at 80,000 kbp) are diluted to 5×10^8 cells/mL (a lower cell density is used due to the large cell size; *see* Note 4). They are then immediately pipetted into a sample mold kept at 0°C and further treated as described in Chapter 8.

This technique allows analysis of chromosome-sized DNA by avoiding mechanical damage to the DNA, and it is appropriate for most protozoan genomes (*see also* the section on sucrose gradient fractionation of high-mol-wt DNA; *see* Note 5).

Optimal separation of the different chromosome-size classes usually requires the use of a range of different PFGE conditions. The conditions described in Table 1 are used with the point electrode, pulsaphor configuration (*see also* Chapter 4), described in materials.

3. Materials

1. Lysis buffer: 0.5M EDTA, pH 9.0, 1% (SLS) N-lauroylsarcosine, 0.5 mg/mL proteinase K (*see* Chapter 8).
2. LMP agarose for preparation of DNA agarose blocks: INCERT agarose, FMC Corp., or Ultrapure agarose, BRL.
3. Sample mold: Dimensions, $2 \times 6 \times 8$ mm; 100 µL/sample.
4. Phosphate buffered saline (PBS): 140 mM NaCl, 2.7 mM KCl; 10 mM Na$_2$PO$_4$; 1.7 mM KH$_2$ PO$_4$; adjust pH to 7.4 with HCl.
5. Agarose for gels (Ultrapure, BRL); LMP agarose (Ultrapure, BRL) for preparative gels (*see* Note 6).
6. For yeast and lambda concatemer DNA size standards, *see* Chapter 8.
7. All size-separations were performed in 20 cm^2 1% agarose gels in a Pharmacia LKB Pulsaphor apparatus at 15°C (*see also* Chapter 4). This unit can be equipped with: (a) an array of movable point electrodes positioned on the North and West sides (cathodes; we used three electrodes positioned at 11, 44, and 78%, respectively, from the North/East and South/West corners of the electrophoresis unit), and single electrodes at the South and East corners (anodes; one electrode positioned at 22%, so that the angle of reorientation of DNA in the gel is about 120°); or (b) a CHEF electrode array (*60, see also* Chapter 2).
8. PFGE running buffer: 89 mM Tris, 90 mM boric acid, 0.2 mM EDTA, prepared and stored at room temperature as a 10X stock.
9. Solutions and materials for Southern blotting are as described in Maniatis (*61*). All Southern blotting procedures are preceded by an acid hydrolysis step of 20 min in 0.25 M HCl at room temperature.
10. Ethidium bromide is prepared as a stock solution of 5 mg/mL in water. Stored at room temperature (**NB**: ethidium bromide is toxic and gloves should be used at all times when handling solutions or gels containing it).
11. Restriction enzyme buffers are those designated by the manufacturer for each enzyme.
12. DNA size-markers for small restriction enzyme fragments are 1-kb size standards (Bethesda Research Lab.) and or wildtype lambda digested with an appropriate restriction enzyme.

Table 1
Optimal Separation of Chromosome-Sized DNA
Using Different PFGE Conditions

Size range	Field strength	Pulse time	Duration of run
50–600 kb	10 V/cm	50 s	40 h
600 kb–2 Mb	5 V/cm	900 s	120 h
2–3 Mb	3 V/cm	2400 s	168 h
3–6 Mb	3 V/cm	3600 s	168 h

13. Sucrose gradients are made by mixing 5 and 20% sucrose solutions in two communicating chambers, in 25 mM Tris-HCl, pH 7.5, 100 mM EDTA, 1% sodium lauroylsarcosine, proteinase K at 50 µg/mL.
14. PMSF, phenylmethyl sulfonyl fluoride (**NB:** this is toxic, wear a face mask when weighing). Dissolve in isopropanol to give a 100 mM PMSF stock solution; it inactivates in water with time). Removal of the solution incubated with blocks is achieved by placing a pipet on the bottom of the tube, with the tip touching the bottom, sucking up the solution. The blocks can be manipulated with glass Pasteur pipets with sealed, rounded off "shephard's crock ends."
15. RNAse A is heat-inactivated for 15 min at 65°C prior to its use to inactivate other nucleases.
16. Phenol: Saturated with 100 mM Tris-HCl, pH 8.0, containing 0.1% 8-hydroxyquinoline. (**NB:** Phenol can cause severe burns and is toxic. Gloves should be worn at all times.)
17. PCI: Phenol:chloroform:isoamylalcohol; 25:24:1; PCI solutions are equilibrated with 10 mM Tris-HCl pH 7.5 and contain 0.1% 8-hydroxyquinoline.
18. Chloroform isoamylalcohol (24:1) for final extraction after phenol chloroform extractions.

4. Methods

4.1. Purification of Chromosome-Specific DNA

4.1.1. Chromosome-Specific DNA from PFGE Gels

The genomes of many protozoa have a relatively low complexity. Obtaining single-copy DNA sequences in clone has therefore been simple, since small libraries have a high probability of containing a clone of interest. However, the genomes of many protozoa encode numerous repetitive elements. The analysis of the transcriptional control of specific repetitive protein coding genes in the protozoan *T. brucei* required the construction of chromosome-specific libraries, facilitating the physical mapping of these chromosomes and the isolation of chromosome-specific clones. Initially, we separated chromosome-sized DNA by PFGE and purified the DNA by electroelution from the PFGE agarose gels, followed by CsCl fractionation *(6,48)*. These methods suffered from relatively low yields and from the fact that the DNA was frequently of insufficient quality for further enzymatic manipulation. We developed additional procedures for the reliable fractionation and purification of large, chromosome-specific DNA *(49)*:

1. Prepare chromosome-sized DNA for PFGE analysis as described above and in Chapter 8.
2. Size-separate chromosome-sized DNA at optimal PFGE conditions (*see* Note 7 and Section 2.5.).
3. Stain the gel with ethidium bromide (30 min in H_2O with 2.5 µg/mL ethidium bromide) and cut the desired band *en bloc* from one lane of the gel with minimal UV exposure. Significant UV exposure of DNA that has been stained with ethidium bromide must be avoided in analyzing large DNA molecules (*see* Note 8).
4. Trim the block to $5 \times 5 \times 2$ mm, and wash 5X in a larger vol (10–20 mL) of 10 m*M* Tris-HCl, pH 8.0 for 2 h for each wash, with gentle shaking.
5. This DNA is a suitable substrate for enzymatic treatment.

4.1.2. Chromosome-Specific DNA from Sucrose Gradients

Pure, minichromosome-sized DNA could also be obtained by sucrose-gradient fractionation *(13)*; a modification of the method by Sloof et al. (*62; see also* Note 9).

1. Make a 35 mL linear, 5–20% sucrose gradient in 100 m*M* EDTA, 25 m*M* Tris-HCl, pH 7.5, 1% SLS, and 50 µg/mL proteinase K in SW 28 ultracentrifuge tubes.
2. Layer 2 mL of a solution containing 200 m*M* EDTA, 1% SLS, and 1 mg/mL proteinase K on top of this gradient.
3. Layer 4×10^8 trypanosomes (*T. brucei*, stock 427-60) in 1 mL PBS on top of this gradient.
4. Allow the trypanosomes to lyse by incubating the gradient at 25°C for 3 h, after which the DNA is size-separated by centrifugation for 16 h at 14,000*g*.
5. Mini-chromosomal DNA of 50–150 kbp is taken from the top of the gradients in 1-mL aliquots with a sterile 1-mL syringe (omitting the needle).

6. Assay the fractions directly for the presence of minichromosomal DNA by PFGE. Minichromosomal DNA is normally observed in fractions 6–10 of these gradients.
7. Concentrate the minichromosomal DNA by ethanol precipitation and confirm its size by PFGE electrophoresis (*see* Note 10).
8. Treat the DNA with RNAse A (final concentration of 10 μg/ml for 15 mins at 37°C).
9. Following RNAse A treatment the MC are precipitated with ethanol (*see* Note 11).
10. Recovery of MC DNA is usually 2–4 μg/gradient. This DNA was of sufficient quality for further enzymatic digestion or electronmicroscopic analysis of the chromosome-sized DNA.

4.2. Manipulation of Chromosome-Specific DNA

4.2.1. Restriction Enzyme Digestion of Chromosome-Specific DNA in Agarose Blocks

Chromosome-specific DNA can be restriction enzyme-digested, and then run in a second PFGE gel, providing "chromosome-specific" PFGE gels for physical mapping of the chromosome (frequently too little DNA is available for direct ethidium bromide visualization):

1. Wash each block containing purified, chromosome-sized DNA that was cut from the LMP agarose gels, twice for 1 h in 10 m*M* Tris-HCl, pH 7.5, 1 m*M* EDTA, 1 m*M* PMSF.
2. Wash each block three times in 10 m*M* Tris-HCl, pH 7.5, 1 m*M* EDTA for 1-1/2 h each.
3. Incubate the blocks in an equal vol of 2X restriction enzyme buffer solution and 30 U of the restriction endonuclease required, for a period of 12 h at the appropriate temperature.
4. After restriction endonuclease digestion, the blocks are removed, inserted into the well of a 1% regular agarose PFGE gel, and subjected to electrophoresis as described above. The gel is then ready for Southern blotting.

4.2.2. Preparation of Chromosome-Specific Libraries

Chromosome-specific DNA is purified for chromosome-specific libraries as described above, except that the DNA is isolated from multiple lanes (approx 20 lanes for the libraries made from the 1.5-Mbp chromosome-specific DNA of *T. brucei;* total recovery of <500 ng of pure 1.5-Mbp DNA; *49*).

1. Incubate the washed blocks (as described above) with the appropriate enzyme chosen for library construction. Using *Bal* 31 digestion followed by digeston with a restriction enzyme, fragments can also be generated to prepare chromosome-specific telomere libraries.
2. Following restriction enzyme digestion, the LMP agarose blocks are melted at 68°C in 2 vol of 10 mM Tris-HCl, pH 8.0, 50 mM NaCl for 15 min.
3. Extract the mixture twice with phenol.
4. Extract once with phenol:chloroform:isoamylalcohol (PCI) and once with chloroform:isoamylalcohol (24:1).
5. Precipitate the DNA with ethanol and wash once with 70% ethanol.
6. Resuspend the pellet in 10 mM Tris-HCl, pH 8.0.
7. For the generation of chromosome-specific libraries, the DNA is redigested with a tenfold excess of the restriction endonuclease required, for 2 h at 37°C (to ensure complete digestion).
8. Extract the DNA with PCI and coprecipitate with 100 ng of calf-intestinal-phosphatase-treated, restriction endonuclease-digested vector DNA.
9. Dissolve the DNA in 10 mM Tris-HCl, pH 7.5, and ligate the suspension for 14 h at 16°C in a final vol of 20 μL.
10. Approximately 1 μL (1–5 ng of DNA) of the ligation mixture is used to transform an aliquot of MAX Efficiency DH5α competent *E. coli* bacteria (Bethesda Research Lab.).

4.2.3. PCR Amplification of Chromosome-Specific DNA

Sequences that are markedly underrepresented in recombinant libraries can make chromosome-walking difficult. We adapted PCR amplification to enrich for these DNA sequences in our chromosome-specific libraries. This allowed us to walk past those short, "unclonable" DNA sequences that had prevented us from obtaining larger clones from our libraries.

1. Purify chromosome-specific DNA from PFGE gels and digest with a restriction endonuclease, and ligate into an appropriate vector as described above. This material is used to generate a recombinant library.
2. Sequences that are underrepresented in the recombinant library are obtained by amplifying an aliquot of the ligated material. For PCR amplification, use one synthetic primer from the nucleotide sequences from the most 5' end of previously cloned sequences; PCR amplification is then performed as described below, using this oligonucleotide

and a second oligonucleotide from vector-derived sequences (*see* Fig. 4; *see also* Note 12).
3. Purify the amplified DNA on LMP agarose gels and then ligate the fragments into the appropriate vector, after which the DNA is transformed into *E. coli* DH5α. After PCR amplification, these "unclonable" recombinant clones could be isolated, thus walking past short unclonable regions.

5. Notes

1. MC fractions that were enriched in replicating molecules appeared to migrate abnormally in PFGE, being trapped near the well of the gels.
2. Extensive purification of the organisms preceding their embedding in agarose is recommended but may not always be possible. The amounts of contaminating DNA from other sources (tissues) should, however, be limited. The tolerance of the system for contamination is dependent on the complexity and quality of the contaminating DNA (i.e., relatively nondegraded and distributed over large chromosomes). The reason for this is that the more complex genomes of higher eukaryotes do not give rise to specific banding patterns in the size range where protozoan chromosomes migrate in PFGE. An initial overview of the size distribution of protozoan chromosomes can therefore still be obtained with impure material. For instance, reliable PFGE banding patterns can be obtained with trypanosomes embedded in the presence of whole blood. However, in general, we prefer purified cells, since this will avoid the complications resulting from contaminating nucleases, the presence of a background of nonspecifically degraded contaminating DNA, and potential cross-hybridization.
3. The blocks are cut, and in general, about 1/10 of a block is loaded per lane. Blocks are sliced by placing them on parafilm and using a large cover glass to cut the agarose.
4. For many parasitic protozoa, the genome complexity is not known. In addition, cell size can be a limiting factor in forming agarose blocks with the correct DNA concentration, and it may not be possible to embed the cells at sufficiently high concentrations to allow ethidium bromide visualization of chromosomes. Having determined a cell density, either based on knowledge of the genome complexity or by trial and error, one can further adjust the DNA concentration in each lane by slicing off a small rectangular piece of a block and inserting this into the well of a PFGE gel (we normally load 1–5 μg of DNA/lane). DNA agarose blocks are cut and transferred with a cover glass.

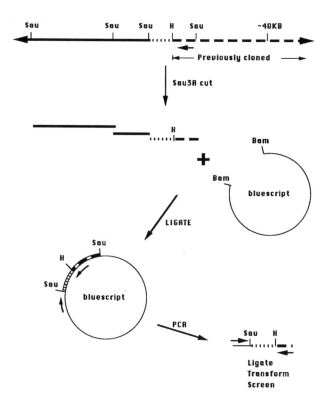

Fig. 4. Schematic presentation of the PCR amplification procedure used in constructing chromosome-specific libraries. The plasmid vector used is Bluescript; previously cloned DNA is marked by heavy dashed lines; sequences that are underrepresented in recombinant libraries are marked with light dashed lines.

The agarose block is then overlaid with 1% LMP agarose, avoiding trapping of air bubbles.

Adjustments in the concentration of DNA per lane are important to allow optimal separation of specific size classes of chromosomes. The actual DNA concentration in a lane in any particular mol wt range is dependent on the size and distribution of chromosomes in the specific sample and on the amount of total DNA loaded. "Local" overloading of a lane can thus affect the ability to visualize individual chromosomes in that specific size class, and the amount of DNA loaded may need to be adjusted depending on the chromosome size-class for which optimal separation is required.

5. DNA can be stored in the lysis solution (*see* Chapter 8) at 4°C for long periods. We have maintained DNA in agarose blocks for up to 8 years without deterioration in the quality of the chromosome-sized DNA.
6. Gels must be poured perfectly horizontal.
7. All size-separations described have been performed in 20 cm² 1% low-melting-point agarose (Bethesda Research Lab.) gels in a Pharmacia LKB Pulsaphor apparatus at 15°C. LMP agarose gels and regular agarose gels achieve equivalent chromosome separations and adjustments in running conditions or agarose concentration are not necessary. For instance, for separation of *T. brucei* chromosomes between 1 and 2 Mbp we performed PFGE for 5 days at 5 V/cm at a pulse time of 900 s, which allowed excellent separation of *T. brucei* band 10 (*see* Fig. 1, second panel from the left) from chromosomes in the adjacent bands 9 and 11.
8. UV exposure in the presence of ethidium bromide will nick the DNA, followed by potential breaking of the DNA, leading to loss of signal. In some cases (for the analysis of restriction fragments that were larger than 150 kbp), we circumvented this problem by staining adjacent lanes and then cutting unstained chromosomes from the PFGE gel.
9. These gradients are easily disturbed and should be handled carefully.
10. Chromosomal breakage could not be detected, and chromosomes up to 300 kbp have been fractionated.
11. Phenol extractions must be omitted since they will lead to substantial shearing of the DNA.
12. We have thus far amplified only relatively short sequences of up to 1 kbp in length. Melting of DNA hybrids is performed for 1 min at 94°C, annealing at 52°C for 2 min, and *Taq* polymerase-mediated elongation for 3 min at 72°C for 40 cycles.

Acknowledgments

We thank all colleagues for critical reading of the manuscript. This work was supported by a grant from the John D. and Catherine T. MacArthur Foundation and by NIH Grants AI 21784 and AI 26497 to LHTVDP. LHTVDP is a Burroughs Wellcome Scholar in Molecular Parasitology.

References

1. Schwartz, D. C., Saffran, W., Welsh, J., Haas, R., Goldenberg, M., and Cantor, C. R. (1983) New techniques for purifying large DNAs and studying their properties and packaging. *Cold Spring Harbor Symp. Quant. Biol.* **XLVII,** pp. 189–195.

2. Borst, P., Bernards, A., Van der Ploeg, L. H. T., Michels, P. A. M., Liu, A. Y. C., De Lange, T., Sloof, P., Schwartz, D. C., and Cantor, C. R. (1983) The role of mini-chromosomes and gene translation in the expression and evolution of VSG genes. *Gene Expression, UCLA Symposia on Molecular and Cellular Biology. New Series, vol. 8*, (Hamer, D. and Rosenberg, M., eds.) Alan R. Liss, New York. pp. 414–435.

3. Schwartz, D. C. and Cantor, C. R. (1984) Separation of yeast chromosome-sized DNAs by pulsed field gradient gel-electrophoresis. *Cell* **37**, 67–75.

4. Van der Ploeg, L. H. T., Schwartz, D. C., Cantor, C. R., and Borst, P. (1984) Antigenic variation in *Trypanosoma brucei* analyzed by electrophoretic separation of chromosome sized DNA molecules. *Cell* **37**, 77–84.

5. Van der Ploeg, L. H. T., Cornelissen, A. W. C. A., Barry, J. D., and Borst, P. (1984) Chromosomes of kinetoplastida. *EMBO J.* **3**, 3109–3115.

6. Van der Ploeg, L. H. T., Cornelissen, A. W. C. A., Michels, P. A. M., and Borst, P. (1984) Chromosome rearrangements in *Trypanosoma brucei*. *Cell* **39**, 213–221.

7. Van der Ploeg, L. H. T., Smits, M., Ponnudurai, T., Vermeulen, A., Meuwissen, J. H. E. Th., and Langsley, G. (1985) Chromosome-sized DNA molecules of *Plasmodium falciparum*. *Science* **229**, 658–661.

8. Van der Ploeg, L. H. T. (1987) Separation of chromosome-sized DNA molecules by pulsed field gel-electrophoresis. *Am. Biotechnol. Lab.* **Jan/Feb.**, 8–16.

9. Abelson, J. N., Simon, M. I. (1990) *Pulsed Field Gel-Electrophoresis*, vol. 1, no. 2. Academic, New York.

10. Detke, S., Chaudhuri, G., Kink, J. A., and Chang, K. -P. (1988) DNA amplification in tunicamycin-resistant *Leishmania mexicana*. *J. Biol. Chem.* **263**, 3418–3424.

11. Ouellete M., Fase-Fowler, F., and Borst, P. (1990) The amplified H circle of methotrexate-resistant *Leishmania tarentolae* contains a novel P-glycoprotein gene. *EMBO J.* **9**, 1027–1033.

12. Beverly S. M., Codere, J. A., Santi, D. V., and Schimke, R. T. (1984) Unstable DNA amplifications in methotrexate-resistant Leishmania consist of extrachromosomal circles which relocalize during stabilization. *Cell* **38**, 431–439.

13. Weiden, M., Osheim, Y. N., Beyer, A. L., and Van der Ploeg, L. H. T. (1991) Chromosome structure: DNA nucleotide sequence elements of a subset of the mini-chromosomes of the protozoan *Trypanosoma brucei*. *Mol. Cell. Biol.* **11**, 3829–3834.

14. Spithill, T. W. and Samaras, N. (1985) The molecular karyotype of *L. major* and mapping of α and β tubulin gene families to unlinked chromosomal loci. *Nucleic Acids Res.* **13**, 4155–4169.

15. Comeau, A. M., Miller, S. I., and Wirth, D. F. (1986) The chromosome location of four genes in Leishmania. *Mol. Biochem. Parasitol.* **21**, 161–169.

16. Galindo, I. and Ramirez Ochoa, J. L. (1989) Study of *Leishmania mexicana* electrokaryotype by clamped homogeneous electric field electrophoresis. *Mol. Biochem. Parasitol.* **34**, 245–252.

17. Gomez-Eichelman, M. C., Holtz, G., Beach, D., Simpson, A. M., and Simpson, L. (1988) Comparison of several lizard Leishmania species and strains in terms of kinetoplast minicircle and maxicircle DNA sequences, nuclear chromosomes and membrane lipids. *Mol. Biochem. Parasitol.* **27,** 143–158.

18. Sternberg, J., Tait, A., Haley, S., Wells, J. M., Le Page, R. W. F., Schweizer, J., and Jenni, L. (1988) Gene exchange in trypanosomes: Characterization of a new hybrid genotype. *Mol. Biochem. Parasitol.* **27,** 191–200.

19. Masake, R. A., Nyambati, V. M., Nantulya, V. M., Majiwa, P. A. O., Moloo, S. K., and Musoke, A. J. (1988) The chromosome profiles of *Trypanosoma congolense* isolates from Kilifi, Kenya and their relationship to serodeme identity. *Mol. Biochem. Parasitol.* **30,** 105–112.

20. Henriksson, J., Aslund, L., Macina, R. A., Franke de Cazullo, B. M., Cazullo, J. J., Frasch, A. C., and Petterson, U. (1990) Chromosomal localization of seven cloned antigen genes provides evidence of diploidy and further demonstration of karyotype variability in *Trypanosoma cruzi. Mol. Biochem. Parasitol.,* **42,** 213–232.

21. Shirley, M. W., Kemp, D. J., Pallister, J., and Prowse, S. J. (1990) A molecular karyotype of *Eimeria tenella* as revealed by contour-clamped homogeneous electric field gel electrophoresis. *Mol. Biochem. Parasitol.* **38,** 169–173.

22. Lundgren, B., Cotton, R., Lundgren, J. D., Edman, J. C., and Kovacs, J. A. (1990) Identificaton of *Pneumocystis carinii* chromosomes and mapping five genes. *Infect. Immun.* **58,** 1705–1710.

23. Morzaria, S. P., Spooner, P. R., Bishop, R. P., Musoke, A. J., and Young, J. R. (1990) *Sfi* I and *Not* I polymorphisms in Theileria stocks detected by pulsed field gel electrophoresis. *Mol. Biochem. Parasitol.* **40,** 203–211.

24. Inselberg, J. W., Bzik, D. J., and Li, W. B. (1990) *Plasmodium falciparum:* Analysis of chromosomes separated by contour-clamped homogenous electric fields. *Exp. Parasitol.* **71,** 189–198.

25. Sheppard, M., Thompson, J. K., Anders, R. F., Kemp, D. J., and Lew, A. M. (1989) Molecular karyotyping of the rodent malarias *Plasmodium chabaudi, Plasmodium berghei and Plasmodium vinckei. Mol Biochem. Parasitol.* **34,** 45–52.

26. Triglia, T. and Kemp, D. J. (1991) Large fragments of *Plasmodium falciparum* DNA can be stable when cloned in yeast artificial chromosomes. *Mol. Biochem. Parasitol.* **44,** 207–212.

27. Korman, S. H., Le Blancq, S. M., Deckelbaum, R. J., and Van der Ploeg, L. H. T. (1991) Investigation of human giardiasis by karyotype analysis. *J. Clin. Invest.* Submitted.

28. Adam, R. D., Nash, T. E., and Wellems, T. E. (1988) The *Giardia lamblia* trophozoite contains sets of closely related chromosomes. *Nucleic Acids Res.* **16,** 4555–4567.

29. Le Blancq, S. M., Korman, S. H., and Van der Ploeg, L. H. T. (1991) Frequent chromosome rearrangements in *Giardia lamblia. Nucleic Acids Res.* **19,** 4405–4412.

30. Giannini, S. H., Schittini, M., Keithly, J. S., Warburton, P. E., Cantor, C. R., and Van der Ploeg, L. H. T. (1986) Karyotype analysis of Leishmania species and its applicability to classification and clinical diagnosis. *Science* **232**, 761–765.
31. Giannini, S. H., Curry, S. S., Tesh, R. B., and Van der Ploeg, L. H. T. (1990) Size-conserved chromosomes and stability of molecular karyotype in cloned stocks of *Leishmania major. Mol. Biochem. Parasitol.* **39**, 9–22.
32. Gibson, W. and Garside, L. (1991) Genetic exchange in *Trypanosoma brucei:* Variable chromosomal location of housekeeping genes in different trypanosome stocks. *Mol. Biochem. Parasitol.* **45**, 77–90.
33. Bishop, R. P. (1990) Extensive homologies between *Leishmania donovani* chromosomes of markedly different size. *Mol. Biochem. Parasitol.* **38**, 1–13.
34. Bastien, P., Blaineau, C., Taminh, M., Rioux, J. A., Roizes, G., and Pages, M. (1990) Interclonal variations in molecular karyotype in *Leishmania infantum* imply a mosaic strain structure. *Mol. Biochem. Parasitol.* **40**, 53–62.
35. Shirley, M. W., Biggs, B. A., Forsyth, K. P., Brown, H. J., Thompson, J. K., Brown, G. V., and Kemp, D. (1990) Chromosome 9 from independent clones and isolates of *Plasmodium falciparum* undergoes subtelomeric deletions with similar breakpoints in vitro. *Mol. Biochem. Parasitol.* **40**, 137–146.
36. Lopes, A. H. de C. S., Iovannisci, D., Petrillo-Peixoto, M., McMahon-Pratt, D., and Beverley, S. M. (1990) Evolution of nuclear DNA and the occurrence of sequences related to new small chromosomal DNAs in the trypanosomatid genus *Endotrypanum. Mol. Biochem. Parasitol.* **40**, 151–162.
37. Ponzi, M., Janse, C. J., Dore, E., Scotti, R., Pace, T., Reterink, T. J. F., Van der Berg, F. M., and Mons, B. (1990) Generation of chromosome-size polymorphism during in vivo mitotic multiplication of *Plasmodium berghei* involves both loss and addition of subtelomeric repeat sequences. *Mol. Biochem. Parasitol.* **41**, 73–82.
38. Cowman, A. F. and Lew, M. (1990) Chromosomal rearrangements and point mutations in the DHFR–TS gene of *Plasmodium chabaudi* under antifolate selection. *Mol. Biochem. Parasitol.* **42**, 21–30.
39. Beverley, S. M. and Coburn, C. M. (1990) Recurrent de novo appearance of small linear DNAs in *Leishmania major* and relationship to extrachromosomal DNAs in other species. *Mol. Biochem. Parasitol.* **42**, 133–142.
40. Pages, M., Bastien, P., Veas, F., Rossi, V., Bellis, M., Zincker, P., Rioux, J.-A., and Roizes, G. (1989) Chromosome-size and number polymorphisms in *Leishmania infantum* suggest amplification/deletion and possible genetic exchange. *Mol. Biochem. Parasitol.* **36**, 161–168.
41. Sharkey, A., Langsley, G., Patrapotikul, J., Mercereau-Puijalon, O., McLean, A. P., and Walliker, D. (1988) Chromosome size-variation in the malaria parasite of rodents, *Plasmodium chabaudi. Mol. Biochem. Parasitol.* **28**, 47–54.
42. Van der Ploeg, L. H. T., Smith, C., Polvere, R. I. P., and Gottesdiener, K. (1989) Improved separation of chromosome-sized DNA from *Trypanosoma brucei* stock 427-60. *Nucleic Acids Res.* **17**, 3217–3227.

43. Gottesdiener, K., Garcia-Anoveros, J, Lee, G.-S. M., and Van der Ploeg, L. H. T. (1990) Chromosome organization in the protozoan *Trypanosoma brucei. Mol. Cell. Biol.* **10**, 6079–6083.
44. Pologe, L. G. and Ravetch, J. V. (1986) A chromosomal rearrangement in a *Plasmodium falciparum* histidine-rich protein gene is associated with the knobless phenotype. *Nature* **322**, 474–477.
45. Corcoran, L. M., Forsyth, K. P., Bianco, A. E., Brown, G. V., and Kemp, D. (1988) Chromosome size polymorphisms in *Plasmodium falciparum* can involve deletions and are frequent in natural parasite populations. *Cell* **53**, 807–813.
46. Kemp, D. J., Corcoran, L. M., Coppel, R. L., Stahl, H. D., Bianco, A. E., Brown, G. V., and Anders, R. F. (1985) Size variation in chromosomes from independent cultured isolates of *Plasmodium falciparum. Nature* **315**, 347–350.
47. Borst, P., Van der Ploeg, M., Van Hoek, J. F. M., Tas, J., and James, J. (1982) On the DNA content and ploidy of trypanosomes. *Mol. Biochem. Parasitol.* **6**, 13–23.
48. Shea, C., Glass, D. J., Parangi, S., and Van der Ploeg, L. H. T. (1986) VSG gene expression site switches in *T.brucei. J. Biol. Chem.* **261**, 6056–6063.
49. Gottesdiener, K., Chung, H. M., Brown, S., Lee, M. G.-S., and Van der Ploeg, L. H. T. (1991) Characterization of VSG gene expression site promoters and identification of promoter associated DNA rearrangement events. *Mol. Cell. Biol.* **11**, 2467–2480.
50. Bellafatto, V. and Cross, G. A. M. (1989) Expression of a bacterial gene in a trypanosomatid protozoan. *Science* **244**, 1167–1169.
51. Laban, A. and Wirth, F. D. (1989) Transfection of *Leishmania enrietii* and expression of chloramphenicol acetyl transferase gene. *Proc. Natl. Acad. Sci.* **86**, 9119–9123.
52. Clayton, C. E., Fueri, J. P., Itzhaki, J. E., Sherman, D. R., Wisdom, G. S., Vijayasarathy, S., and Mowatt, M. R. (1990) Transcription of the procyclic acidic repetitive protein genes of *Trypanosoma brucei. Mol. Cell. Biol.* **10**, 3036–3047.
53. Rudenko, G., Le Blancq, S. M., Smith, J., Lee, G.-S. M., Rattray, A., and Van der Ploeg, L. H. T. (1990) Alpha-amanitin resistantly transcribed PARP genes located in an unusually small polycistronic transcription unit: PARP promoter analysis by transient DNA transfection of *Trypanosoma brucei. Mol. Cell. Biol.* **10**, 3492–3504.
54. Lee, M. G.-S. and Van der Ploeg, L. H. T (1990) Homologous recombination and stable transfection in the parasitic protozoan *Trypanosoma brucei. Science* **250**, 1583–1589.
55. Lee, M. G.-S and Van der Ploeg, L. H. T. (1991) The hygromycin B resistance gene as a selectable marker for stable DNA transfection of *Trypanosoma brucei. Gene* **105**, 255–257.
56. Cruz, A. and Beverley, S. (1990) Gene replacement in parasitic protozoa. *Nature* **348**, 171–173.

57. Tobin, J. F., Laban, A., and Wirth, D. F. (1991) Homologous recombination in *Leishmania enriettii. Proc. Natl. Acad Sci.* **88,** 864–868.
58. ten Asbroek, A. L. M. A., Quelette, M., and Borst, P. (1990) Targeted insertion of the neomycin phosphotransferase gene into the tubulin cluster of *Trypanosoma brucei. Nature* **348,** 174–175.
59. Eid, J. and Sollner-Webb, B. (1991) Stable integrated transformation of *Trypanosoma brucei* that occurs exclusively by homologous recombination. *Proc. Natl. Acad Sci.* **88,** 2118–2121.
60. Chu, G., Volrath, D., and Davis, R.W. (1986) Separation of large DNA molecules by contour-clamped homogeneous electrical fields. *Science* **234,** 1582–1585.
61. Maniatis, T., Fritsch, E. F., and Sambrook, J. (1982) *Molecular Cloning. A Laboratory Manual.* Cold Spring Harbor Laboratory, Cold Spring Harbor, NY.
62. Sloof, P., Henke, H. H., Caspers, M. P. M., and Borst, P. (1983) Size fractionation of *Trypanosoma brucei* DNA: Localization of the 177 bp repeat satellite DNA and variant glycoprotein gene in a mini-chromosomal DNA fraction. *Nucleic Acids Res.* **11,** 3889–3901.
63. Boothroyd, J. C., Wang, A., Campbell, D. A., and Wang, C. C. (1987) An unusually compact ribosomal DNA repeat in the protozoan *Giardia lamblia. Nucleic Acids Res.* **15,** 4065–4084.

CHAPTER 16

Construction of Yeast Artificial Chromosome Libraries by Pulsed-Field Gel Electrophoresis

Anthony P. Monaco, Zoia Larin, and Hans Lehrach

1. Introduction

Yeast artificial chromosomes (YACs; *1*) are rapidly becoming the major cloning system to study eukaryotic genomes by physical mapping and chromosome walking projects *(1–6)*. The main advantages of YACs over prokaryotic-based cloning systems are their large insert capacity (100–1500 kb), and their ability to maintain sequences that are unstable or not well represented in bacteriophage or cosmid genomic libraries.

YAC libraries have been constructed by preparing and size fractionating high-mol-wt DNA in solution using sucrose gradients *(1–3)* or in agarose by pulsed field gel electrophoresis (PFGE; *4–7*). When DNA is prepared in agarose, YAC insert sizes are larger on average for the reason that shear forces seen with DNA in solution are minimized. However, partial degradation of DNA occurs when melting agarose containing high-mol-wt DNA, perhaps owing to metal ion contamination or denaturation *(7)*. The presence of polyamines protects DNA in agarose from degradation at the melting step. By incorporating polyamines in the cloning procedure, mouse and human YAC libraries were constructed with average insert sizes of 700 and 620 kb, respectively *(7)*.

From: *Methods in Molecular Biology, Vol. 12: Pulsed-Field Gel Electrophoresis*
Edited by: M. Burmeister and L. Ulanovsky
Copyright © 1992 The Humana Press Inc., Totowa, NJ

In this chapter we describe in detail the protocols used to construct large insert YAC libraries. This includes preparation of pYAC4 vector *(1)*, partial digestion of genomic DNA in agarose blocks *(7)*, size fractionation by PFGE both before and after ligation to vector *(6,7)*, and transformation of the yeast host AB1380 *(1,8)*. The transformation is essentially done as described by Burgers and Percival *(8)* and yeast media and regeneration plates are described in Rothstein *(9)*.

2. Materials

1. Preparation of vector. All library construction protocols in this chapter are based on the pYAC4 vector *(1)*. Vector DNA is prepared by large scale plasmid extractions and purification by CsCl gradient centrifugation *(10)*.
2. Restriction digest buffers. For most restriction digests, buffers recommended by the manufacturer are adequate. We recommend 10X T4 polymerase buffer *(10)* when digesting vector DNA because it works with almost all restriction enzymes and calf intestinal alkaline phosphatase (CIP; Boehringer Mannheim, Germany, 1 U/μL), thus eliminating precipitation of DNA and buffer changes between enzyme reactions. 10X T4 polymerase buffer: 0.33M Tris-Acetate, pH 7.9, 0.66M potassium acetate, 0.10M magnesium acetate, 0.005M dithiothreitol (DTT), 1 mg/mL bovine serum albumin (BSA). Store frozen at $-20°C$ in small aliquots.
3. Preparation and lysis of cells in agarose blocks. High-mol-wt DNA from fibroblast or lymphoblastoid cell lines, whole blood or fresh mouse spleen tissue is prepared in low melting point (LMP) agarose blocks as described by Barlow (*see* Chapter 8) and Herrmann et al. *(11)*, with $2–5 \times 10^6$ cells/block (approx 15–40 μg DNA).
4. *Eco*RI partial digestion reaction buffer: 1 agarose block with DNA 80–100 μL, BSA (5 mg/mL) 50 μL, 10X *Eco*RI methylase buffer 50 μL, Spermidine (0.1M) 13 μL, *Eco*RI 1 U, *Eco*RI methylase (New England Biolabs [NEB]) 50–200 U, distilled water to 500 μL final vol.
5. 10X *Eco*RI methlyase buffer: 800 μM S-adenosyl-methionine (SAM, NEB); 0.02M MgCl$_2$, 1.0M NaCl, 0.5M Tris-HCl, pH 7.5, 0.01M DTT. Store frozen at $-20°C$ in small aliquots.
6. 100X polyamines: 0.075M spermidine-(HCl)$_3$, 0.030M spermine-(HCl)$_4$. Store frozen at $-20°C$ in small aliquots.
7. 10X ligase buffer: 0.5M Tris-HCl, pH 7.5, 0.1M MgCl$_2$, 0.3M NaCl, 10X polyamines.
8. YPD: 1% yeast extract, 2% bactopeptone, 2% dextrose.

9. Regeneration plates: 1.0 *M* sorbitol (Sigma, St. Louis, MO), 2% dextrose, 0.67% yeast nitrogen base w/o amino acids, (Difco, Detroit, MI, add as filter-sterile after autoclaving of agar), 1X amino acid supplements (-uracil), 2% agar.

10. 10x amino acid supplements: adenine (200 µg/mL), arginine (200 µg/mL), isoleucine (200 µg/mL), histidine (200 µg/mL), leucine (600 µg/mL), lysine (200 µg/mL), methionine (200 µg/mL), phenylalanine (500 µg/mL), tryptophan (200 µg/mL, light sensitive, filter sterilize, and store at 4°C), valine (1.5 mg/mL), tyrosine (300 µg/mL), uracil (200 µg/mL, omit in regeneration and selective plates).

11. SCE: 1.0 *M* sorbitol, 0.1 *M* sodium citrate, pH 5.8, 0.01 *M* EDTA pH 7.5, 0.03 *M* β-mercaptoethanol or 0.01 M DTT (add fresh).

12. STC: 1.0 *M* sorbitol, 0.01 *M* Tris-HCl, pH 7.5, 0.01 *M* CaCl₂.

13. PEG: 20% Polyethylene glycol 6000 (PEG, Serva, Heidelberg, Germany), 0.01 *M* Tris-HCl, pH 7.5, 0.01 *M* CaCl₂. Make fresh and filter sterilize.

14. SOS: 1.0 *M* sorbitol, 25% YPD, 0.0065 *M* CaCl₂, 10 µg/mL tryptophan, 1 µg/mL uracil. Make fresh and filter sterilize.

15. YAC selective media and plates: 2% dextrose, 0.67% yeast nitrogen base w/o amino acids (add filter sterile), 1X amino acid supplements (-uracil, -tryptophan), 2% agar for plates.

16. Contour-clamped homogeneous electric field (CHEF) apparatus. We recommend the Bio-Rad (Richmond, CA) system.

17. Small horizontal gel electrophoresis apparatus. Use to check restriction enzyme digests of vector and test ligations of vector and genomic DNA.

18. Electrophoresis buffer. For both CHEF and horizontal gels we recommend TBE. 10X TBE: 0.89 *M* Tris-borate, 0.89 *M* boric acid, 0.016 *M* EDTA.

19. Agarose. We recommend regular (SeaKem) and LMP (Seaplaque GTG) agarose from FMC Bioproducts (Rockland, ME). Most gels will be 1% w/v (aqueous).

20. Yeast and/or lambda concatamer size markers (Bio-Rad or make as described by Barlow, Chapter 8).

21. Agarase (Calbiochem, San Diego, CA or Sigma) dissolved in 50% v/v glycerol in water and store at 10 U/µL at –20°C.

22. T4 DNA ligase (NEB) at 400,000 U/mL.

23. T4 polynucleotide kinase (NEB) at 10 U/µL.

24. 1X TE: 0.01 *M* Tris-HCl, pH 7.5, 0.001 *M* EDTA, pH 7.5.

25. Phenylmethylsulfonylfluoride (PMSF, Sigma). Prepare at 40 mg/mL in ethanol or isopropanol and heat for several min at 68°C to dissolve. Caution: Use gloves, it is toxic.

26. Proteinase K (Boehringer). Dissolve in water at 10 mg/mL and store in small aliquots at –20°C.
27. 0.5*M* EDTA, pH 8.0.
28. Lyticase (Sigma). Weigh out fresh prior to spheroplast formation (1000 U/20 mL SCE) and dissolve in SCE or water. Lyticase is difficult to get in solution and will need extensive vortex mixture.
29. β-mercaptoethanol (BDH, Poole, England). Open in hood and use gloves.
30. For the yeast transformation, a spectrophotometer, a student microscope (10X, 25X, and 40X objectives and phase contrast) and a hemocytometer cell counter are needed.
31. Phenol equilibrated with 0.1*M* Tris-HCl, pH 8.0. Caution: Wear gloves since phenol burns.
32. Chloroform.
33. Ethanol, 100%.
34. Trinitriloacetic acid (BDH). Dissolve in water at 0.15*M* and store frozen in small aliquots at –20°C. Used to inactivate CIP.

3. Methods
3.1. Preparation of pYAC 4 Vector

1. Before preparing pYAC4 arms for ligation to genomic DNA, test plasmid preps for deletions of telomere sequences during propagation in *E. coli*. Digest 0.5 µg of the pYAC4 plasmid with *Hind*III and check on a 1% agarose gel. Four bands should be visualized: 3.5 kb, 3.0 kb, 1.9 kb, and a 1.4 kb doublet.
2. If there is an additional smaller fragment below the 1.4 kb doublet, then telomere sequences have been deleted from the plasmid and another preparation should be attempted.
3. For preparative vector arms, digest 100–200 µg of pYAC4 with *Eco*RI and *Bam*HI to completion in 500 µL 1X T4 polymerase buffer and check on a 1% agarose gel. Three bands should be visualized: 6.0 kb, 3.7 kb, and 1.7 kb.
4. Heat kill the *Eco* RI and *Bam*HI at 68°C for 10 min.
5. Add directly 0.03–0.06 U/µg vector of calf intestinal phosphatase (CIP) and incubate at 37°C for 30 min.
6. Inactivate the CIP with trinitriloacetic acid to 0.015*M* at 68°C for 15 min.
7. Extract twice with an equal volume of phenol, once with an equal volume of chloroform, and precipitate with ethanol.
8. Resuspend the vector arms at a concentration of 1 µg/µL in 0.01*M* Tris-HCl, pH 7.5 and 0.001*M* EDTA (1X TE).

9. Check the efficiency of dephosphorylation of vector ends and the ability of these ends to ligate after phosphorylation. Set up two 20-μL ligation reactions (2 μL 10X ligase buffer without polyamines, 0.5 μg of digested and CIP-treated pYAC4 vector, 1 U T4 DNA ligase) one with and one without 1 U of T4 polynucleotide kinase.
10. Check ligations on a 1% agarose gel:
    a. Without kinase; 3 bands should be visualized as after digestion.
    b. With kinase; The 1.7 kb *Bam*HI fragment can ligate to itself and form several supercoiled forms below 1.7 kb. The upper arms (6.0 and 3.7 kb) should ligate together by their *Eco*RI and *Bam*HI sites and form several larger fragments.

3.2. Partial Digestion of Genomic DNA

1. Partial digestion reactions. Prior to enzyme digestion, wash the blocks containing genomic DNA in 1X TE with 40 μg/mL phenylmethyl-sulfonylfluoride (PMSF) at 50°C to inactivate the proteinase K and twice in 1X TE to remove the PMSF.
2. Perform partial *Eco*RI digestions by incubating blocks with a combination of *Eco*RI and *Eco*RI methylase. To determine the best mixture of the two enzymes set up analytical reactions of 1 U of *Eco*RI with 0, 20, 40, 80, 160, 320, and 640 U of *Eco*RI methylase.
3. Place individual blocks in *Eco*RI partial digestion buffer (*see* Step 4 in Materials) with the various combinations of *Eco*RI and *Eco*RI methylase and incubate on ice for 1 h.
4. Transfer the reactions to 37°C for 4 h.
5. Add EDTA and proteinase K to 0.02M and 0.5 mg/mL respectively to terminate the reactions, and incubate at 37°C for 30 min.
6. Check partial digests on a 1% agarose gel in a CHEF apparatus with yeast chromosomes as size markers to see which combination of enzymes gives most DNA in the range of 200–2000 kb.
7. Then digest many (6–12) blocks preparatively for the library construction using several of the best enzyme combinations (usually 1 U *Eco*RI and 50–200 U *Eco*RI methylase).

3.3. First Size Fractionation by PFGE

1. Pool blocks containing partially digested DNA in a 50-mL Falcon tube and wash once in 0.01M Tris-HCl, pH 7.5, and 0.05M EDTA.
2. Place blocks adjacent to each other in a trough in a 1% LMP agarose gel in 0.5X TBE, preset for 1 h at 4°C. Place a genomic DNA block in the adjacent gel slot on either side of the trough and place yeast chromosome size markers in the outside gel slots.

3. Overlay the gel slots and trough with 1% LMP agarose. Subject the gel to electrophoresis at 160 V (4.7 V/cm), using a switch time of 30 s (which selects fragments 400 kb and above) for 18 h at 15°C in a CHEF apparatus.
4. Remove the gel from the CHEF apparatus. Cut away only the outside lanes, including one lane each of partially digested genomic DNA and yeast chromosome size markers, and stain with ethidium bromide (1 µg/mL) for 45 min. Keep the central portion of the preparative gel in 0.5X TBE plus 0.02M EDTA at 4°C.
5. Under UV light notch the marker lanes at the edges of the limiting mobility (above 400 kb), and take a photograph. Place adjacent to the central portion of the preparative gel, cut out the limiting mobility using the notches in the outside lanes as a guide, and place in a 50-mL Falcon tube. Stain all of the remaining preparative gel with ethidium bromide and take a photograph.

3.4. Ligation to Vector

1. Equilibrate the gel slice (1–2 mL) containing the limiting mobility of size selected DNA four times (30 min each) in 1X ligase buffer (*see* Section 2., Step 7).
2. Place the gel slice equilibrated in 1X ligase buffer in an Eppendorf tube (<1 mL agarose/tube) and melt at 68°C for 10 min together with digested and CIP-treated pYAC4 vector (*see* Section 3.1.) in a ratio of 1:1 by weight of genomic DNA.
3. Stir the vector and genomic DNA in molten agarose slowly with a pipet tip and incubate at 37°C for 1–2 h.
4. Add T4 DNA ligase to 400 U/µL, ATP pH 7.5 and DTT to 0.001M each in 1X ligase buffer by slow stirring at 37°C. Incubate the reaction at 37°C for an additional 0.5–1 h and then overnight at room temperature. For ligation efficiency controls, *see* Note 2.
5. Terminate the reaction by adding EDTA, pH 8.0, to 0.02M (Optional).

3.5. Second Size Fractionation by PFGE

1. Melt the ligation reaction at 68°C for 10 min and cool to 37°C.
2. Carefully pipet the molten agarose with a tip of bore diameter >4 mm into a trough in a 1% LMP agarose gel in 0.5X TBE, preset for 1 h at 4°C. Place some molten agarose ligation mix in the gel slots adjacent to the trough on each side and place yeast chromosome size markers in the outside gel slots. Overlay the gel slots and trough with 1% LMP agarose.

3. Subject the gel to electrophoresis in a CHEF apparatus using the same conditions as described earlier for the first size fractionation (*see* Section 3.3.).

4. Excise the limiting mobility as described earlier for the first size fractionation (*see* Section 3.3.). If any degradation of DNA is seen at this step, *see* Note 1.

5. Equilibrate the gel slice (about 2–3 mL), containing the limiting mobility from the second size fractionation, 4X (30 min each) in 0.01 M Tris-HCl, pH 7.5; 0.03 M NaCl; 0.001 M EDTA; and 1X polyamines.

6. Score the equilibrated gel slice with a sterile scalpel and place less than 1 mL of agarose into individual Eppendorf tubes. Melt at 68°C for 10 min, cool to 37°C and add agarase (150–200 U/mL of molten agarose). Incubate at 37°C for 2–6 h prior to transformation.

3.6. Transformation

Transformation is carried out according to Burgers and Percival (8) using lyticase (Sigma) to spheroplast cells from the *S. cerevisae* strain AB1380 (1).

1. Streak a fresh YPD plate with AB1380 from a frozen glycerol stock. Grow at 30°C for 2–3 d. Inoculate a single colony into 10 mL of YPD. Let sit overnight at 30°C.

2. The next evening, inoculate 200 mL of YPD in a 1-L flask with 200 µL of the 10 mL overnight AB1380 culture. Shake at 30°C overnight for 16–18 h.

3. When the OD_{600} nm of a 1/10 dilution of the AB1380 culture is between 0.1 and 0.2, split the culture into 50-mL Falcon tubes. Check some of the culture under the microscope for bacterial contamination.

4. Spin the tubes at 400–600 g (3000 rpm on table top centrifuge) for 5–10 min at 20°C. Decant media and resuspend pellets in 20 mL of distilled, sterile water for each tube.

5. Spin 400–600 g for 5–10 min at 20°C. Decant water and resuspend pellets in 20 mL of 1.0 M sorbitol.

6. Spin 400–600 g for 5–10 min at 20°C. Decant sorbitol and resuspend pellets in 20 mL SCE.

7. Add 46 µL of β-mercaptoethanol and take 300 µL from one tube for a prelyticase control. Add 1000 U Lyticase (Sigma), mix gently, and incubate at 30°C.

8. At 5, 10, 15, and 20 min test the extent of spheroplast formation of one tube by two independent methods:

    a. Using a spectrophotometer, measure OD_{600} nm of a 1/10 dilution in distilled water. When the value is 1/10 of the prelyticase value, spheroplast formation is 90% complete.

    b. Mix 10 μL of cells with 10 μL 2% SDS and check under the microscope using phase contrast. When cells are dark ("ghosts") they are spheroplasted.

9. Take the spheroplast formation to 80–90%. This should take 10–20 min. Then spin cells at 200–300g (1100 rpm on table top centrifuge) for 5 min at 20°C.

10. Decant SCE, resuspend pellets gently in 20 mL of 1.0M sorbitol. Spin 200–300g for 5 min at 20°C. Decant sorbitol, resuspend pellets in 20 mL STC.

11. Take a cell count of one tube by making a 1/10 to 1/50 dilution in STC and count on a hemocytometer.

12. Spin cells at 200–300g for 5 min at 20°C and then resuspend in a vol of STC calculated for a final concentration of $4.0–6.0 \times 10^8$ cells/mL when added to genomic DNA.

13. Add approx 0.5–1.0 μg of DNA in digested agarose solution (50–75 μL) to 150 μL of spheroplasts in 15-mL conical polystyrene Falcon tubes. For transformation controls use:

    a. No DNA.

    b. 10 ng supercoiled YCp50 *(12)*.

    c. 100 ng restricted and CIP-treated pYAC4. Let DNA and spheroplasts sit for 10 min at 20°C.

14. Add 1.5 mL PEG, mix gently by inverting tubes. Let sit for 10 min at 20°C. Spin at 200–300g for 8 min at 20°C.

15. Carefully pipet off PEG solution and do not disturb pellets. Gently resuspend pellets in 225 μL of SOS. Place at 30°C for 30 min.

16. Keep molten top regeneration agar at 48°C. If using small plates, add 5 mL of (-uracil) regeneration top agar to each 225 μL of SOS and cells. If you are using large (22 × 22 cm) plates, pool 10 tubes of 225 μL of SOS and cells to a 50-mL Falcon tube and add 50 mL of (-uracil) regeneration top agar. Mix gently by inverting the tube and pour quickly onto the surface of a prewarmed (-uracil) regeneration plate and let sit. Incubate plates upside down at 30°C for 3–4 d.

17. YAC analysis (*see* Chapter 17) and replication of transformants. Good transformation efficiencies are between $2–8 \times 10^5$ clones/μg YCp50 and 100–1000 clones/μg genomic DNA. For low transformation efficiencies, *see* Note 3. Pick YAC clones individually onto selective plates (-uracil, -tryptophan, *see* Section 2., Step 14) to test for both

vector arms. When using minimal adenine, visualize red color in YAC colonies containing inserts. Grow YAC clones in selective media and make agarose blocks containing chromosomes (*see* Chapter 7,8,and 17) and check the size of YAC clones by PFGE. To replicate clones for library screening, pick YAC clones individually into microtiter dishes for screening of pools by polymerase chain reaction (PCR) amplification *(13)* or by colony hybridization after spotting onto filters using manual devices. A multipin transfer device, containing 40,000 closely spaced pins, has been used to efficiently replicate YAC clones from the supportive agar matrix of regeneration plates to the surface of selective plates, for colony hybridization and picking into microtiter dishes *(7)*.

4. Notes

1. Degradation of DNA. If anywhere in the cloning procedure you encounter complete or partial degradation of high-mol-wt DNA, use yeast chromosomes in a series of control reactions to pinpoint the problem. Since yeast chromosomes can be visualized as distinct bands on PFGE, degradation can be detected much easier than in partial digests of genomic DNA. Test all buffers and enzymes (*Eco*RI methylase, T4 DNA ligase, proteinase K, agarase) for nuclease activity in mock cloning experiments using yeast chromosomes. Also, melt agarose blocks containing yeast chromosomes in buffers with and without 1X polyamines to test for partial degradation.

2. Ligation controls for vector and genomic DNA. As in Section 3.1., Steps 9 and 10, test the efficiency of ligation of vector arms to partially digested genomic DNA by incubating a small sample of the ligation reaction with and without 1 U T4 polynucleotide kinase. Melt the samples and load them on a small 1% agarose gel to check for no change of vector arms without kinase and disappearance of vector arms to larger sized fragments when incubated with kinase.

3. Transformation efficiency. If your transformation efficiencies are routinely lower than expected, check the following:
   a. Always streak AB1380 onto a fresh YPD plate before setting up cultures. Cultures grown from old plates (>2 wk) seem to transform less well although they will appear to spheroplast normally.
   b. Try different concentrations of Lyticase and percent spheroplast formation for optimum efficiency.

c. Try various batches of sorbitol and PEG to see if there is any difference in transformation efficiency.

d. Always use distilled, deionized water to guard against heavy metal ion contamination which can degrade DNA or decrease transformation efficiency.

e. Check the temperature of room. Transformation is best at 20–22°C and decreases dramatically at temperatures around 30°C.

References

1. Burke, D. T., Carle, G. F., and Olson, M. V. (1987) Cloning of large DNA segments of exogenous DNA into yeast by means of artificial chromosome vectors. *Science* **236,** 806–812.
2. Coulson, A., Waterston, R., Kiff, J., Sulston, J., and Kohara, Y. (1988) Genome linking with yeast artificial chromosomes. *Nature* **335,** 184–186.
3. Garza, D., Ajioka, J. W., Burke, D. T., and Hartl, D. L. (1989) Mapping the Drosophila genome with yeast artificial chromosomes. *Science* **246,** 641–646.
4. Anand, R., Villasante, A., and Tyler-Smith, C. (1989) Construction of yeast artificial chromosome libraries with large inserts using fractionation by pulsed-field gel electrophoresis. *Nucleic Acids Res.* **17,** 3425–3433.
5. McCormick, M. K., Shero, J. H., Cheung, M. C., Kan, Y. W., Hieter, P. A., and Antonarakis, S. E. (1989) Construction of human chromosome 21-specific yeast artificial chromosomes. *Proc. Natl. Acad. Sci. USA* **86,** 9991–9995.
6. Albertsen, H. M., Abderrahim, H., Cann, H. C., Dausset, J., Le Paslier, D., and Cohen, D. (1990) Construction and characterization of a yeast artificial chromosome library containing seven haploid human genome equivalents. *Proc. Natl. Acad. Sci. USA* **87,** 5109–5113.
7. Larin, Z., Monaco, A. P., and Lehrach, H. (1991) Yeast artificial chromosome libraries containing large inserts from mouse and human DNA. *Proc. Natl. Acad. Sci. USA* **88,** 4123–4127.
8. Burgers, P. M. J. and Percival, K. J. (1987) Transformation of yeast spheroplasts without cell fusion. *Anal. Biochem.* **163,** 391–397.
9. Rothstein, R. (1985) Cloning in yeast, in *DNA Cloning, Vol. II* (Glover, D. M., ed.), IRL Press, Oxford, pp. 45–65.
10. Maniatis, T., Fritsch, E. F., and Sambrook, J. (1982) *Molecular Cloning: A Laboratory Manual,* Cold Spring Harbor Laboratory, Cold Spring Harbor, NY.
11. Herrmann, B. G., Barlow, D. P., and Lehrach H. (1987) An inverted duplication of more than 650 Kbp in mouse chromosome 17 mediates unequal but homologous recombination between chromosomes heterozygous for a large inversion. *Cell* **48,** 813–825.
12. Hieter, P., Mann, C., Snyder, M., and Davis, R. W. (1985) Mitotic stability of yeast chromosomes: A colony color assay that measures nondisjunction and chromosome loss. *Cell* **40,** 381–392.
13. Green, E. D. and Olson, M. V. (1990) Systematic screening of yeast artificial chromosome libraries by use of the polymerase chain reaction. *Proc. Natl. Acad. Sci. USA* **87,** 1213–1217.

CHAPTER 17

Analysis of Yeast Artificial Chromosome Clones

Settara C. Chandrasekharappa, Douglas A. Marchuk, and Francis S. Collins

1. Introduction

The technology of cloning large fragments of DNA in yeast as yeast artificial chromosomes (YACs) *(1)*, combined with that of electrophoretic separation of large fragments by pulsed-field gel electrophoresis (PFGE), has revolutionized research in molecular genetics. In addition to the large insert size, the YAC cloning system offers relatively unbiased genome representation compared to the conventional bacterial cloning systems based on plasmids or bacteriophages *(2,3)*. The enormous impact of YAC cloning and PFGE techniques on physical mapping is being realized with the building of multimegabase contigs (contiguous cloned stretches of DNA). Some examples include mapping and cloning of a large gene like cystic fibrosis (CF) *(4)*, of a chromosomal band like that of 18q21.3 *(5)*, of a substantial portion of an entire human chromosome, Xq24-28 *(3)*, and even an entire organism, *C. elegans* *(2)*. Similarly, the use of YAC clones in the identification of disease-related genes like neurofibromatosis (NF1) is well documented *(6)*. PFGE techniques play a major and critical role in the construction (*see* Chapter 16) and characterization (later in this chapter) of YAC clones.

From: *Methods in Molecular Biology, Vol. 12: Pulsed-Field Gel Electrophoresis*
Edited by: M. Burmeister and L. Ulanovsky
Copyright © 1992 The Humana Press Inc., Totowa, NJ

This chapter deals with the characterization of YAC clones once isolated from a YAC library. Various procedures described in this chapter include (1) growing yeast in culture to prepare DNA embedded in agarose blocks or in liquid form, and for storage of YAC clones; (2) determination of the size of the insert, and detection of multiple YACs as a preliminary characterization of YAC clones; and (3) construction of restriction maps of the YAC clones. Stability of YAC clones and methods available for analysis of chimeric YAC clones are also discussed.

2. Materials

2.1. General Reagents

1. YPD medium: Prepare 1 L of medium by dissolving 10 g yeast extract and 20 g peptone, adjust to pH 5.8 with HCl. Autoclave for 20 min. When cooled to 65–80°C, add 50 mL of sterile 40% glucose. If needed for plates, add 17 g of agar per liter before autoclaving.
2. AHC medium and plates: Prepare 1 L of medium as follows: Dissolve 1.7 g yeast nitrogen base without amino acids and without ammonium sulfate (cat. #0335-15, Difco Laboratories, Detroit, MI), 5 g ammonium sulfate, 10 g casein hydrolysate-acid (cat. #12852, U.S. Biochemicals, Cleveland, OH), 100 mg adenine hemi-sulfate (cat. #A-9126, Sigma Chemical Co., St Louis, MO). Adjust to pH 5.8 with about 50–60 µL of concentrated (12N) HCl. For plates add 17–20 g of bacto agar. Autoclave for 20 min. Cool to 65–80°C, add 50 mL of sterile 40% glucose. AHC is a selective medium for YAC-containing clones and cells grown in AHC containing 20 mg/L of adenine hemi-sulfate will produce a red precipitate as a result of the color screen for insert-containing YACs (7). To avoid having this red precipitate being carried through during DNA preparation, 100 mg/L of adenine hemi-sulfate is included in the medium.
3. 40% Glucose solution: Dissolve 40 g of glucose (dextrose) in water, sterilize by filtration through sterile filter.
4. SCE: 1 M sorbitol, 100 mM trisodium citrate, pH 5.8, and 60 mM EDTA, pH 8. Prepare 1 L of SCE by mixing 100 mL of 1 M trisodium citrate, pH 5.8, 120 mL of 0.5 M EDTA and dissolving 182 g of D-sorbitol in water to a final volume of 1 L. Sterilize the solution by filtration through a sterile filter.
5. N-Lauroylsarcosine (sodium salt) (cat. #L-5125, Sigma Chemical Co., St. Louis, MO).

6. Proteinase K solution (10 mg/mL): Prepare 10 mg/mL solution of proteinase K (Boehringer Mannheim Biochemicals, Indianapolis, IN) in water. Store frozen at –20°C.

7. Yeast lytic enzyme (Zymolyase) (cat. #152270, ICN Biomedicals, Costa Mesa, CA, supplied as 70 U/mg).

8. 1M Dithiothreitol (DTT).

9. Low-melting point (LMP) agarose: InCert agarose or LGT agarose (SeaPlaque) (FMC Bioproducts, Rockland, ME).

10. 20% glycerol in YPD: Add 20 mL of glycerol to 50 mL of 2X YPD medium and 30 mL of water. Mix and sterilize the solution by filtration.

11. 0.5X TBE buffer: 45 mM Tris, 45 mM boric acid, and 2 mM EDTA. Dilute with water from 5X TBE stock. Prepare 1 L of 5X stock as follows: 54 g of Tris base, 27.5 g of boric acid, and 40 mL of 0.5M EDTA.

12. 0.5M EDTA, pH 8.0.

13. TE: 10 mM Tris-HCl, pH 8, 1mM EDTA pH 8.

14. Restriction endonucleases: Enzymes and the buffers are available from many vendors. The enzymes needed are indicated in the Materials and in the Methods section.

15. 1M trisodium citrate, pH 5.8: Prepare by dissolving trisodium citrate salt in water and adjust pH to 5.8 with concentrated (12N) HCl.

16. Freezing vials: NuNC-style flat-bottomed screw-cap vial (Irvine Scientific, Santa Ana, CA).

17. Glycerol.

18. Ampicillin: Prepare 25 mg/mL solution in water. Store frozen at –20°C.

19. 1M MgCl$_2$ solution.

20. Yeast and lambda concatamer DNA size markers (*see* Chapters 7 and 8 for preparation of the markers). The markers are also commercially available from suppliers such as New England Biolabs, Inc., Beverly, MA, and Bio-Rad, Richmond, CA.

21. Left-arm vector DNA probe: A 1130 bp *Bam*HI /*Pst*I fragment of pBR322 serves as a left-arm probe. Digest pBR322 with *Pst*I and *Bam*HI. Separate the products by electrophoresis, cut out the 1130 bp band. Isolate the fragment from the gel piece using Geneclean® kit (Bio 101, La Jolla, CA).

22. Right-arm vector DNA probe: A 1050 bp *Bam*HI /*Ava*I fragment of pBR322 serves as a right-arm probe. Digest pBR322 with *Ava*I and *Bam*HI. Separate the fragments by electrophoresis. Cut out the gel piece containing the 1050 bp fragment, and elute the DNA out of the gel using the Geneclean® kit (Bio 101).

23. Human DNA probe: Any one of the sources like circulating leucocytes from blood, placenta, or a cell line can be used to prepare total human DNA. A repetitive sequence such as *Alu* sequences can substitute for human DNA as probe in all the situations mentioned in this chapter.

24. Ethidium bromide solution: prepare a stock solution containing 10 mg/mL in water. CAUTION: carcinogenic. Always use gloves while handling this reagent.

25. DNase-free RNase A (10 mg/mL): Prepare a 10 mg/mL of RNase A (cat. #109142, Boehringer) and boil *(8)* to inactivate DNase.

26. D-Sorbitol (cat. #S-6021 Sigma Chemical Co., St. Louis, MO).

27. 100 m*M* phenylmethanesulfonyl flouride (PMSF): Prepare 100 m*M* solution of PMSF (cat. # 236608, Boehringer) in isopropanol. Store at 4°C in dark.

28. Molds for agarose blocks: Molds are available from most suppliers of the PFGE equipment. The size of the wells is not important. If molds are not available, the cell-agarose mix can be poured into a small tissue culture dish. Treatment with different reagents can be carried out in the same dish.

2.2. Materials for the Preparation of Agarose Blocks

Lysis solution: 0.45*M* EDTA, pH 8, 10m*M* Tris-HCl, pH 8, 1% *N*-lauroylsarcosine, 1mg/mL proteinase K. Prepare 100 mL, by mixing 90 mL of 0.5*M* EDTA, 1 mL of 1*M* Tris-HCl, 1 g of *N*-lauroylsarcosine, and 9 mL of water. Just before use, add proteinase K from stock solution to the final concentration of 1 mg/mL.

2.3. Materials for the Preparation of Liquid DNA (see Section 3.1.2.)

1. 40 m*M* EDTA/90 m*M* β-mercaptoethanol: Prepare the mixture in water using 0.5*M* EDTA stock solution and β-mercaptoethanol (14*M*).

2. Sorbitol reagent: 1*M* D-sorbitol, 1 m*M* EDTA, pH 8, 3 m*M* DTT.

3. Lysis solution A: 100 m*M* NaCl, 10 m*M* Tris-HCl, pH 7.4, 10 m*M* MgCl$_2$, 5 m*M* β-mercaptoethanol.

4. Lysis solution B: 50 m*M* Tris-HCl, pH 7.4, 1*M* NaCl, 50 m*M* EDTA, and 2% SDS.

   All these reagents can be prepared and stored in a refrigerator (4°C). Lysis solution B needs to be warmed before use since SDS will precipitate in the cold. Reagents containing mercaptoethanol or DTT cannot be stored for more than 2–3 mo.

2.4. Materials for the Preparation of High-Mol-Wt DNA (see Section 3.1.3.)

1. Lysis buffer: 0.5 M Tris-HCl, pH 9, 3% N-lauroylsarcosine, 0.2 M EDTA.
2. Sucrose solutions: prepare 15%, 20%, and 50% sucrose in 0.8 M NaCl, 20 mM Tris-HCl, pH 8, and 10 mM EDTA. Dissolve 15, 20, or 50 g of sucrose to a final volume of 100 mL in this buffer.

3. Methods

3.1. Preparation of DNA from YAC Clones

Yeast containing YAC clones can be grown either in the selective medium (AHC) or in an enriched medium (YPD). Although they grow slowly in AHC compared to YPD medium, AHC is the medium used for growing cells containing YAC clones, unless cells are being grown in order to freeze them away for long-term storage. In this section, preparation of DNA in agarose blocks and in liquid form is presented. DNA embedded in agarose blocks serves as a source of DNA for almost any kind of experiment. PFGE of agarose blocks is used in the characterization of the YAC clones, in terms of insert size, stability, presence of multiple YACs, and construction of restriction maps. Other enzymatic reactions such as PCR can be carried out with agarose blocks, after melting them by heating to 65°C and diluting with water to an appropriate concentration. If needed, DNA of high quality and in liquid form can be squeezed out of the agarose blocks very easily (*see* Note 1). A procedure for the extraction of yeast DNA in liquid is given below (*see* Section 3.1.2.). This procedure produces DNA of adequate quality for most routine uses, such as Southern blots and PCR reactions. High-mol-wt chromosomal quality DNA is required for introducing YACs into cells or to prepare a cosmid library from the YAC DNA. A protocol for the isolation of high-mol-wt DNA, which involves a sucrose stepgradient procedure, is given here (*see* Section 3.1.3.) *(9,10)*.

3.1.1. Preparation of DNA Embedded in Agarose Blocks

1. Inoculate a YAC clone into 100 mL of AHC medium containing ampicillin (50 µg/mL) in a 250-mL flask. Grow the cells for 2–3 d at 30°C with shaking at 300 rpm (*see* Note 2).

2. Pour the culture into two 50-mL tubes and harvest the cells by centrifugation for 5 min on a table-top centrifuge. Pour off the supernatant and resuspend the cells in an equal volume of 50 mM EDTA.

3. Centrifuge for 5 min and pour off the supernatant. Resuspend the cells and pool them from both tubes into 40 mL of 50 mM EDTA. Make sure the cells are evenly suspended. Dilute 10 µL of cells with 990 µL of water and read the absorbance at 600 nm in a spectrophotometer using water as blank. If the optical density (OD) is too low (<0.05) or too high (>0.5), dilute appropriately to get an OD in this range. Calculate the total OD of cells in the suspension based on the dilution and the total volume of cells present. Multiply the OD units by 2×10^7. Each OD is equivalent to ~2×10^7 cells.

4. Centrifuge the cells, pour off the supernatant, and resuspend the cells in SCE solution so that the final concentration of cells in the SCE solution is 2×10^9 cells/mL (*see* Note 3). Make sure to add a much smaller initial volume of SCE and resuspend cells first, because the large pellet of cells will take up a substantial portion of the volume. Transfer the suspension to a smaller tube (Falcon 2059-round bottom) so that the final volume can be more accurately measured. Then add the remainder of the SCE to the correct volume.

5. Add DTT from 1M stock solution to a final concentration of 70 mM (this will be 0.07 vol of the cell suspension).

6. Add yeast lytic enzyme (Zymolyase) to 280 U of enzyme per mL of cell suspension. Wrap the tube with parafilm and invert the tube for 1–2 min until it looks as if most or all the solid enzyme has gone into solution. Incubate the suspension at 37°C for 5 min with occasional inverting. It is critical not to let the spheroplasting go much longer than this in order to avoid complete cell lysis and shearing of the DNA in the subsequent steps.

7. Add an equal volume of 1% LMP agarose prepared in SCE and maintained at 50°C. Invert the tube several times until the cell suspension and agarose are completely mixed.

8. Replace the tube to 50°C and begin pipeting the suspension into alcohol-rinsed molds using a P-1000 pipetman. Keep the cell-agarose suspension at 50°C and pull out only as much as you can pipet into plugs without the agarose solidifying inside the pipet tip.

9. Cool the plugs for about 20 min in the cold (4°C). Pop the plugs into a 25-mL SCE containing 280 U/mL of yeast lytic enzyme (100 mg) using an alcohol-soaked cotton swab or Texswab (Tx 709, Texwipe, Upper Saddle River, NJ).

10. Incubate the blocks at 37°C for 2–4 h in a 37°C incubator shaker with the shaking at very low speed (about 60 rpm). Watch occasionally. If you find that the shaking is breaking the blocks, then incubate without shaking, but mix them carefully every 15–20 min.
11. Carefully remove the solution with a sterile 10-mL pipet : Pour off as much of the solution as you can and then place the pipet tip at the bottom of the tube and slowly remove the solution.
12. Add 40 mL of lysis solution to the tube. (This lysis solution contains 1 mg/mL proteinase K.)
13. Incubate overnight at 50°C.
14. On the following day, cool the blocks to room temperature. Carefully pour off the supernatant and remove the remainder as in Step 11. Replace with fresh lysis solution (40 mL) with proteinase K. Incubate at 50°C for 4 h.
15. Cool the blocks to room temperature, and carefully remove the solution as in Step 11. Replace with 30 mL of 50 mM EDTA and place blocks on a rocking platform in cold. Repeat washing the blocks (4–5 times over a period of 12–24 h) with fresh EDTA solution. Store most of the blocks in 50 mM EDTA except for those that you may need for PFGE right away (*see* Note 4).
16. These blocks are NOT ready for restriction enzyme digestion or for any enzymatic reaction. Wash the blocks in TE at least 4–5 times in the cold over a period of 1–2 d before enzyme digestion. Include 1 mM PMSF (from the 100 mM stock solution) in the first TE wash.

3.1.2. Preparation of Liquid DNA

1. Inoculate 250 mL of AHC medium with a colony and grow the cells for 2–3 d.
2. Pour cells into a centrifuge bottle and centrifuge at 1000g for 10 min.
3. Suspend the pellet in 12.5 mL of 40 mM EDTA/90 mM β-mercaptoethanol solution, transfer it to a 50-mL polypropylene tube, and centrifuge at 1000g for 10 min.
4. Resuspend the pellet in 12.5 mL of sorbitol reagent. Add 25 mg of yeast lytic enzyme to a final concentration of 2 mg/mL. Place the tubes in a test tube rack and place the rack on its side in a shaking incubator for no less than 1 h at 30°C.
5. Centrifuge the tubes at 1000g for 10 min.
6. Resuspend the pellet in 5 mL of lysis solution A. Add 5 mL of solution B. Mix the contents by inverting the tube several times until the pellet is suspended. Incubate at 45°C for 10 min. Mix by inverting once or twice during incubation.

7. Extract with 10 mL of phenol/chloroform (1:1). Repeat extraction of the upper aqueous layer with phenol/chloroform until the interphase is clear. Finally, extract the aqueous layer with a mixture of chloroform: isoamyl alcohol (24:1). At each step of extraction, mix the contents of the tube and centrifuge for 15 min at $3000g$.

8. Precipitate the final aqueous layer by adding an equal volume (10 mL) of ice-cold isopropanol. Invert the tube gently and incubate at $-20°C$ for at least 20 min. Centrifuge at $3000g$ for 15 min. Pour off the supernatant and rinse the pellet with 70% ethanol. Drain the liquid by inverting the tubes on a paper towel for 2 min, taking care that the pellet does not drop off the tube, and air-dry the pellet for about 20 min.

9. Resuspend the pellet in 2.5 mL of TE, and add 12.5 µL of 10 mg/mL of DNase-free RNase A. Incubate at $37°C$ for at least 4 h or overnight. The pellet may not go into solution readily. Tap the tube occasionally during incubation, and within 10 min the pellet will dissolve.

10. Reprecipitate the DNA by adding 0.1 vol of $3M$ sodium acetate (0.25 mL) and 2.5 vol of ice-cold 95% ethanol (6.25 mL). Keep in a $-20°C$ freezer for 20 min. Centrifuge at $3000g$ for 15 min. Rinse the pellet with 70% ethanol.

11. Dissolve the pellet in TE (0.5 mL). If the DNA does not dissolve readily, incubate at $37°C$ until the DNA dissolves.

12. Quantitate the DNA by reading the absorbance at 260 and 280 nm. Usually 20 µL of DNA in 1 mL of water will give an absorbance of about 0.1–0.3 OD at 260 nm. Calculate the ratio of absorbance at 260/280. The ratio should be between 1.6–2.0. If it is over 2.0, repeat extraction with phenol/chloroform, then with chloroform: isoamylalcohol, and precipitate the DNA as in steps 7 and 8.

3.1.3. Preparation of High-Mol-Wt Liquid DNA

1. Inoculate a single colony into 5-mL AHC media. Grow at $30°C$ for 2 d. Inoculate 400-mL AHC with 0.5 mL of the aforementioned liquid culture. Grow at $30°C$ with shaking for 2–3 d.

2. Centrifuge the cells at $1000g$. Resuspend the pellet in 40 mL of water, transfer to a 50-mL tube, and centrifuge again. (Weigh the wet pellet; the weight should be around 3 g.)

3. Resuspend the pellet in 3.5-mL SCE, and add 160 µL of $1M$ DTT and 20 mg yeast lytic enzyme.

4. Incubate at $37°C$ for 1 h or until spheroplasting is complete. Mix gently every 15 min. An increase in viscosity occurs as the cells lyse;

this is a good monitor for the extent of spheroplasting. It is important to get a fairly viscous solution at this step.

5. Pour the spheroplasts directly into 7 mL of lysis buffer containing 100 µg/mL proteinase K, placed in a 25-mL flask. Make sure that the mixture is homogenous by mixing gently before proceeding. No clumps should be visible and some clearing may be apparent.

6. Heat the lysate at 65°C for 15 min with occasional gentle mixing, then cool quickly to room temperature in water. The solution will become clearer.

7. Prepare a sucrose gradient during the 65°C incubation period. First, place 11 mL of the 20% sucrose solution in the centrifuge tube (36-mL tubes for a Beckman SW 28 rotor), then add 11 mL of the 15% sucrose solution on top. Then, carefully underlay 3 mL of the 50% sucrose solution.

8. Pour the lysate on top of the crude gradient; if pipeting is necessary, use a wide bore pipet. Spin the gradient at $131,000g$ for 3 h at 20°C in a SW 28 rotor (*see* Note 5).

9. Aspirate the top 25–30 mL of the gradient with vacuum and discard (*see* Note 6). When nearing the bottom, watch out for viscous material (DNA) trying to enter the suction setup. As soon as you feel that, stop siphoning.

10. Collect the remaining 3–6 mL with a large bore 10-mL pipet. Start very close to the white polysacharide pellet but do not touch it, and slowly remove the solution.

11. Transfer the DNA solution to a dialysis bag and dialyze overnight at 4°C against TE with one change. Leave just enough air space in the bag to float. The volume will increase significantly if you leave more space in the bag.

12. Surround the bag with solid sucrose, replacing it with fresh sucrose when saturated. During this concentration step, bring the clamps closer together until the volume is 0.5–1 mL. Redialyze with the clamps as close to each other as possible for 10 h or longer at 4°C with two changes of TE.

13. DNA yields can be quantitated by spectroscopic analysis, and usually are 100–200 µg per gradient.

3.2. Storage of YAC Clones

1. Grow the desired YAC clone on an AHC plate at 30°C for 2–3 d.
2. Remove a swab of cells with the flat end of a sterile toothpick and add to 1 mL 20% glycerol in YPD medium in a NuNC-style flat-bottomed screw-cap vial.

3. Vortex briefly to suspend the cells. Freeze and store at –80°C. The frozen cells must be stored at –55°C or lower.

Alternatively, inoculate 5 mL of YPD medium with the YAC clone and incubate in a shaker at 30°C overnight. Mix 3 mL of this culture with 1 mL of 80% glycerol in water. Aliquot 1 mL of culture into three freezing vials and store at –80°C. The stored YAC needs to be repurified on a selective medium (AHC plate) after long storage. Even if a white colony appears during the repurification step, it is usually a petite variant and retains the original YAC *(11)*.

3.3. Preliminary Characterization

The artificial chromosome containing the cloned DNA will be present as an additional chromosome among the host yeast chromosomes. Comparison of chromosomes from the host yeast cells and the cells carrying the YAC, after separation by PFGE and staining with ethidium bromide, will indicate the location of the artificial chromosome(s). The size of the YAC can be obtained by comparing its mobility with those of the markers of known sizes. Phage lambda multimers (multiples of approx 50 kb) and yeast *S. cerevisiae* chromosomes (245–1500 kb) serve as useful markers for the analysis of YAC clones by PFGE. If the size of the YAC(s) is indistinguishable from one of the host chromosomes, it may not be detected from the ethidium bromide stained gel. Therefore, it is necessary to transfer the separated chromosomal DNA to a membrane by Southern transfer and to hybridize it to a pBR322 probe. Since the vector sequences (both left and right arms) contain sequences from pBR322, all the YACs will be detected by hybridization to this probe. Although each clone is likely to contain a single YAC, about 10% of the clones in most libraries contain multiple YACs *(3)*. In this section, characterization of YAC clones for the size of the insert and for the presence of multiple YACs is presented.

Various versions of PFGE are available for separation of large molecules of the size of yeast chromosomes (*see* Chapters 1–7). The procedure described here is for a CHEF apparatus (BioRad).

1. Prepare 100 mL of 1% agarose in 0.5X TBE. Pour most of the gel (about 95–98 mL) into the gel tray (14 cm × 12.7 cm). Save the remaining gel solution in a 50°C water bath.
2. Cut an appropriate sized piece of agarose block, which is approx 100 µL for the 10-well comb provided with CHEF apparatus. Load the

agarose blocks in the wells, and use two wells for size markers, one for phage lambda multimers and another for yeast chromosome markers. Seal the wells with the 1% agarose (stored at 50°C). Do not submerge the gel in the electrophoresis buffer before loading the gel. Keep at 4°C for about 2–3 min for the gel to solidify.

3. Perform electrophoresis for 24 h using 200 V at 14°C with pulse time ramping from 50–90 s. These conditions will give good separation from 50–900 kb.

4. Stain the gel with ethidium bromide (5 mL stock reagent per 100 mL of water) and destain in water. Take a photograph of the gel with a ruler on the gel (*see* Note 7).

5. Set up the gel for Southern transfer using a method of your choice.

6. Prepare radioactively labeled pBR322 and hybridize it to the membrane using a method of your choice.

7. An autoradiograph shows the location of the YAC. More than one signal is seen if multiple YACs are present in the clone (*see* Notes 8 and 9).

8. Determine the size of the YACs with the help of a graph constructed using the mobilities of the known size markers.

3.4. Stability of YACs

Yeast strains used in the construction of the library maintain most artificial chromosomes, and replicate and transfer them faithfully to the daughter cells through many generations. This has been an important factor in the development of the YAC cloning system. The utility of the YAC clones in biological research largely depends on the stability of the YAC clones. Brownstein et al. *(11)* have demonstrated that through passage of 60 generations, two independent Factor IX clones, one of 60 kb and the other 650 kb, remained intact. Growing the cells either in selective medium (AHC) or enriched medium (YPD) did not alter the stability of these YACs. This appears to be true for most of the YAC clones analyzed so far *(12,13)*. On the other hand, Gaensler et al. *(14)* in their analysis of YAC clones for the beta-globin region, found that leaving the clones on a plate for a month introduced artifacts in some YAC clones, resulting in the appearance of new smaller YACs. Because of the significant homology between various regions of the beta-globin locus, the smaller derivatives were thought to have originated as a result of recombination and/or deletion in the original YAC. Similarly, YAC clones with inserts of tandemly repeated sequences have been found to be unstable *(15)*, and so have the inserts that contain alpha satellite regions near the centromere (H. Willard, personal communication). Therefore, it appears that the stability of a YAC clone depends primarily on the sequence that it contains. Stability can be tested by preparing agarose blocks from cells grown for a different number of generations, or from plates stored for various lengths of time (1–2 mo), and testing them by PFGE. If YAC clones are unstable, an additional smaller YAC may appear along with the original YAC (*see* Note 9). On the other hand, if two or more independent, stable YACs are present in a single clone, then the YAC of interest can be identified by hybridization to a specific probe that was used to screen the YAC library.

3.5. Restriction Mapping of YAC Clones

Unlike in the case of plasmid or phage clones where cloned DNA is smaller in size and can easily be isolated away from the host bacterial DNA, YAC clones contain larger inserts that make the isolation of YAC from host yeast chromosomal DNA difficult. Therefore, restriction mapping of YACs is not as straightforward as it is for plasmid or phage clones. In addition, more than 20% of the YAC clones isolated so far from general human libraries have been found to contain chimeric inserts, i.e., the insert is derived from two unrelated pieces of DNA joined together: A description of the chimeric clones and their identification is given in Section 3.6. However, restriction maps are very useful for subsequent analysis and use of the clone:

1. Alignment of restriction maps of different YACs obtained for the same probe will aid in placing the clone in an overlapping map, and subsequently in building large contigs or in linking contigs (*see* Introduction).
2. Identification of the regions containing clusters of rare-cutting enzymes indicates the location of CpG islands (*see* Section 3.5.3.) often situated at the 5'-ends of genes (*see also* Chapter 18).
3. Comparison of the restriction map of a YAC clone with that of the genomic DNA, if available, will show the authenticity of the YAC clone(s) isolated for that given region. Nonconcordant regions between the two maps may suggest that the YAC is chimeric. More importantly, however, differences do not always indicate anomalous clones. There are differences in methylation between the native genomic and the YAC DNA (YAC DNA is unmethylated) that will affect digestion of DNA by many of the rare-cutting restriction enzymes. Another cause for differences in the maps can be polymor-

phic variations between the source of DNA used to prepare the YAC library and the DNA used to construct the genomic restriction map (*see* Note 10).

In order to construct a long-range map of a YAC clone, rare-cutting enzymes are needed. There are several such enzymes available. The frequency of restriction sites depends on the GC/AT ratios and methylation status of the DNA source. A list of rare-cutting enzymes can be obtained from catalogs of the many vendors (*see also* Chapters 12, 14, and 18). For mammalian DNA, among the rare-cutting enzymes, most are sensitive to CpG methylation including *Not*I, *BssH*II, *Nru*I, *Mlu*I, *Nae*I, *Sac*II, *Eag*I, *Sal*I, *Cla*I, *Xho*I, *Pvu*I; some are not affected by methylation, such as *Sfi*I and *Pac*I. In order to construct a restriction map of a YAC clone with a mammalian DNA insert, choose at least six from the aforementioned list of methylation-sensitive enzymes and include either one or both of the methylation-insensitive enzymes. The former will be useful in locating the position of CpG islands, and the latter will help in aligning the genomic DNA and the YAC restriction maps. Therefore, an ideal list of enzymes would consist of *Not*I, *Nru*I, *BssH*II, *Nae*I, *Mlu*I, *Sac*II, and should definitely include *Sfi*I and/or *Pac*I.

In this section, we describe a strategy and give details about constructing a restriction map of a YAC clone with rare-cutting enzymes.

3.5.1. Strategy

From the vector used for construction, each YAC contains distinct sequences at the right and left ends. A Southern blot containing the fragments generated by partial digestion of DNA and separated by PFGE is made. This blot is sequentially hybridized to the left (Trp)-arm probe and the right (Ura)-arm probe to enable the localization of the restriction sites from each end. A complete map for an enzyme can then be assembled from this information. This is similar in concept to the Smith and Birnstiel *(16)* strategy used for restriction mapping of smaller clones like plasmid or phage clones. To account for all the sites for a given enzyme in a YAC, a Southern blot containing the DNA fragments separated after complete digestion with the same enzyme is also hybridized to total genomic DNA to identify all the fragments that are generated. In the case of human DNA inserts, a repetitive sequence probe such as an *Alu* sequence can be used as a probe instead of total human DNA.

3.5.2. Partial and Complete Enzyme Digestion

Agarose blocks should be dialyzed against TE thoroughly before digestion.

1. Transfer an agarose block of about 25 μL to a 1.5-mL Eppendorf tube by cutting with a razor blade on a glass slide, cleaned and wiped with ethanol. Wear gloves whenever handling blocks to avoid degradation of DNA.
2. Add 100 μL of 10X restriction buffer, 4 μL of 1 *M* spermidine (final concentration, 4 m*M*), 100 μL of 1 mg/mL BSA, and water to make up the total volume of 1 mL. Incubate the tubes on ice for about 1 h. Remove 700 μL of the buffer, For partial digestions, add 15 U of enzyme, and digest for 1 h. For complete digestions, add 50 U of enzyme, and incubate for 3 h at 37°C (*see* Note 11). At the end of the incubation, keep the tubes on ice until ready to load them on a gel.
3. Run the digests on FIGE (*see* Chapter 1). For YAC clones of about 200 kb, forward pulse ranging from 6–30 s and reverse pulse from 2–10 s are used. For clones of about 400 kb and larger, use pulses ranging from 6–60 s in the forward direction and 2–30 s in the reverse direction. After the run, take a photograph of the ethidium bromide stained gel and set up for Southern transfer of the gel by the method of your choice.

3.5.3. Construction of the Map

1. Hybridize the membrane containing completely digested DNA to the total human DNA probe.
2. Hybridize the membrane containing the partially digested DNA , sequentially, to the left-arm vector probe and to the right-arm vector probe.
3. Hybridize both the blots to the single-copy probe, if available, from the gene or gene region used to isolate YAC clone.

Determine the size of the fragments generated for each enzyme by partial and complete digestion. Start from one end and build a linear map for one enzyme at a time. Similarly, build from the other end, one enzyme at a time. Look for concordance in the map for the same enzyme from the other end. This gives a map for each enzyme. Now place them all on the same map. While constructing a map for a single enzyme the number of sites should match the number of fragments generated by complete digestion, although some sites might be missed because of comigrating bands. In addition, sites that are closer yield smaller fragments and therefore might be missed. For most practical purposes, however, such missed locations might not be of great consequence. While constructing such a map you may notice that as you move away from the end, the accuracy of the size determination may not be as good. Therefore, the map locations will be more accurate toward the ends compared to the middle of the clone. Because size determinations are made on the fragments generated by several enzymes, digested at the same time and present on the same blot, the sizes of the fragments will be more reliable, and easier to compare from one enzyme digest to the other. Furthermore, the order of sites should be quite reliable.

In the case of partial digests, it is important that a signal for the undigested clone is detected. If the undigested band is not seen, the digestion needs to be repeated at a lower enzyme concentration or shorter time in order to ensure that no sites are missed. An example of a membrane containing partially digested DNA from a single YAC clone, hybridized to a left-arm vector probe, is shown in Fig. 1. A restriction map for one enzyme (*Eag*I) is shown, constructed from the information obtained from this hybridization. Clustering of the restriction sites for several rare-cutting enzymes identifies a CpG island, which can be easily identified from the autoradiogram as a signal at the same mobility for many enzymes (Fig. 1).

3.6. Chimeric YAC Clones

Ideally, YAC clones should contain a DNA insert of contiguous DNA. However, clones with inserts made up of DNA from noncontiguous regions are often encountered. In a study of YAC clones for the cystic fibrosis gene, spanning a region of more than 1500 kb, 6 of 30 clones (20%) contained inserts that were noncontiguous *(4)*. This figure appears to hold for the analysis for YAC clones for the entire Xq24-28 *(3)*, although as high as 60–65 % of the YACs have been found to be chimeric in the analysis of the clones for MHC region *(17)*. In general, YAC libraries made from total human DNA have a higher proportion of chimeric clones compared to those prepared from human-rodent hybrid cell lines *(3)*. Although coligation events are thought to be mainly responsible, recombination events at homologous regions between two incipient YACs have been shown to contribute to the origin of chimeric clones *(18)*. The presence of chimeric inserts limits the use of the clone in applications like walking

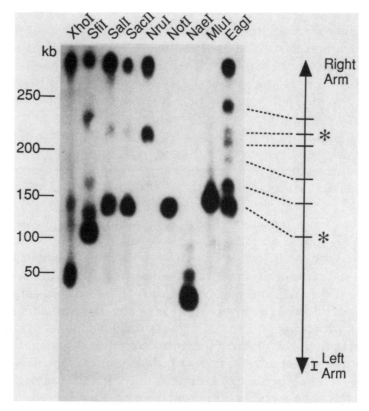

Fig. 1. Partial restriction mapping from the left arm of a YAC clone. Agarose blocks were digested partially with the enzymes listed, blotted onto a GeneScreen membrane, and hybridized with the left-arm vector probe (indicated by a bar). The map for the enzyme, *Eag*I is shown. The locations of two CpG islands are shown by asterisk.

across the ends of the clone; yet, such inserts may still be useful in building sequence tagged sites (STS) contig maps. Unfortunately, there are no simple methods for the detection of chimeric YAC clones. Some approaches for the detection of chimeric inserts are discussed in this section.

One of the methods is to use the fluorescent *in situ* hybridization (FISH) technique, which involves hybridization of biotinylated total yeast DNA as a probe to a spread of metaphase chromosomes, and identification of the chromosome(s) and subchromosomal bands to which the probe has hybridized *(19,20)*. Chimeric clones will be identified by their binding to regions of chromosomes in addition to the expected region. The origin of DNA in the chimeric inserts is also obtained by this method. However, since the resolution of metaphase chromosomes is of one band width, any chimeric clones formed from noncontiguous regions from within this band region will not be identified. The fluorescent signals on metaphase chromosomes for DNA probes smaller than 20 kb are weaker and therefore chimeric fragments smaller than 20 kb may go undetected. Also, inserts that contain related genes or pseudogenes from another chromosomal region will hybridize to this region and might be mistaken for chimeric clones.

Another strategy to distinguish chimeric inserts from normal inserts is to obtain end fragments of the YAC inserts, and prove that both ends hybridize to the expected chromosome. This can be determined by hybridizing radiolabeled end fragments to a panel of somatic cell hybrids. If both ends of the YAC insert hybridize to the same chromosome, then it is very likely that the insert is contiguous. Somatic cell hybrid panels can only point out the chromosome the probes are hybridizing to, therefore chimeric clones consisting of different parts of the same chromosome will be mistaken as nonchimeric unless a panel of hybrids representing different portions of that chromosome is also available.

There are several approaches to isolate end fragments of a YAC insert. Many of these methods are based on PCR. There are two strategies that are not based on PCR. One of them allows the isolation of only the left-end fragment that takes advantage of the ampicillin resistance marker present in the left-arm vector. This method involves digestion of DNA from the YAC clone with either *Nde*I or *Xho*I, circularization of the digestion products, transformation of competent bacterial cells with the circularized DNA, and identification of the ampicillin-resistant colonies *(1)*. Another method involves homologous recombination in yeast that allows the rescue of both the ends *(21,22)*. The PCR-based methods include *Alu*-vector PCR *(23)*, primer-ligation-mediated PCR *(24)*, capture-PCR *(25)*, and inverse PCR *(26)*. Although each of these methods in itself can generate the end fragments of any of the YAC inserts, a combination of these methods may be needed. Although the homologous recombination approach is likely to yield larger end fragments, PCR-based methods are simpler and faster. A detailed protocol for the isolation of end fragments by inverse PCR is given here.

1. Digest about 2 μg of DNA, either in liquid form or in agarose blocks, overnight with 20 U of three different enzymes, *Hae*III, *Taq*I, and *Sau*3AI, in a total volume of 100 μL, (*see* Note 12). At the end of the reaction, inactivate the enzymes by heating at 70°C for 20 min.

2. Set up a ligation reaction overnight in a total volume of 50 μl at 14°C with 20 U of T4 DNA ligase and 5 μL of the aforementioned digest.

3. Set up two sets of PCR reactions with 1 μL of each of the digested/ligated DNA in a total volume of 20 μL, with the required amounts of buffer (50 mM KCl, 10 mM Tris-HCl, pH 8.3, 1.5 mM MgCl$_2$, 0.01% gelatin), dNTPs (200 mM each), and Taq polymerase (0.1 μL; 0.5 U). Carry out one set of reactions with 1 μM each of right (Ura)-end vector primers (5'-TTT CAA GCT CTA CGC CGG AC-3' and 5'-CCG ATC TCA AGA TTA CGG AA-3') and another set with 1 μM each of left (Trp)-end vector primers (5'- CCT TCC AAG ATG GT T CAG AG -3' and 5'-AGC CAA GTT GGT TTA AGG CG-3'). For both sets of primers, do PCR for 30 cycles using the same amplification conditions that is (94°C/1 min; 56°C/2 min; 72°C/2 min).

4. Separate the PCR products by agarose gel electrophoresis, and stain the gel with ethidium bromide. Once a distinct band is identified, set up a large-scale PCR reaction (100–300 μL) with the appropriate enzyme digested/ligated DNA to prepare large amounts of this product. Isolate the fragment from the agarose gel using the Geneclean kit (Bio 101).

In addition to the aforementioned strategies, fingerprinting of overlapping YAC clones could provide information as to the common region shared by the YAC clones, allowing identification of the region that is not chimeric. For mammalian DNA inserts, fingerprinting can be carried out by comparing the *Alu*-PCR *(23)* products. Also, hybridization of total genomic DNA (or *Alu* repeat sequence) probe to a membrane that contains digestions of YAC clones with enzyme such as *Hind*III, E *co*RI, or *Pst*I, will generate patterns that can be compared *(27)*. However, fingerprinting is only useful for comparing the extent of overlap among the clones. The nonoverlapping regions could be extensions beyond the contiguous region or could come from an entirely different region of the genome (chimeric).

4. Notes

1. DNA in agarose blocks can be extracted by centrifugation through a glass-wool plug. Punch a pinhole at the bottom of a 0.5-mL Eppendorf tube and place glass-wool at the bottom. Add agarose blocks to the

tube on top of the glass-wool. Place this tube in a 1.5-mL Eppendorf tube suspended at its rim. Centrifuge at full speed in a microfuge for about 10 min. The DNA will collect at the bottom of the 1.5-mL Eppendorf tube in liquid form, and the agarose gel will stay behind in the 0.5-mL tube. The size range of the eluted DNA is about 50 kb.

2. Yeast grow slowly in the AHC medium, and therefore the cells need to be grown for 2–3 d. The ampicillin will reduce the chance that a faster growing bacterial contaminant will overtake the culture. The yeast are unaffected by the ampicillin. Bacterial contamination of the culture is likely to produce thick whitish material after an overnight period. Therefore, it is necessary to examine the culture after 1 d of incubation. If contaminated, discard the culture after treating with a bleaching agent like clorox. If the bacterial contamination cannot be avoided, then streak your YAC clone stab on an AHC plate containing 20 mg/L of adenine hemi-sulfate. Isolate a single red colony away from the bacterial colony and use that colony to prepare a new culture.

3. The number of cells in a 100-mL culture, grown for 3 d, is usually about $4–8 \times 10^9$. Therefore, the yield of agarose blocks will be about 4–8 mL. This should be more than sufficient for analysis of the clone including restriction mapping. If needed, the entire procedure can be scaled up or down for larger or smaller cultures.

4. Treatment of agarose blocks with DNase-free RNase is recommended. Incubate the blocks with DNase-free RNase at a final concentration of 5 µg/mL for 45 min at 37°C.

5. Centrifugation for a shorter or longer period is not advisable, as it results in DNA either not sedimenting completely or sticking to the polysaccharide pellet, respectively.

6. Aspiration is better than removal by pipeting; less protein, RNA, and membranous debris are carried down the gradient. Viscosity throughout the gradient indicates overloading.

7. Chromosomes show up as sharp bands. If the ethidium bromide stained gel shows a diffuse background in the lane, this is a result of undigested RNA that can be removed by treatment with RNase. On the other hand, if the autoradiogram shows a smear instead of discrete signals, this means that the chromosomes are degraded.

8. More often than not, screening of the YAC libraries with a given probe generates more than one YAC clone. Therefore, the clones with multiple YACs could be discarded without much of a loss. However, there are ways to generate a YAC clone with a single YAC from the ones with multiple YACs. Sometimes the needed YAC can be isolated

away from the other YAC as well as from the host chromosomes by preparative PFGE (*see* Chapters 15 and 21 for the procedure). Such isolated YAC DNA can be used to retransform spheroplasts to generate a clone containing the single YAC *(13)*. Alternatively, one can retransform competent yeast spheroplasts with the total yeast DNA that contains multiple YACs *(7)*. Some of the retransformed cells will contain a single YAC. Yet another way is by mating the yeast strain that contains multiple YACs with the compatible strain and separating the tetrads in the following generation *(4)*.

9. In addition, if the multiple YACs in a YAC clone originated from a single YAC, the larger YAC is likely to be unstable, and could disappear in subsequent generations by converting to a smaller stable one. Therefore, growing the YAC clone on a AHC plate and testing the DNA from cells grown for different number of generations by PFGE, should reveal the disappearance of a YAC. However, the stable smaller YAC may have deletions or rearrangements in the insert DNA.

10. The restriction map generated for genomic DNA may not agree with that of the YAC clones. For mammalian DNA, this is mainly owing to cytosine methylation of CpG sequences in genomic DNA, and most of the rare-cutting enzymes that recognize and cleave at such sequences are sensitive to methylation. But in cases where methylation does not inhibit the digestion by enzymes, absence of a restriction site in a YAC insert that is present in genomic restriction map could still be a result of either a polymorphism or be an error in either of the maps. The differences resulting from a polymorphism can be resolved by testing for the presence or absence of the site by digesting the human DNA used to construct the library. For YAC clones isolated from Washington University library, this cell line is available from the Center for Genetics in Medicine of Washington University. Use of this cell line as a control DNA in mapping experiments will assist in clarifying the ambiguities that might arise during the construction of a restriction map of a given genomic region.

11. Partials have to be generated by trial and error using different amounts of enzyme. There are other approaches to obtain partials, such as using magnesium concentration in the buffer lower than the optimal level *(27)*, or competition with a methylase that recognizes and methylates the same site as that recognized by the enzyme (*see* Chapters 12 and 16).

12. The enzymes chosen are those that digest the vector near the *Eco*RI cloning site but beyond the region where the opposing PCR Primers are placed. The enzymes that meet these criteria are *Hae*III, *Sau*3AI, *Fnu*DII, *Taq*I, *Nae*III for the left end and *Hae*III, *Sau*3AI, *Hha*I, *Hae*II,

*Aha*I, *Nla*IV for the right end. Similarly, primers can be chosen for each vector arm to accommodate the enzyme sites used to digest the YAC clone. *Taq*I digestion of a YAC clone does not yield a right-end fragment with the primers chosen here, as the enzyme cuts the vector, preventing these primers from amplifying.

Acknowledgments

We thank Roxanne Tavakkol for providing the autoradiograph used in Fig. 1; Laura Gross for optimizing the procedure for liquid DNA preparation, and Louise Hallet for assistance in preparing the manuscript.

References

1. Burke, D. T., Carle, G. F., and Olson, M. V. (1987) Cloning of large segments of exogenous DNA into yeast by means of artificial chromosome vectors. *Science* **236**, 806–812.
2. Coulson, A., Kozono, Y., Lutterbach, B., Shownkeen, R., Sulston, J., and Waterson, R. (1991) YACs and the C.elegans genome. *BioEssays* **13**, 413–417.
3. Schlessinger, D., Little, R. D., Freue, D., Abidi, F., Zucchi ,I., Porta, G., Pilia, G., Nagaraja, R., Johnson, S. K., Yoon, J-Y., Srivastava, A., Kere, J., Palmieri, G., Ciccodicola, A., Montanaro, V., Romano, G., Casamassimi, A., and D'Urso, M. (1991) Yeast artificial chromosome-based genome mapping: Some lessons form Xq24–q28. *Genomics* **11**, 783–793.
4. Green, E. D. and Olson, M. V. (1990) Chromosomal region of the cystic fibrosis gene in yeast artificial chromosomes: A model for human genome mapping. *Science* **250**, 94–98.
5. Silverman, G. A., Jockel, J. I., Domer, P. H., Mohr, R. M., Taillon-Miller, P., and Korsmeyer, S. J.(1991) Yeast artificial chromosome cloning of a two-megabase-size contig within a chromosomal band 18q21 establishes physical linkage between BCL2 and plasminogen activator inhibitor type-2. *Genomics* **9**, 219–228.
6. Wallace, M. R., Marchuk, D. A., Andersen, L. B., Letcher, R., Odeh, H. M., Saulino, A. M., Fountain, J. W., Brereton, A., Nicholson, J., Mitchell, A. L., Brownstein, B. H., and Collins, F. S. (1990) Type 1 neurofibromatosis gene: Identification of a large transcript disrupted in three NF1 patients. *Science* **249**, 181–186.
7. Burke, D. T. and Olson, M. V. (1991) Preparation of clone libraries in yeast artificial chromosome vectors. *Methods Enzymol.* **194**, 251–270.
8. Maniatis, T., Fritsch, E. F., and Sambrook, J. (1982) *Molecular cloning: A laboratory manual.* Cold Spring Harbor Laboratory Press, Cold Springs Harbor, NY.
9. Olson, M. V., Loughney, K., and Hall, B. D. (1979) Identification of the yeast DNA sequences that correspond to specific tyrosine-inserting nonsense suppressor loci. *J. Mol. Biol.* **132**, 387–410.

10. Carle, G. F. and Olson, M. V. (1984) Separation of chromosomal DNA molecules from yeast by orthogonal-field-alternation gel electrophoresis. *Nucleic Acids Res.* **12**, 5647–5664.

11. Brownstein, B. H., Silverman, G. A., Little, R. D., Burke, D. T., Korsmeyer, S. J., Schlessinger, D., and Olson, M. V. (1989) Isolation of single-copy human genes from a library of yeast artificial chromosome clones. *Science* **244**, 1348–1351.

12. Little, R. D., Porta, G., Carle, G., Schlessinger, D., and D'Urso, M. (1989) Yeast artificial chromosomes with 200-to 800-kilobase inserts of human DNA containing HLA, V_k, 5S, and Xq24-28 sequences. *Proc. Natl. Acad. Sci. USA* **86**,1598–1602.

13. Wada, M., Little, R. D., Abidi, F., Porta, G., Labella, T., Cooper, T., Della Valle, G., D'Urso, M., and Schlessinger, D. (1990) Human Xq24-Xq28: Approaches to mapping with yeast artificial chromosomes. *Am. J. Hum. Genet.* **46**, 95–106.

14. Gaensler, K. M. L., Burmeister, M., Brownstein, B. H., Taillon-Miller, P., and Myers, R. M. (1991) Physical mapping of yeast artificial chromosomes containing sequences from the human beta globin gene region. *Genomics* **10**, 976–984.

15. Neil, D. L., Villasante, A., Fisher, R. B., Vetrie, D., Cox, B., and Tyler-Smith, C. (1990) Structural instability of human tandemly repeated DNA sequences cloned in yeast artificial chromosome vectors. *Nucleic Acids Res.* **18**, 1421–1428.

16. Smith, H. O. and Birnstiel, M. L. (1976) A simple method for DNA restriction site mapping. *Nucleic Acids Res.* **3**, 2387–2398.

17. Bronson, S. K., Pei, J., Taillon-Miller, P., Chorney, M. J., Geraghty, D. E., and Chaplin, D. D. (1991) Isolation and characterization of yeast artificial chromosome clones linking the HLA-B and HLA-C loci. *Proc. Natl. Acad. Sci. USA* **88**, 1676–1680.

18. Green, E. D., Riethman, H. C., Dutchik, J. E., and Olson, M. V. (1991) Detection and characterization of yeast artificial chromosome clones. *Genomics* **11**, 658–669.

19. Rowley, J. D., Diaz, M. O., Espinosa III, R., Patel, Y. D., van Melle, E., Ziemin, S., Taillon-Miller, P., Lichter, P., Evans, G. A., Kersey, J. H., Ward, D. C., Domer, P. H., and LeBeau, M. M. (1990) Mapping chromosome band 11q23 in human acute leukemia with biotinylated probes: Identification of 11q23 translocation breakpoints with a yeast artificial chromosome. *Proc. Natl. Acad. Sci. USA* **87**, 9358–9362.

20. Chandrasekharappa, S. C., Stock., W., Neuman, W. L., LeBeau, M. M., Brownstein, B., and Westbrook, C. A. Characterization of yeast artificial chromosomes containing interleukin genes on human chromosome 5. (manuscript in preparation).

21. Hermanson, G. G., Hoekstra, M. F., McElligot, D. L., and Evans, G. A. (1991) Rescue of end fragments of yeast artificial chromosomes by homologous recombination in yeast . *Nucleic Acids Res.* **19**, 4943–4948.

22. Marchuk, D. A., Tavakkol, R., Wallace, M., Brownstein, B., Taillon-Miller, P., Fong, C., Legius, E., Andersen, L., Glover, T., and Collins, F. S. A yeast artificial chromosome contig encompassing the type 1 neurofibromatosis (NF1) gene. (submitted).

23. Nelson, D. (1990) Current methods for YAC clone characterization. *Gene Anal. Tech. Appl.* **7**, 100–106.

24. Riley, J., Ogilvie, D., Finnier, R., Jenner, D., Powell, S., Anand, R., Smith, J. C., and Markham, A. F. (1990). A novel, rapid method for the isolation of terminal sequences from yeast artificial chromosome (YAC) clones. *Nucleic Acids Res.* **18**, 2887–2890.

25. Lagerstrom, M., Parik, J., Malmgrem, H., Stewart, J., Patterson, U., and Landegren, U. (1991) Capture PCR: Efficient amplification of DNA fragments adjacent to a known sequence in human and YAC DNA. *PCR Methods and Applications* **1**, 111–119.

26. Silverman, G. A., Ye, R. D., Pollock, K. M., Sadler, J. E., and Korsmeyer, S. J. (1989) Use of yeast Artificial chromosome clones for mapping and walking within human chromosome segment 18q21.3. *Proc. Natl. Acad. Sci. USA* **86**, 7485–7489.

27. Grootscholten, P. M., Den Dunnen, J. T., Monaco, A. P., Anand, R., and Van Ommen, G-J. B. (1991) YAC mapping strategies applied to the DMD-gene. *Technique* **3**, 41–50

28. Albertsen, H. M., Le Paslier, D., Abderrahim, H., Dausset, J., Cann, H., and Cohen, D. (1989) Improved control of partial DNA restriction enzyme digest in agarose using limiting concentrations of Mg. *Nucleic Acids Res.* **12**, 808.

CHAPTER 18

Strategies for Mapping Large Regions of Mammalian Genomes

Margit Burmeister

1. Introduction

In many instances, pulsed-field gel electrophoresis (PFGE) is used to construct a map of a large region from the human or mouse genome. Sometimes, a gene or a gene-complex are very large, as in the case of the Duchenne muscular dystrophy gene *(1)* or the major histocompatibility complex (MHC) *(2,3)*. Frequently, such a study is initiated because there is evidence that a disease-causing gene is localized in a particular region. Before starting such elaborate molecular techniques as chromosome walking, YAC-cloning (*see* Chapter 16), cutting bands out of PFGEs (Chapter 21), and so on, it is important to know the size of the region of interest. In addition, PFGE is used to determine the order of markers more precisely than possible with genetic linkage analysis, and with a PFGE map available, new markers can be quickly localized within that region. Since the construction of a PFGE map is not trivial, this chapter will not so much deal with the techniques to generate the data for mapping, but focus on the construction of a PFGE map and on strategies how to generate data efficiently. Whenever possible, I refer to the relevant chapters in which the techniques are discussed in detail.

In general, a physical map is constructed by cleaving genomic DNA with restriction enzymes that cut infrequently in the mammalian genome, then separating the fragments by any of the PFGE versions,

From: *Methods in Molecular Biology, Vol. 12: Pulsed-Field Gel Electrophoresis*
Edited by: M. Burmeister and L. Ulanovsky
Copyright © 1992 The Humana Press Inc., Totowa, NJ

blotting the DNA fragments to nylon membranes, called filters, and hybridizing the filters with radioactively labeled probes. The hybridization patterns of different probes on the same filter are then compared: e.g., when two probes hybridize to the same 500-kb fragment, they cannot be more than 500 kb apart. The restriction enzymes used and size ranges separated will determine the resolution of the map.

This chapter deals with the construction of a physical map, the problems that frequently arise during the construction of such maps, and the ways in which these problems can be approached. In the materials section, the necessary probes, cell lines, and restriction enzymes are discussed. For the construction of a rough map, probes are placed on identical fragments by hybridization to a set of filters, on which DNA fragments were resolved by PFGE in three different size ranges. When probes have been successfully linked to identical large PFGE fragments, more precise mapping, which is addressed in the second part, can be performed by cleaving the DNA more frequently and by using combinations of restriction enzymes (double digests). The combination of single- and double-digest patterns renders a PFGE mapping approach similar to the mapping of plasmids or cosmids, except that hybridization rather than ethidium bromide staining is used to visualize DNA fragments. In the third part, we discuss reasons why markers sometimes cannot be linked by PFGE even though they are close to each other, and strategies to overcome this problem. The fourth part deals with the localization of CpG-rich islands, which are often found by PFGE mapping. Since such islands are indicative of 5' ends of genes, they are often the first landmark investigated when looking for expressed sequences. The fifth part addresses the identification of structural rearrangements, such as inversions, translocation breakpoints, or deletions in DNA from patients as compared to normal DNA.

2. Materials

1. One of the most critical prerequisites for the construction of a PFGE-map of a genomic region is the availability of markers spaced on average every 500 kb in the region to be mapped. Only in rare instances is it possible to construct a complete map when probes are spaced more than 1000 kb apart. For comparison, an average size cytogenetic band is about 10 Mbp in size, and a high-resolution (prophase) band 2–3 Mbp.

2. It is very useful to have an approximate idea where your probes map—
at least within the cytogenetic range. The approximate positions of
markers can be determined by using a panel of somatic cell hybrids
with broken chromosomes, e.g., from translocations *(4)*, or by *in situ*
hybridization *(5)*, genetic linkage, or radiation hybrid mapping *(6)*.
The reason why a rough map is useful is that in most people's hands,
PFGE gels do not run completely reproducible. Therefore,
comigration of bands is best identified by comparing hybridizations
to the same filter. When mapping more than about 10 probes, group
those probes together that are likely to map close to each other, and
use identical filters for each group.

3. Carefully plan the DNA sources. For constructing a map of human
chromosomes, it is best to have four different DNA sources available:
blood, sperm, a somatic cell hybrid containing only the chromosome
of interest on a hamster or mouse background, and a lymphoblastoid
cell line. Some consideration should be given to the individual whose
blood and sperm are used for making agarose blocks: Do not use
many different blood samples, especially not from a source such as a
blood bank, because you will inevitably run out of material when you
can afford it least. One of the best donors in terms of availability and
convenience is probably yourself (*but see* Note 1). The donor for blood
and sperm DNA need not necessarily be the same person, but if you
are interested in studying methylation or CpG-rich islands in the
region you are mapping, it is often useful to have blood and sperm
available from the same donor. When constructing a map from mouse
or other rodent genomes, obvious fresh tissue sources are spleen and
liver (*see also* Chapter 8), a fast-growing cell line such as A9 cells, and
if possible a somatic cell hybrid containing the chromosome of interest on a nonrodent, e.g., canine or primate, background. Use only
one strain of mice for the first round of experiments, and preferably
do not mix different tissues.

For a small fee, some useful somatic cell hybrid lines are available
from American Type Culture Collection (ATCC, 12301, Parklawn
Drive, Rockville, MD 20852, (301) 231-5585). ATCC can also provide
cell lines derived from many different tissues from many different
species. Similarly, the Coriell Institute for Medical Research
(Copewood and Davis St., Camden, NJ 08103, (609) 966-7377) can
also provide human lymphoblastoid or fibroblast cell lines, and
somatic cell hybrids containing nearly every human chromosome
as the only human chromosome on a mouse or chinese hamster

background. Preparation of agarose blocks from human and mouse tissues and their digestion with restriction enzymes are described in Chapter 8.

4. Restriction enzymes (*see* Note 2): *Bss* HII, *Cla* I, *Eag* I, *Mlu* I, *Not* I, *Nru* I, *Pvu* I, *Sal* I, *Sfi* I, *SnaB* I are recommended for initial studies. *Rsr* II (*Csp* I), *Sgr* AI, *Spl* I, *Sse*83871, *Sst* II may also be useful, but I have little experience with them myself. *Ksp* I, *Nae* I, *Pac* I, *Sfu* I, *Sma* I, *Sph* I, *Spe* I, and *Xho* I may be useful for fine mapping if probes are found to be very close to each other. All of these enzymes are available from one or more of the following companies: Amersham, Boehringer, New England Biolabs, and Promega. Take extreme care not to contaminate enzymes, and keep enzymes used for PFGE separate from other enzyme stocks. This point cannot be overstressed, insofar as it is very easy to contaminate enzymes with a small amount of other restriction enzymes, which may cause the DNA in the blocks to appear degraded. In addition to infrequently-cutting restriction enzymes, combinations of methylases and restriction enzymes can be used to generate rare-cutter sites (*see* Chapter 12)

3. Methods

3.1. Preparation of a Rough Map

1. To optimize conditions for separation of DNA fragments in the appropriate size ranges, *see* Chapters 1–7, and/or the instruction manual from the supplier of your PFGE apparatus.
2. Prepare a first set of filters with a resolution in two to three size ranges: high (50–700 kb), medium (800–2000 kb), and possibly a low resolution (1000–5000 kb) filter (*see* Note 3). Since new restriction enzymes come constantly on the market, *see* Note 2 for general rules to find infrequently cutting enzymes. I found the following enzymes useful for mammalian DNA for the first set of filters:
   a. high resolution: *Cla* I, *Sal* I, *Sfi* I, *Bss* HII, *Eag* I;
   b. medium resolution: *Cla* I, *Nru* I, *Mlu* I, *Not* I, *Sal* I;
   c. low resolution: *Nru* I, *Mlu* I, *Not* I, *Sna* BI, *Sal* I.

   Note that *Sal* I-digests can result in fragments in many different size ranges. Prepare these filters initially from two different DNA sources each, e.g., blood DNA and the somatic cell hybrid DNA. Using just a few enzymes, you can usually fit digests from both sources on one gel. Load the digests of the two sources on different parts of the gel (i.e., load all five digests from blood DNA, then size marker, then all five digests from the cell line DNA), and cut the gel into these two halves after blotting (*see* Note 4). Plan to use about 10 probes on these first filters and prepare more filters if you want to map more probes.

3. Hybridize those probes that are expected to be close to each other consecutively to the same filter. Ensure that the previous probe has been stripped prior to subsequent hybridizations by exposing the stripped filter overnight, especially when a strong or repetitive signal had been detected before.

4. Align autoradiographs to detect probes that might hybridize to the same restriction fragment (*see also* Chapter 8 for alignments). Estimate the size, and tabulate the results. At this point, do not worry too much about exact sizes, but pay attention to the shape of fragments: Identical fragments should have the same shape. In addition, irregularities such as air bubbles and artifactual hybridization spots or hybridization to the loading slot can often help to align blots when comparing hybridizing fragments.

5. A major problem in any PFGE analysis at this stage is to distinguish between truly identical fragments and fragments that are comigrating by coincidence. A common PFGE pattern between two probes can never be considered an absolute proof for physical linkage of markers. However, several considerations can improve the likelihood for physical linkage: When two probes have three or more different comigrating hybridizing fragments in common, this is usually not coincidence. In addition, the results should be checked for logical consistency: Often, partial methylation and/or digestion results in many different fragments (*see* Note 5). An identical fragment has to be of the same relative intensity with both probes, whereas coincidentally comigrating fragments may have the same size, but often have different relative intensities (*see* Fig. 1A–C). If many different bands are hybridizing to two probes, at least a fraction of the fragments larger than the comigrating fragment also has to be shared. An example is shown in Fig. 1D, E: The smaller *Sal* I fragments to which the two probes hybridize are of different sizes, but all larger fragments are in common, suggesting true linkage. A third consideration is the pattern from the different DNA sources (*see* Note 6), which has to be consistent, and often can confirm physical linkage. An example is shown in Fig. 1F–I. For mapping purposes, it is usually irrelevant whether additional bands in different tissues are a result of partial methylation, partial digestion, or polymorphisms. Either one can provide additional evidence in favor of physical linkage.

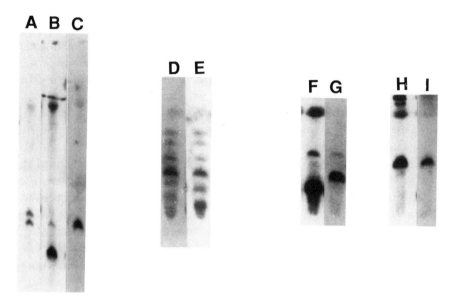

Fig. 1. Partial digestion products, problems, and use: **A–C**: three probes from human chromosome 21 were hybridized to *Sal* I-digested DNA separated by PFGE. Although a fragment of the same size is recognized, it is not identical, because either a large partial fragment is not shared (lane **A**) or the intensity ratio is different (lane **B**). **D** and **E**: hybridization of two probes to *Sal* I-digested DNA shows close physical linkage on partially digested fragments. Although the smallest *Sal* I fragments are different, the larger ones are identical in the size distribution pattern and the intensity ratios between the two probes. **F–I**: **F** and **G** show hybridization of two probes to *Mlu* I-digested blood DNA. The larger partial digest product suggests identity, especially since the sizes of the two smaller fragments approximately add up to the size of the larger one (data not shown). In contrast, when hybridized to cell line DNA (**H** and **I**), the site between the two markers is methylated or absent. Taken together, the two DNA sources give complementing information: Physical linkage on the large DNA fragment, and the position of a polymorphic site between the probes gives a finer resolution map. Lanes **D–I** reproduced with permission from ref. *(37)*.

6. Once several large fragments that are shared between some of your probes have been established, a first rough map can be drawn (*see* Fig. 2). First, consider the enzyme that results in the largest fragments. Draw these fragments to scale on a piece of graph paper, assuming that there is no space between sites. If larger fragments were detected that are owing to partial methylation and that are not recognized by

any of your other probes, include them in this map as well. Otherwise, the true size of the region will be underestimated. This map gives you the minimal size of the region you are analyzing.

7. With other fragments that are corecognized by some of the probes, you can usually order some of the large fragments. At this point, you will have maps for several enzymes, but no information to position the enzyme sites relative to each other (*see* Fig. 2). Often, it will be impossible to detect whether two probes that hybridize to an identical restriction fragment are very close to each other or on opposite sites of that fragment. Therefore, at this stage, conclusions such as "probe A and B are between 0 and 1500 kb apart" can be made. Fine mapping is necessary to get more exact positional information. In cases where no restriction fragments are found to be shared between various probes, proceed as described in Section 3.3.

3.2. Fine Mapping

1. Prepare double digests with combinations of four or five of the enzymes that were the most useful in linking your probes into a map. However, avoid combinations with enzymes that result in more than three fragments already in single digests. Partial digests are often too complicated to interpret, and digesting them with a second enzyme is often even more confusing. However, if you have a hypothesis about the arrangement of several partially digested fragments, you can test it by using double digests. At this stage, you will know the size range of the analyzable fragments, and thus which PFGE conditions will separate them best. For these gels, use yeast chromosomes as well as lambda ladders as size markers, since you now want to size fragments as accurately as possible.

2. Hybridize as before and, in addition, visualize the size markers on the filters by hybridization with labeled yeast or lambda DNA. Do not mix the labeled marker into the mammalian probes, but rather hybridize the markers separately. Sometimes, you may find a probe in your collection that cross-hybridizes to your markers anyway. Since autoradiographs can be overlaid easily, but autoradiographs and pictures of ethidium-bromide stained gels are usually not of the same size, this is a better strategy than trying to compare the results of an ethidium-bromide stained gel picture with autoradiographs by photographing and aligning rulers. If you are storing your information from both ethidium-bromide-stained gels and autoradiographs electronically (*see* Chapter 13), exact size determination is easy and there is no need to visualize size markers.

locus	Not I	Mlu I	Nru I
S3	700, 1800	700, 1600	600
S101	1400	1000, 1600	1400, 2000
S15	1400	1250, 1400	1400, 2000
S39	700	1250, 1400	600, 2000
S51	700	200, 1400	500

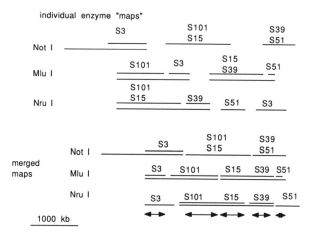

Fig. 2. Constructing a map. The table shows the size of hybridizing fragments (in kb) revealed by PFGE using five different probes from human chromosome 21. As typical for the arithmetic in such experiments, 700 + 1000 add up to 1600 kb, and 200 + 1250 to 1400 kb. The map for each enzyme is shown below. By combining the *Not* I with the *Mlu* I map (or the *Mlu* I with the *Nru* I map), the order of markers can be determined unambiguously, as shown in the combined map below. The arrowheads show the position of each marker, which is known only within a several hundred kb. Fine mapping using other enzymes, and double digests, are needed at this stage to further narrow down the distances and positions. Data are derived from ref. *(37)*.

3. When double digests result in fragments of the same size as single digests, the sites are probably clustered in CpG-rich islands flanking your probe. In Fig. 3, a probe is shown to result in the same size fragments with a variety of combinations of enzymes, with the exception of *Sfi* I. When this is the case, enzymes cleaving preferentially outside of CpG-rich islands, such as *Sfi* I, should be used for fine mapping (*see* Note 2).
4. When the sizes of fragments from double digests are compared to the sizes of fragments in the single enzyme digest, very often the arith-

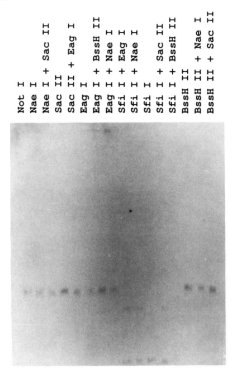

Fig. 3. A probe flanked by two CpG-rich islands. DNA was digested with the enzymes and combinations of enzymes shown above each lane. All of these except for *Sfi* I are known to cut in CpG-rich islands. The identical size of about 700 kb for each digest indicates that the probe (SF85/D21S46) is located between two CpG-rich islands, and that both islands contain sites for *Not* I, *Nae* I, *Sac* II, *Eag* I, and *BssH* II.

metic does not work perfectly well. To improve the size determination, always estimate the size at the bottom of a strongly hybridizing band. If the cleaved fragments add up to less than the original fragment, there is more than one restriction site between the two probes in question. Most often, however, the cleaved fragments will add up to *more* than the original fragment, e.g., a 500-kb fragment is cleaved into a 350- and a 250-kb fragment. This phenomenon is owing to the fact that bands will be shifted up in gels that are overloaded. If this is the case, assume that the lower sizes are probably correct (e.g., in reality, the fragments may be 260, 170, and 430 kb in size). An extreme example, in which the double-digest product actually ran slower than the single digest, is shown in Fig. 4. Other such examples are shown in ref. 7. The reason for this phenomenon is local over-

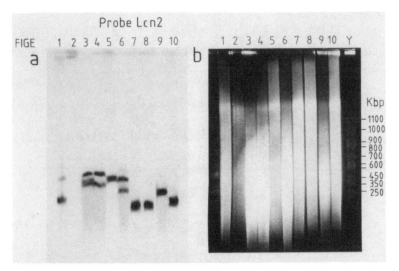

Fig. 4. Artifacts resulting from local overloading of lanes. Probe Lcn2.1 is hybridized to blood DNA that was separated by FIGE (panel a). The *Sal* I + *Sfi* I double-digest product (lane 9) actually run *slower* than the *Sal* I single-digest (lane 7). This is owing to the fact that *Sfi* I digestion moves more DNA fragments into the 200- to 350-kb range (*see* ethidium bromide staining in panel b), therefore overloading the 200-kb region relative to the digest with *Sal* I alone. Reproduced by permission from ref. *(15)*, copyright 1987 by the AAAS.

loading, i.e., the second enzyme moves more DNA into the region of the hybridizing band, thereby overloading that size range of the gel. Such artifacts are more likely in FIGE than in OFAGE or CHEF. Because some overloading of lanes is a very frequent problem, agarose blocks with smaller amounts of DNA should be prepared when exact size determination is important, e.g., when cosmid walking is planned. Generally, the apparent size will decrease when less DNA is loaded. However, for many mapping studies, the focus is more on the order of markers rather than exact size determination. For the determination of the order of markers, it is more important to have sufficient DNA on a filter to compare many different probes on the same filter. Thus, you have to weigh the trade-off between preparing many filters for reprobing when using less DNA, and having less accurate size determinations when using more DNA on your filters.

5. When you construct the final map, it is important to ask whether the final map is just compatible, or the *only* map that can be constructed from the results *(7)*. Sometimes, several different maps can be con-

structed, and it is important to be aware of any still existing ambiguities. In that case, design combinations of digests to prove or discard a particular hypothetical map.

6. When probes are closer than about 300 kb to each other, a different set of enzymes can be used for more precise mapping. For example, *Ksp* I, *Nae* I, *Pac* I, *Sfi* I, *Sma* I, *Sph* I, *Spe* I, *Stu* I and *Xho* I often result in fragments smaller than 300 kb. As a general rule, choose enzymes that contain no more than one CpG-site in their recognition sequence for such high-resolution mapping, and use shorter switching intervals to resolve these smaller fragments.

3.3. Closing Gaps in the Map

There are many possible reasons why no overlapping restriction fragments are found while constructing a map, and the remedy for this will depend on the cause of the problem.

3.3.1. Hybridization to the Slot

One problem is that the only hybridization signal detected is in the loading slot, or in the slot and in the limiting mobility. The most trivial explanation is that the restriction enzyme did not completely digest the DNA. In this case, most of the ethidium bromide staining pattern will look similar to that of undigested DNA. A better control is to use a probe that is known to hybridize to a small, completely digested fragment on the blot to make sure that the digest is complete. Refer to Chapter 8 to trouble-shoot incomplete digestions. When you are sure that the digests are complete, hybridization to the slot means that all DNA fragments with the enzymes chosen are too large to enter the gel. This problem is most often encountered when the genomic region under investigation is poor in CpG-rich islands, and when restriction enzymes that cut nearly exclusively in CpG-rich islands are used (*see* Note 2). This type of result is expected most often with probes from a G-band (black band in standard banding pictures; *see* Note 7 for the general features of different cytogenetic bands). Figure 5 shows such a result with D21S1 and D21S18. Another possible explanation is that the region contains tandem repeats of some kind, such as in the centromeric region (Fig. 5). Try a different set of enzymes in this situation (*see* Note 2). For special cases such as centromeres, even otherwise frequently cleaving enzymes such as *Eco*RI or *Bam*HI can behave like rare-cutters (*see also* Chapter 20). In addition, a PFGE gel that separates fragments in the very large size-range (*see* Chapters 2, 5, and 7) can be run using those enzymes whose

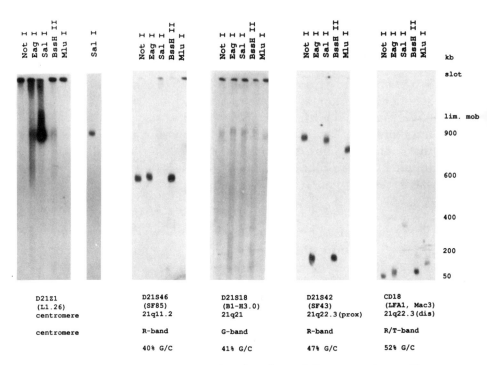

Fig. 5. Hybridization pattern of probes from different regions of human chromosome 21. A filter was prepared from DNA digested with the enzymes indicated above each lane, and hybridized successively to the probes indicated below each panel. The GC-content around each probe are from ref. *(34)*. The centromeric region shows unusual behavior because of tandem repeats (*see* Chapter 20). With the exception of *Sal* I, all enzymes used are likely to cleave in CpG-rich islands. A lower exposure of the Sal I lane is shown next to the pattern from probe L1.26. Note that most of the hybridization of probe B1-H3.0 is to the loading wells, typical for a G-band probe. Further experiments showed that the hybridizing fragments with this probe are between 2 and 5,000 kb in size *(6)*. The pattern with probe CD18 is typical of a very GC-rich T-band, whereas the pattern with SF43 indicates that several CpG-rich islands flank the probe, but not all of the enzymes cleave in each of the islands.

fragments could initially not be resolved. With that, you may be able to take advantage of the scarcity of restriction sites and link up very large genomic regions into a map.

3.3.2. Too Small Fragments

Another problem are relatively small, completely cleaved restriction fragments with all restriction enzymes result. This is a frequent problem when probes are derived from the relatively gene-rich regions of the genome that are often located near telomeres of chromosomes, called T-bands (*see* Note 7). An example is shown in Fig. 5 with probe CD18. Small fragments, often of equal size, are most likely the result of a large number of CpG-rich islands in that region of the genome. Therefore, use enzymes that cleave outside of CpG-rich islands to construct the map from such a region (*see* Note 2). Since CpG-rich islands are often methylated in cell lines (*see* Note 6 and ref. *8*), use cell lines rather than fresh tissue as DNA sources in this situation. In addition, you might try to link fragments by using partial digests. Set up digests with different dilution of enzymes, an example of which is shown in Fig. 6. Alternatively, inhibiting intercalating dyes (*9*) or partial methylation (*see* Chapter 12) can be used. The larger, partially digested fragments may be in common between probes that do not hybridize to identical digested fragments.

3.3.3. Fragments Cannot Be Linked by PFGE

If your probes recognize different size fragments of 500 kb or larger that are not identical, the most likely explanation is that the probes that you are using are not close enough to each other. Very often, the cytogenetic estimates of band sizes, and derived physical distances, are very inaccurate. Alternatively, even if you are not in an extremely CpG-rich region of the genome, one CpG-island that is cleaved by all enzymes that you are using may disrupt the mapping effort. With only one such CpG-rich island, restriction fragments are often not of the same size. By using combinations of enzymes (double digests) you can find out whether there is such a cluster of sites in one place. If that is the case, try to use enzymes that cleave outside of CpG-rich islands (*see* Note 2) or try partial digests (*see* Section 3.3.2.). In addition, test DNA from several different sources. Often some DNA sources result in larger fragments than others. However, none of these techniques is easy to perform, and it may be more efficient to generate more probes from the region before proceeding with your mapping effort.

3.4. Localization of CpG-Rich Islands

Although for many mapping efforts CpG-rich islands are not very helpful because restriction enzymes cluster in these regions, they are a very useful tool for biological reasons. CpG-rich restriction sites cluster near the 5' ends of genes (*10*), and have been used to find genes on cosmids (*11*). Physical maps can be used to locate such CpG-rich

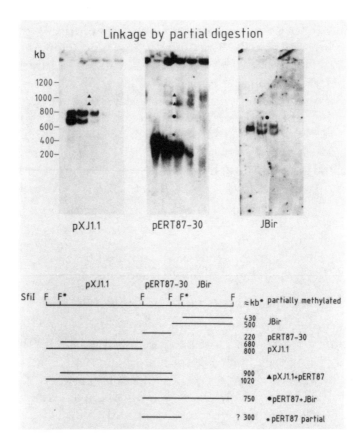

Fig. 6. Partial digests to link up fragments. *Sfi* I digests with dilutions of enzymes were used to generate partially digested DNA. These were hybridized to probes that were known to be close to each other. The identical partial digestion products are marked with dots. The map, including the fragments seen, is shown below. This figure also illustrates a difficulty in using partial digests: To reveal higher order partial digest products, very high signal intensities are needed. Note that the lanes containing completely digested DNA look severely overloaded, although only weak signals can be detected from the partial digestion products.

islands and thereby help in the identification of candidate genes in a small region *(12–16)*. As shown in Fig. 3, the localization of CpG-rich islands often emerges from the mapping effort. To specifically search for CpG-islands, cleave several blocks containing blood or sperm DNA (but not DNA from a cell line) with an enzyme that generates large fragments. Possible good candidates are *Sal* I, *Sfi* I, and *Cla* I, and sometimes *Not* I, *Nru* I, or *Mlu* I. Then redigest all but one of these agarose blocks with enzymes that cleave frequently in CpG-rich islands (*see* Note 2 for such enzymes). Separate the double digest products and the single digest control DNA by PFGE, and hybridize the filters to the probes in question. The presence of a CpG-rich island is detected when the single-digest fragment is cleaved by many different enzymes into a smaller fragment of identical size (*see* Fig. 7). The closer the marker is to an island, the more frequently cleaving restriction enzymes can be used for this type of analysis.

3.5. Detection of Rearrangements

Most studies in which a gene was successfully cloned using its chromosomal position as the start point took advantage of patients with rearrangements such as translocations or deletions. Finding and cloning markers close to such breakpoints is therefore often a very high priority. When a patient is known to have a translocation along the chromosomal region studied, or is suspected to show a deletion, it is expected that these rearrangements will alter the PFGE map. When the breakpoint is associated with a disease gene, one would like to detect the rearrangement breakpoint as altered PFGE fragments, and ultimately clone the breakpoint site. Several studies have successfully used this approach (*1, 17–22*). In principle, the approach is straightforward: Take patient DNA and DNA from control persons, digest with a number of restriction enzymes, and look for fragments of altered size by PFGE using a probe assumed to be close to the breakpoint.

Practically, however, there are many potential pitfalls (*23*).

1. When dealing with a translocation, there are two chromosomes involved. Although only one chromosome contains the gene of interest, consider using markers from both chromosomes. Often, on one side of the translocation many more markers have been characterized or mapped. Once you find a breakpoint from one side, it will be helpful even if the aim is to clone the gene located on the other chromosome. As an example, one group that cloned the Duchenne muscular dystrophy gene used a translocation into the ribosomal genes, for which probes were available (*24*). Similarly, use markers from either side of a deletion when possible.
2. Define which markers might be closest to the breakpoint. In the case of a translocation, if possible use somatic cell lines containing only one of the chromosomes involved to determine on which side of the

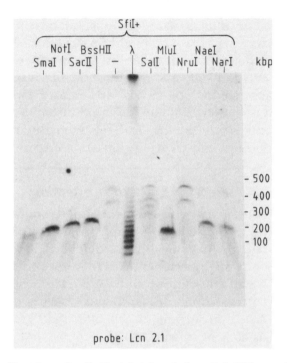

Fig. 7. Localization of a CpG-rich island. Lcn 2.1 *(15)* was hybridized to a filter containing DNA fragments separated by OFAGE after digestion with *Sfi* I, and double digests of *Sfi* I with the enzymes indicated above each lane. Fragments of identical size with *Sma* I, *Not* I, *Sac* II, *BssH* II, *Nae* I, and *Nar* I indicate the presence of a CpG-rich island within about 250 kb of the *Sfi* I site.

translocation breakpoint your probes map. In the case of a deletion, determine which probes are deleted.

3. Construct a map of the region near the translocation breakpoint using the approaches mentioned before. After that, you should be left with only one or very few markers that might be close to the breakpoint. At least you should be able to exclude a number of markers that way: If the order of markers is A-B-C-D, and all of these are on one side of the translocation breakpoint, probes B and C cannot be the closest to the breakpoint, and should not be used in further analysis.

4. Prepare agarose blocks from many different control DNAs. If blood DNA is used, for this purpose a blood bank might be a possible source. However, because of the risks of AIDS, hepatitis, and so forth associated with using blood from unknown sources, prefer-

ably use blood that is known to be free of these pathogens. Most people prefer using donors they personally know. In any case, take extra precautions assuming the blood might be contaminated. The preparation of the control DNAs is one of the most crucial steps. To avoid methylation differences between different tissues (*see* Note 6), use the same tissue as a DNA source for controls and your patient sample. If available, a very important control are samples from the parents of the patient (or the strain on which the mutation arose in the case of mouse DNA). These should be included even if the tissue differs, because the parental DNA can be used to rule out restriction site polymorphisms. If available, use more than one DNA source from the patient *(23)*. Take great care to accurately determine the number of cells used for the preparation of the agarose blocks (*see* Chapter 8). Differences in the amount of DNA can easily mimic different fragment sizes (*see* Fig. 8).

5. Prepare filters using those enzymes that result in large fragments in the region you are analyzing. If possible, use enzymes that are not sensitive to methylation such as *Sfi* I. However, currently there are not many such enzymes available (*see* Note 2 for rules to find suitable enzymes). Load all samples that were cleaved with the same enzyme next to each other, and the most important controls (e.g., parental DNA samples) next to the patient DNA sample.

6. Hybridize the filter to the probe(s) closest to the breakpoint. Compare with the control lanes. Very often at this stage you will get ambiguous results, i.e., altered fragments in the patient lane that may or may not be an indication of the breakpoint. With the controls at hand, consider the alternative explanations for new bands: polymorphisms, partial digest, differential methylation, and differences in the amount of DNA loaded. The best positive control is the detection of the same band with a probe from the other side of the deletion or translocation. Since most regions of the genome are not densely enough covered with markers, this is rarely possible. Compare the intensities carefully and adjust the amount of DNA if necessary. A probe that hybridizes to a fragment of the same size as the truncated fragment, but which is not near the deleted region, provides another good control (*see* Fig. 8). A more difficult problem is the presence of rare polymorphisms *(20)*. Except when DNA from both parents of the patient are available, test a large number of control samples to ensure that a change in fragment size is not a polymorphism. Another frequently observed problem at this stage is differential and/or partial methylation. Cultured cell lines can show

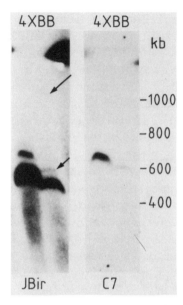

Fig. 8. Localization of a breakpoint: control for loading artifacts. Hybridization of a probe, JBir, that detects a deletion in the Duchenne muscular dystrophy gene is shown. The deletion is only detected in a partial-digest product. Since the amount of DNA loaded is not the same in the two lanes, it is not immediately obvious from the hybridization with JBir alone whether the downshift in the lane containing DNA from the patient BB is because less DNA was loaded, or indicating a rearrangement. The relative amounts of DNA loaded cannot be inferred from the signal intensity, since the control DNA (4X) contains four X chromosomes per genome equivalent, whereas the patient DNA contains only one X chromosome per genome equivalent. Control hybridization with a probe, that hybridizes outside of the deleted region, resolves that problem: The C7 hybridizing fragment runs slightly faster than the JBir fragment in the control lane, but definitely slower than the JBir fragment in the BB lane. Reproduced with permission from ref. *1.*

additional bands because they loose their control of methylation, and start methylating random sites partially *(1)* as well as the otherwise unmethylated CpG-rich islands *(8).* If possible, analyze DNA from fibroblasts, blood cells, and lymphoblastoid cell lines as well as the DNA from the parents of a patient with the presumed deletion/translocation. A breakpoint is probably real when it is observed with several different enzymes. Exclude new bands owing to partial digest by digesting with more enzyme, hybridization with a probe that results in a completely digested fragment. If one of the altered PFGE fragments is smaller than the controls, a partial digest cannot be the problem. On the other hand, a partial digest can be used to discriminate between true breakpoints, polymorphisms and changes in methylation patterns: A partial digest across a breakpoint will result in several altered fragments.

7. If you cannot find a breakpoint, you probably need more probes from the region. Once a translocation breakpoint is found, use other probes in the region to rule out more complex rearrangements such as small inversions or small deletions. Define the smallest fragment that contains the breakpoint. Depending on its size, you may now want to chromosome walk, jump *(25, 26)* or cut out the fragment from PFGE (*see* Chapter 21) in order to actually clone the breakpoint.

4. Notes

4.1. EBV Transformation

Never make a lymphoblastoid (Epstein–Barr-virus-transformed) cell line from your own blood. You will not have an immunological reaction that would otherwise protect you against accidental infection by these potentially tumorigenic cell lines.

4.2. Restriction Enzymes

Since more restriction enzymes are constantly being introduced, it is worthwhile checking catalogs (e.g., Amersham, Boehringer, New England Biolabs, and Promega) from time to time for new, infrequently-cutting restriction enzymes. These are enzymes with a 7- or 8-bp recognition sequence. (Currently, the following 8-bp cutters are available: *Asc* I, *Not* I, *Pac* I, *Pme* I, *Sfi* I, *Sse* 83871, and *Swa* I; *see also* Chapter 14) *See* Chapter 12 for how to determine the frequencies of restriction sites in species other than mammals. In addition, 6-bp cutters cut rarely in mammalian DNA if they contain 1 or 2 CpG sequences in their recognition sequence. Those with 2 CpGs usually cut less frequently. Determine whether the enzymes are sensitive to CpG methylation, i.e., will not cut CpG-methylated DNA (for most enzymes this can be found in standard cloning manuals such as ref. *27*). For example, *Ksp* I is an isoschizomer of (i.e., it has the same recognition sequence as) *Sac* II and *Sst* II, but unlike the latter two is not sensitive to CpG methylation. *Ksp* I therefore cuts mammalian DNA much more frequently than *Sst* II. Most enzymes that contain CpG in their recognition sequence are methylation sensitive (*see below*, Note 5). There is no need to worry about any other methylation, because mammalian DNA is not methylated at any other base. Note, however, that an enzyme such as *Sfi* I, although not containing CpG in its recognition sequence, can still be methylation sensitive (the *Sfi* I recognition sequence is ggccnnnnnggcc, but it may be sensitive to methylation at the last C, which can happen when the recognition sequence is followed by a G). To some extent, trial and error may be necessary to determine the frequency of cleavage of new restriction enzymes. Although *Sal* I (gtcgac) and *Xho* I (ctcgag) are expected to cleave mammalian DNA with similar frequency (they have the same GC content and the same number of CpGs in recognition sequence, and both are methylation sensitive), empirically *Xho* I cuts about 5–10 times more frequently. Usually, enzymes that contain only one CpG in their recognition sequence and that are methylation sensitive have a cleavage frequency of once every 100–300 kb, whereas those with two CpG residues in their recognition sequence cleave about every 400–1500 kb.

Most unmethylated, i.e., cleavable, restriction sites that contain CpG in their recognition sequence will be clustered in so-called CpG-rich islands, often found at the 5'-end of genes *(10)*. Enzymes that are most likely to cleave in CpG islands are those that contain two CpG residues in their recognition sequence and whose site is G + C only (e.g., *Eag* I [cggcgg], *Not* I [ccggccgg], *BssH* II [gcgcgc], or *Sst* II [ccgcgg]). In contrast, enzymes with recognition sequences containing an A + T component are also present outside of islands (e.g., *Mlu* I [acgcgt], *Nru* I [tcgcga], *Cla* I [atcgat] and *Sal* I [gtcgac]) (*see also* ref. *28*). Rare-cutting enzymes without CpG in their recognition sequence should be completely independent of CpG-rich islands, but currently, there are only five such enzymes available, namely *Sfi* I (ggccnnnnnggcc), *Pac* I (aattaatt), *Swa* I (atttaaat), *Pme* I (gtttaaac), and *Sse* 83871 (cctgcagg). The latter four are new on the market and have not yet been extensively tested for PFGE. *Pac* I, *Pme* I, and *Swa* I may be good as a more frequently cutting enzyme, whereas *Sse* 83871 is expected to cut outside of CpG-rich islands and would be expected to cut with a similar frequency as *Sfi* I.

An additional problem can be a specificity for the environment around a restriction site. This means that the same amount of enzyme will cleave one site completely, and another site only partially. However, digestion with 20 to 50 times more enzyme can cleave all sites completely. Without actually performing the often expensive overdigestion experiment, site specificity is hard to discriminate from partial digests due to methylation. Several enzymes are known to show site specificity on plasmid DNA. These are *Nar* I, *Nae* I, *Sac* II, and *Xma* III (as stated in the New England Biolabs catalog). Although they might be useful for mapping, I prefer not to use them for PFGE mapping because of the irreproducible results often obtained. *Sfi* I occasionally also shows such site-specificity. *Sst* II is an isoschizomer of *Sac* II that is less affected by such site-specificity. Similarly, *Eag* I should be used instead of *Xma* III.

4.3. Different Tissues

For the initial screening of probes it is more useful to have different tissues and several size ranges available and relatively few enzymes rather than a large collection of different enzymes. Many enzymes often give similar information, and having several different tissue types often can confirm a tentative linkage.

4.4. Crossreactivity of Probes

Certain probes contain repeats that only allow unique hybridization to somatic cell hybrids containing a single human chromosome, but result in a smear when hybridized to total human DNA. This can also be the case when cosmids are used as a probe, and repeats are removed by competition with total human DNA *(29)*. If competition is not completely successful, map such probes using only the filter containing somatic cell hybrid DNA. However, sometimes artifact bands appear that disappear when a unique probe is used. If possible, use of probes that need competition should be avoided. On the other hand, cDNA probes that cross-react with rodent DNA should not be used on the somatic cell hybrid filter, because discrimination between rodent and human DNA on PFGE filters is often not possible. Even a control lane with rodent DNA often does not help, because the rodent DNA may have a different hybridization pattern owing to methylation differences. Since PFGE hybridizations often reveal partially cleaved bands, even low-level cross-reactivity can cause confusion, because you cannot rely on intensity ratios between control and sample DNA, as you would in regular Southern blot hybridizations.

4.5. Partial Digests and Partial Methylation

Most of the enzymes discussed in this chapter are sensitive to methylation at CpG residues, meaning that they will not cut methylated DNA. Although partial digests are more difficult to interpret, they often help in confirming a physical linkage between two probes, as shown in

Fig. 2. Enzymes that cleave outside of CpG-rich islands (*see* Note 2) are particularly prone to reveal such partial methylation.

4.6. Methylation Differences Between Tissues

The degree of methylation may vary drastically between different tissues. When using different DNA sources from different donors it may be difficult to discriminate between RFLPs and differential methylation. However, the mechanism is often irrelevant for mapping purposes. Make sure to construct the map initially by using information from only one source, and then to use the additional information from other sources to verify and refine the map. Because the pattern of methylation in different tissues is not well known, the following rules should be taken with some reservations. In general, cell lines have a more "relaxed" methylation pattern than tissues, i.e., they will often show many bands, whereas in fresh tissues especially if there is only one predominant cell type in the tissue, usually a site is either methylated or not. CpG islands are by definition unmethylated at their CpG sites in all tissues, even in nonexpressing tissues *(10)*. However, this generally only holds for fresh tissues. Lymphoblastoid cell lines have been shown to methylate CpG islands *(8)*, and other cell lines are likely to do that as well. In addition, sites outside of CpG islands (e.g., *Sal*I sites) are methylated to a different degree in different tissues. My experience is that blood DNA tends to be more highly methylated than sperm DNA, but that may depend on the genomic region. Cell lines show more partial methylation at such sites, resulting in many PFGE bands. *See* ref. *1* for a comparison of cell line, blood, and sperm DNA methylation in the Duchenne muscular dystrophy region. Lymphoblastoid cell line DNA may often be undermethylated compared to blood DNA, whereas fibroblast cell lines show more methylation the longer they have been kept in culture, i.e., the higher the passage number *(30)*. Somatic cell hybrid cell lines containing human chromosomes are unpredictable in their methylation patterns.

4.7. Genome Organization

An important feature of the genome is that it is organized into regions of different cytogenetic and molecular properties *(31–33)*. The well-known cytogenetic banding pattern with Giemsa results in dark (Giemsa or G) and Giemsa-light or reverse (R) bands. In general, G bands are A/T-rich, and heavily methylated at CpG residues. In contrast, R-bands contain most or all of the CpG-rich islands of the genome, and are usually G/C-rich. However, some A/T-rich R-bands have been reported (*34*). T-bands are a subpopulation of R-bands, and are extremely G/C-rich, very gene-rich and are often located near telomeres, but also in other gene-rich regions such as the MHC complex (*35,36*). In general, the region from which a probe is derived will be reflected in the PFGE-banding pattern (*see* Fig. 5): In the A/T-rich G-bands (e.g., 21q21), CpG-recognizing enzymes cut very infrequently, and sites do not cluster. PFGE fragments with probes from A/T-rich R-bands (such as 21q11.2) will show some clustering of sites, but fragments are relatively large, on the order of several 100 to more than 1000 kb. In the very G/C-rich T-bands, such as distal 21q22.3, CpG sites will cluster, and with most CpG-recognizing restriction sites, fragments of identical size will hybridize to the probes, and the fragments tend to be small.

Acknowledgments

The experiments described here were performed in the labs of Richard M. Myers (University of California, San Francisco) and Hans Lehrach (European Molecular Biology Laboratory, Heidelberg, presently Imperial Cancer Reseach Fund, London). I thank them as well as David R. Cox (University of California, San Francisco) and the members of these labs for encouragement, support, and many useful discussions. I also appreciated the critical comments on this manuscript by Marion Buckwalter (University of Michigan, Ann Arbor), Pragna Patel (Baylor, Houston, Texas), and Chris Vulpe (University of California San Francisco).

Further Reading

Many very large PFGE maps have been published. The following references are reviews or examples that represent a variety of approaches to the construction of maps.

Burmeister, M., Monaco, A. P., Gillard, E. F., van Ommen, G. J., Affara, N. A., Ferguson-Smith, M. A., Kunkel, L. M., and Lehrach, H. (1988) A 10-megabase physical map of human Xp21, including the Duchenne muscular dystrophy gene. *Genomics* **2**, 189-202.

Gardiner, K., Horisberger, M., Kraus, J., Tantravahi, U., Korenberg, J., Rao, V., Reddy, S., and Patterson, D. (1990) Analysis of human chromosome 21: Correlation of physical and cytogenetic maps; gene and CpG island distributions. *EMBO J.* **9**, 25-34.

Poustka, A. (1990) Physical mapping by PFGE. *Methods* 1, 204-211.

Burmeister, M., Kim, S., Price, E. R., de Lange, T., Tantravahi, U., Myers, R. M., and Cox, D. R. (1991) A map of the distal region of the long arm of human chromosome 21 constructed by radiation hybrid mapping and pulsed field gel electrophoresis. *Genomics* 9, 19-30.

Compton, D. A., Weil, M. M., Jones, C., Riccardi, V. M., Strong, L. C., and Saunders, G. F. (1988) Long range physical map of the Wilms' tumor-aniridia region on human chromosome 11. *Cell* 55, 827-36.

Fulton, T. R., Bowcock, A. M., Smith, D. R., Daneshvar, L., Green, P., Cavalli, S. L. L., and Donis, K. H. (1989) A 12 megabase restriction map at the cystic fibrosis locus. *Nucleic Acids Res* 17, 271-84.

Herrmann, B. G., Barlow, D. P., and Lehrach, H. (1987) A large inverted duplication allows homologous recombination between chromosomes heterozygous for the proximal t complex inversion. *Cell* 48, 813-25.

Van Ommen, G. J. B., den Dunnen, J. T., Lehrach, H., and Poustka, A. (1990) Pulsed field gel electrophoresis: Applications to long-range genetics, in *Electrophoresis of Large DNA Molecules: Theory and Applications* (E. Lai and B. W. Birren, eds.), Cold Spring Harbor Laboratory Press, Cold Spring Harbor, NY, pp. 133-148.

Poustka, A. M., Lehrach, H., Williamson, R., and Bates, G. (1988) A long-range restriction map encompassing the cystic fibrosis locus and its closely linked genetic markers. *Genomics* 2, 337-45.

References

1. Burmeister, M., Monaco, A. P., Gillard, E. F., van Ommen, G. J., Affara, N. A., Ferguson-Smith, M. A., Kunkel, L. M., and Lehrach, H. (1988) A 10-megabase physical map of human Xp21, including the Duchenne muscular dystrophy gene. *Genomics* 2, 189–202.

2. Hardy, D. A., Bell, J. I., Long, E. O., Lindsten, T., and McDevitt, H. O. (1986) Mapping of the class II region of the human major histocompatibility complex by pulsed-field gel electrophoresis. *Nature* 323, 453–455.

3. Lawrance, S. K., Smith, C. L., Srivastava, R., Cantor, C. R., and Weissman, S. M. (1987) Megabase-scale mapping of the HLA gene complex by pulsed field gel electrophoresis. *Science* 235, 1387–1390.

4. Gardiner, K., Horisberger, M., Kraus, J., Tantravahi, U., Korenberg, J., Rao, V., Reddy, S., and Patterson, D. (1990) Analysis of human chromosome 21: Correlation of physical and cytogenetic maps; gene and CpG island distributions. *EMBO J.* 9, 25–34.

5. Lichter, P., Tang, C.-J., Call, K., Hermanson, G., Evans, G., Housman, D., and Ward, D. C. (1990) High-resolution mapping of human chromosome 11 by in-situ hybridization with cosmid clones. *Science* 247, 64–69.

6. Cox, D. R., Burmeister, M., Price, E. R., Kim, S., and Myers, R. M. (1990) Radiation Hybrid mapping: A somatic cell genetic method for constructing high-resolution maps of mammalian chromosomes. *Science* 250, 245–250.

7. Poustka, A. (1990) Physical mapping by PFGE. *Methods* **1**, 204–211.
8. Antequera, F., Boyes, J., and Bird, A. (1990) High levels of de novo methylation and altered chromatin structure at CpG islands in cell lines. *Cell* **62**, 503–514.
9. Barlow, D. P., and Lehrach, H. (1990) Partial Not I digests, generated by low enzyme concentration or the presence of ethidium bromide, can be used to extend the range of pulsed-field gel mapping. *Technique* **2**, 79–87.
10. Bird, A. P. (1986) CpG-rich islands and the function of DNA methylation. *Nature* **321**, 209–213.
11. Lindsay, S., and Bird, A. P. (1987) Use of restriction enzymes to detect potential gene sequences in mammalian DNA. *Nature* **327**, 336–338.
12. Brown, W. R., and Bird, A. P. (1986) Long-range restriction site mapping of mammalian genomic DNA. *Nature* **322**, 477–481.
13. Pritchard, C. A., Goodfellow, P. J., and Goodfellow, P. N. (1987) Mapping the limits of the human pseudoautosomal region and a candidate sequence for the male-determining gene. *Nature* **328**, 273–275.
14. Estivill, X., Farrall, M., Scambler, P. J., Bell, G. M., Hawley, K. M., Lench, N. J., Bates, G. P., Kruyer, H. C., Frederick, P. A., Stanier, P., Watson, E. K., Williamson, R., and Wainwright, B. J. (1987) A candidate for the cystic fibrosis locus isolated by selection for methylation-free islands. *Nature* **326**, 840–845.
15. Michiels, F., Burmeister, M., and Lehrach, H. (1987) Derivation of clones close to met by preparative field inversion gel electrophoresis. *Science* **236**, 1305–1308.
16. Poustka, A. M., Lehrach, H., Williamson, R., and Bates, G. (1988) A long-range restriction map encompassing the cystic fibrosis locus and its closely linked genetic markers. *Genomics* **2**, 337–345.
17. Fountain, J. W., Wallace, M. R., Bruce, M. A., Seizinger, B. R., Menon, A. G., Gusella, J. F., Michels, V. V., Schmidt, M. A., Dewald, G. W., and Collins, F. S. (1989) Physical mapping of a translocation breakpoint in neurofibromatosis. *Science* **244**, 1085–1087.
18. Gessler, M., and Bruns, G. A. (1988) Molecular mapping and cloning of the breakpoints of a chromosome 11p14.1-p13 deletion associated with the AGR syndrome. *Genomics* **3**, 117–123.
19. Gessler, M., Simola, K. O., and Bruns, G. A. (1989) Cloning of breakpoints of a chromosome translocation identifies the AN2 locus. *Science* **244**, 1575–1578.
20. Julier, C., and White, R. (1988) Detection of a NotI polymorphism with the pmetH probe by pulsed-field gel electrophoresis. *Am. J. Hum. Genet.* **42**, 45–48.
21. Kenwrick, S., Patterson, M., Speer, A., Fischbeck, K., and Davies, K. (1987) Molecular analysis of the Duchenne muscular dystrophy region using pulsed field gel electrophoresis. *Cell* **48**, 351–357.
22. Meitinger, T., Boyd, Y., Anand, R., and Craig, I. W. (1988) Mapping of Xp21 translocation breakpoints in and around the DMD gene by pulsed field gel electrophoresis. *Genomics* **3**, 315–322.

23. Reilly, D. S., Sosnoski, D. M., and Nussbaum, R. L. (1989) Detection of translocation breakpoints by pulsed field gel analysis: Practical consideration. *Nucleic Acids Res.* **17**, 5414.

24. Bodrug, S. E., Ray, P. N., Gonzalez, I. L., Schmickel, R. D., Sylvester, J. E., and Worton, R. G. (1987) Molecular analysis of a constitutional X-autosome translocation in a female with muscular dystrophy. *Science* **237**, 1620–1624.

25. Poustka, A., Pohl, T. M., Barlow, D. P., Frischauf, A. M., and Lehrach, H. (1987) Construction and use of human chromosome jumping libraries from NotI-digested DNA. *Nature* **325**, 353–355.

26. Collins, F. S., Drumm, M. L., Cole, J. L., Lockwood, W. K., Vande, W. G. F., and Iannuzzi, M. C. (1987) Construction of a general human chromosome jumping library, with application to cystic fibrosis. *Science* **235**, 1046-1049.

27. Sambrook, J., Fritsch, E. F., and Maniatis, M. (1989) *Molecular Cloning: A Laboratory Manual.* Cold Spring Harbor Laboratory, Cold Spring Harbor, NY.

28. Bird, A. P. (1989) Two classes of observed frequency for rare-cutter sites in CpG islands. *Nucleic Acids Res.* **17**, 9485.

29. Sealey, P. G., Whittaker, P. A., and Southern, E. M. (1985) Removal of repeated sequences from hybridisation probes. *Nucleic Acids Res.* **13**, 1905–1922.

30. Shmookler-Reis, R. J., Finn, G. K., Smith, K., and Goldstein, S. (1990) Clonal variation in gene methylation: c-H-ras and alpha-hCG regions vary independently in human fibroblast lineages. *Mutat. Res.* **237**, 45–57.

31. Bernardi, G., Olofsson, B., Filipski, J., Zerial, M., Salinas, J., Cuny, G., Meunier, R. M., and Rodier, F. (1985) The mosaic genome of warm-blooded vertebrates. *Science* **228**, 953–958.

32. Holmquist, G. P. (1987) Role of replication time in the control of tissue-specific gene expression. *Am. J. Hum. Genet.* **40**, 151–173.

33. Bickmore, W. A., and Sumner, A. T. (1989) Mammalian chromosome banding—an expression of genome organization. *Trends Genet.* **5**, 144–148.

34. Gardiner, K., Aissani, B., and Bernardi, G. (1990) A compositional map of human chromosome 21. *EMBO J.* **9**, 1853–1858.

35. Dutrillaux, B. (1973) [New system of chromosome banding: The T bands (author's transl)]. *Chromosoma* **41**, 395–402.

36. Ambros, P. F., and Sumner, A. T. (1987) Correlation of pachytene chromomeres and metaphase bands of human chromosomes, and distinctive properties of telomeric regions. *Cytogenet. Cell Genet.* **44**, 223–228.

37. Burmeister, M., Kim, S., Price, E. R., de Lange, T., Tantravahi, U., Myers, R. M., and Cox, D. R. (1991) A map of the distal region of the long arm of human chromosome 21 constructed by radiation hybrid mapping and pulsed field gel electrophoresis. *Genomics* **9**, 19–30.

CHAPTER 19

Two-Dimensional DNA Electrophoresis (2D-DE) for Mammalian DNA

Michael A. Walter and Diane W. Cox

1. Introduction

Pulsed-field gel electrophoresis (PFGE) has been successfully used to generate physical maps of regions of human chromosomes near genes or single copy probes. However, many regions of the human genome contain gene families that often include pseudogenes. Other regions contain stretches of repeats. Conventional PFGE requires a series of single copy probes, which are sometimes difficult if not impossible to find for clusters of homologous genes or repeats. If probes detect several different genes that are dispersed on a chromosome, standard PFGE techniques will not be useful for the generation of large scale restriction maps. To overcome this problem, we have developed a method of two-dimensional DNA electrophoresis (2D-DE). The 2D-DE method consists of two DNA digestion steps and two electrophoretic separation steps (shown in Fig. 1):

1. The first digestion step involves cleavage of high-mol-wt DNA into large restriction fragments by a rare-cutting (first) restriction enzyme.
2. PFGE is then used as the first electrophoresis step to size separate the digested DNA.
3. The DNA, separated by PFGE and still embedded within the agarose gel, is then digested with a second frequent-cutting restriction endonuclease.

From: *Methods in Molecular Biology, Vol. 12: Pulsed-Field Gel Electrophoresis*
Edited by: M. Burmeister and L. Ulanovsky
Copyright © 1992 The Humana Press Inc., Totowa, NJ

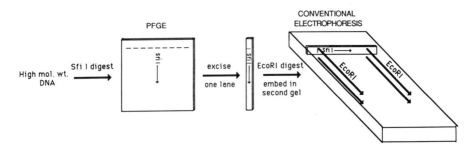

Fig. 1. 2D-DE method. High-mol-wt DNA is digested with a rare-cutting restriction enzyme (e.g., *Sfi* I) and size separated by PFGE. The lane of the LMP agarose PFGE gel is excised and digested with a second, common-cutting restriction enzyme (e.g., *Eco* RI). The lane is then embedded in a second gel and subjected to an electrical field perpendicular to that used in the PFGE apparatus. (The figure is from Walter and Cox *(3)*, *Genomics* **5,** 157–159, reproduced with permission of the publisher; copyright © Academic Press.)

4. The agarose gel slice, now containing DNA digested with both the first and second restriction enzymes, is separated in a conventional gel, in a direction perpendicular to that of the PFGE direction.

After transfer of the DNA to a nylon membrane, DNA probes are then hybridized to the 2D-DE blot. 2D-DE allows the simultaneous mapping, to specific large restriction fragments, of all genes or gene segments detected by a probe, without the need for single copy probes.

The method of 2D-DE has been used to generate a map of the immunoglobulin heavy chain variable (VH) region genes. The VH genes have been subdivided into six families based on DNA homologies *(1,2)*. Each VH gene segment probe detects all members of the relevant VH family. The immunoglobulin VH region is a difficult locus to physically map using conventional PFGE techniques owing to its size—greater than 1400 kbp, and more importantly, to the lack of single copy probes that map within the VH region. Using 2D-DE, a map of 1500 kbp of the human VH region was constructed, indicating the locations of 62 different VH gene segments *(3)*. This example illustrates the potential of 2D-DE for the development of long range restriction maps of gene and repeat families. Our protocol has also been adapted for the study of centromeres, as described in Chapter 20 in this volume.

2. Materials

2.1. Solutions

1. Ficoll-hypaque, room temperature.
2. Phosphate buffered saline (PBS), room temperature.
3. Sodium chloride/EDTA (SE) resuspension buffer—75 mM NaCl, 25 mM EDTA, pH 7.5, room temperature.
4. 1% low melting point (LMP) agarose (Bethesda Research Laboratories, Bethesda, MD), w/v in SE.
5. Lysis buffer (ESP): 0.5M EDTA, 1% N-lauroylsarcosine, 0.5 mg/mL protease K, pH 9.5. (ESP). Make the EDTA / N-lauroylsarcosine solution and sterilize. Add protease K immediately before using.
6. Buffer (TE8): 10 mM Tris-HCl, 1 mM EDTA, pH 8.
7. Sterile water (~0.5 L)
8. 10X loading buffer: 1% Bromphenol blue, 10 mM EDTA, 25% ficoll 400.
9. 1.5 mL of 1X restriction enzyme buffer/2D-DE gel. Owing to the large vol of restriction buffer required for 2D-DE analysis, make 100 mL of the appropriate second enzyme buffer. (Sterilize the solution prior to the addition of β-mercaptoethanol; add the BSA at time of DNA digestion.)
10. 0.1M spermidine trihydrochloride, pH 7. (Add 4 μL/100 mL digest solution.)
11. 20 mg/mL nuclease-free bovine serum albumin (Boehringer, Mannheim).
12. Tris-borate-EDTA (TBE) electrophoresis buffers: 0.5X TBE for the PFGE running buffer; 1X TBE for the conventional electrophoresis running buffer. 10X TBE buffer is 0.09M Tris-borate; 0.002M EDTA.
13. 1% LMP agarose gel in 0.5X TBE (w/v), for a PFGE separation system.
14. 0.8% agarose gel in 1X TBE (w/v), containing 0.8 mg/mL ethidium bromide, for a large conventional electrophoretic system (e.g., Bethesda Research Laboratories, Bethesda, MD).
15. Restriction enzymes. For each 2D-DE gel: rare-cutting (first) restriction enzyme—15–20 U/plug (two plugs; one for the 2D-DE analysis, one for a double digest control lane); frequent cutting (second) restriction enzyme—1000 U to digest PFGE-separated DNA contained within an excised PFGE gel lane, 15–20 U/plug (three plugs; two single digest control lanes, one for the double digest control lane).

16. DNA size markers. Conventional gel size markers: λ phage DNA digested with *Hind* III (e.g., New England Biolabs). PFGE size markers: yeast chromosomes or concatemers of λ phage molecules—*see* Chapter 8 for the methods for construction and isolation of PFGE markers, or purchase commercially available products, e.g., New England Biolabs.

17. 0.5 L of each of the solutions required for DNA transfer (for one 2D-DE gel).
    Acid nicking solution: 0.25 N HCl.
    Denaturing solution: 1.5 M NaCl, 0.4 N NaOH.
    Neutralization solution: 1.5 M NaCl, 0.5 M Tris-HCl, pH 7.6.
    Transfer solution (2X SSC): 300 mM NaCl, 30 mM Na citrate, pH 7.0.

18. Prehybridization and hybridization solutions: 0.5 M phosphate buffer, pH 7.2, 1% BSA (w/v), 7% SDS (w/v), 0.1 M EDTA (4).

19. Reprobing solutions (e.g., for Hybond N nylon membrane).
    Stripping solution: 0.4 N NaOH.
    Neutralization solution: 1.5 M NaCl, 0.5 M Tris-HCl, pH 7.6.

2.2. Equipment

1. Disposable plastic pipet with bulb (one/ DNA sample).
2. Nalgene 8000 tubing (id 0.23 cm), with a 12-mL syringe attached at one end (one/DNA sample), or a PFGE casting mold.
3. Petri dish (one/DNA sample).
4. Sterile glass cover slip (one/DNA sample).
5. 50-mL Falcon tubes (one for each DNA sample).
6. PFGE apparatus (e.g., CHEF electrophoresis system, Bio-Rad, Richmond, CA).
7. Large submarine gel apparatus (e.g., Bethesda Research Laboratories, Bethesda, MD).
8. Paper towels for capillary transfer of DNA (Southern blotting).
9. Nylon membrane (e.g., Hybond N, Amersham, Canada Ltd., Oakville, Ontario).
10. Plastic wrap.
11. 1.5-mL Eppendorf tubes.

3. Method

1. To obtain high-mol-wt DNA from a leukocyte sample, dilute 60 mL of heparinized whole blood approx 1:2 with PBS (60 mL blood : 100 mL PBS) (*see* Note 1 regarding use of leukocyte DNA). Gently layer 40 mL of diluted whole blood onto 10 mL of Ficoll-hypaque in a 50-mL Falcon tube (5). Centrifuge at 2000 rpm (800g), 30 min, at room

temperature. With a plastic pipet, carefully remove the milky white cell layer (10–15 mL) that is between the yellow tinted plasma layer and the red blood cell layer. Be careful not to disturb the red blood cell layer. Go to Step 2. If starting with a lymphoblastoid cell sample, centrifuge cells at 2000 rpm (800g), room temperature, discard supernatant, go to Step 2. (*see* Note 2).

2. Gently add 20 mL of PBS to the cell pellet, resuspend cells, then centrifuge at 2000 rpm for 10 min at room temperature. Do this twice.

3. Resuspend cells in SE, to a concentration of approx 6×10^7 cells/mL SE. Assume an initial concentration of 2×10^6 cells/mL whole blood (*see* Note 3).

4. Mix cell solution with an equal vol of 1% LMP agarose (in SE), to a final concentration of 3×10^7 cells/mL. Attach a 12-mL syringe to the end of the Nalgene 8000 tubing, then draw the agarose/cell solution into tubing. Allow to gel. We find this method of generating PFGE plugs, as suggested by Gardiner et al. (*6*), very useful when dealing with several different DNA samples. However, the use of conventional PFGE molds for plug formation is also suitable.

5. Using the syringe, gently blow the agarose cell solution out of the tubing and into a Petri dish. Cut the agarose into 1/2 cm lengths (plugs) with a sterile glass cover slip. Each plug is approx 30 µL in vol, contains 1×10^6 cells, and is suitable for one lane of a PFGE gel (approx 2 µg DNA).

6. Move the plugs to a 50-mL Falcon tube. Add approx 6 vol of ESP to the plugs in each Falcon tube. Lyse cells for 48 h in ESP at 50°C, in a gently shaking water bath.

7. Remove ESP lysis solution carefully with gentle aspiration. Plugs are very fragile. Add 30 mL of sterile water to rinse plugs. Repeat. Add 30 mL of sterile TE8 twice for 1 h, then once overnight to dialyze plugs. Plugs can now be stored at 4°C in TE8 buffer indefinitely.

8. For the first restriction enzyme digestion of each 2D-DE gel, equilibrate 2 plugs for 1 h in 0.5 mL of 1X restriction endonuclease buffer in an 1.5-mL Eppendorf tube. Remove the equilibration solution and then digest overnight with 15–20 U/plug of a rare-cutting restriction endonuclease (e.g., *Bss*HII, *Eag*I, *Not* I, *Sfi*I) (*see* Note 4). When setting up the digest, take into account the volume of the plug (~ 30 µL) now equilibrated in restriction enzyme buffer. Stop the restriction enzyme digestion of one plug with the addition of 0.1 vol of 10X loading buffer. This plug will be used for the 2D-DE portion of the gel, the second will be digested with the second restriction enzyme (Step 12) and used as a control (*see below*).

9. Embed one plug (the plug that was stopped with loading buffer in Step 8) in a 1% LMP agarose PFGE gel (0.3 cm thick) for each 2D-DE gel required. Multiple lanes in one LMP PFGE gel can be used to generate several 2D-DE gels, using one LMP lane for each 2D-DE gel. Use a PFGE apparatus that results in straight lanes (e.g., the Bio-Rad CHEF electrophoresis system). Place an unloaded second agarose gel on top of the loaded LMP gel during the run to keep the thin LMP gel from floating. *See* Chapters 1–7 for appropriate PFGE separation and running conditions for separation in the desired size range.

10. After PFGE, stain the gel in 2 µg/mL ethidium bromide for 20 min, then destain for 30 min in sterile water.

11. Move the thin gel carefully onto plastic wrap and photograph. Excise the entire lane containing PFGE separated DNA using a UV transilluminator. Each excised lane can be used for one 2D-DE gel (Fig. 1).

12. Equilibrate gel slices for 2 h in 10 mL of 1X restriction enzyme buffer, taking into account the volume occupied by the gel slice. This can be done in a trough made in plastic wrap. Remove 5 mL of buffer and add: 1000 U of the second restriction enzyme (usually a frequent cutting restriction enzyme such as *Eco*RI), 200 µL of 0.1 *M* spermidine trihydrochloride, and 25 µL of 20 mg/mL (100 µg/mL final) nuclease-free bovine serum albumin (*see* Note 5). Do not add spermidine trihydrochloride to low salt restriction enzyme buffers. Digest for 3 h to overnight, then stop with 500 mL of 10X loading buffer.

13. For 2D-DE gel controls, rinse the plug from Step 8 (not stopped with loading dye) in sterile water gently, and then equilibrate this plug and 2 additional plugs (the latter 2 plugs in a separate Eppendorf tube), for 1 h in 0.5 mL of 1X restriction endonuclease buffer. Remove the equilibration solution and then digest the plugs overnight with 15–20 U/plug of the selected frequent cutting restriction endonuclease. When setting up the digests, take into account the volume of each plug (~ 30 µL), now equilibrated in restriction enzyme buffer.

14. Embed the gel slice in a 0.5-cm deep slot in a 20 cm x 20 cm 0.8 % agarose gel containing 0.8 µg/mL ethidium bromide (*see* Fig. 1). Load control lanes of plugs digested with the second enzyme alone, a plug digested with both restriction enzymes (*see* Note 6), and *Hind* III-digested λ DNA size marker. Carry out standard electrophoresis at 50 V for 20 h in a large submarine gel apparatus (*see* Note 7).

15. Photograph gel, then acid nick the gel for 10–20 min in a 1:50 dilution of HCl. Briefly rinse in sterile water, denature for 30 min, and

neutralize 1 h. Transfer to a nylon membrane (e.g., Hybond N, Amersham) in 2X SSC, for >18 h.

16. Bake the membrane for 2 h at 80°C, then crosslink with UV light (365 nm, intensity 7000 $\mu W/cm^2$ at the filter surface) for 1 min.

17. Prehybridize for 1 h and hybridize at 65°C for 18–24 h with 1 x 10^6 cpm of random primed probe/mL of hybridization solution.

18. Wash the membrane in 2X SSC for 30 min, and once for 1 h in 0.1X SSC, 0.1% SDS at 50–65°C. Expose to film at –70°C for up to 2 wk, with two Dupont Lightning Plus intensifying screens.

19. Develop autoradiogram. Use positions of the fragments in the control lanes to determine the identity of dots in the 2D-DE portion of the autoradiogram (*see* Note 8). A typical result is shown in Fig. 2.

20. To reprobe the 2D-DE filter, agitate the nylon filter (e.g., Hybond N) in stripping solution, 20–30 min at room temperature. Neutralize for 30 min at room temperature. Prehybridize, then reprobe.

21. Generate a map:
    a. Determine the size of PFGE fragments by comparison with the markers (*see* Note 9).
    b. Localize dots on the final autoradiogram to specific PFGE fragments (*see* Note 10). An example of map generation is shown in Fig. 2 (*see* Note 11).
    c. Use partial digestion to confirm assignment of the genes or repeat segments (*see* Notes 12 and 13).

4. Notes

1. The compression zone appears to contain DNA of two types: large restriction fragments not separated by the PFGE conditions used in the first electrophoretic separation step, and fragments that are the result of partial digestion of restriction enzyme sites. The latter type appears to be particularly common in DNA isolated from peripheral blood, possibly as a result of the mixture of different cell types contained in this DNA source. Therefore, 2D-DE analysis of DNA from peripheral blood will often result in the apparent discovery that all gene or repeat segments observed in conventional electrophoresis map to a fragment within the compression zone, as well as to different large restriction fragments.

2. Appropriate DNA sources should be selected for 2D-DE analysis to avoid problems caused by genetic polymorphism. A common problem complicating the interpretation of results in PFGE mapping is heterozygosity of restriction enzyme sites within a DNA source. Genetic polymorphism is a concern especially in 2D-DE analysis since

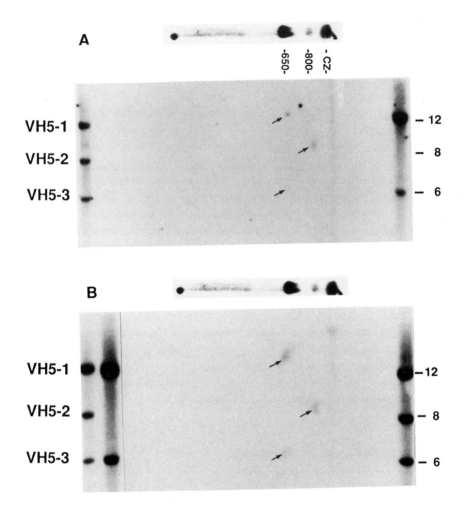

Fig. 2. Hybridization of a VH5 probe to an *Sfi* I/*Bgl* II 2D-DE gel, showing the effects of increased DNA concentration on hybridization intensity.
A: 2 μg of DNA was analyzed by 2D-DE. **B**: 4 μg of DNA was similarly
analyzed by 2D-DE. PFGE direction of migration was right to left, conventional electrophoresis direction was top to bottom. CZ - compression zone.
Control lanes of *Bgl*II-digested DNA were loaded to the left of the 2D-DE
lanes for comparison of fragment migration. A lane of *Sfi* I-digested, and
PFGE-separated DNA, hybridized with the VH5 family probe (2-V) was
aligned at the top of each 2D-DE gel for comparison. VH gene segments
are identified to the left. Numbers at the top and to the right of the 2D-DE
gel indicate the sizes, in kbp, of DNA fragments in PFGE and conventional
electrophoresis, respectively. Arrows indicate hybridization signals.

large regions of DNA are often surveyed in the generation of restriction maps of complex gene or repeat families. DNA sources which minimize genetic polymorphism are those from individuals or cell lines hemizygous in the region of interest, or from hydatidiform mole cell lines *(7)*.

3. Using more DNA in the PFGE direction will increase the hybridization signals (and decrease exposure times) upon probing a 2D-DE filter, but the dots will often become larger as well as darker (*see* Fig. 2). For precise matching of the dots detected in 2D-DE separated DNA with the fragments in the control lanes, use 2–3 µg of DNA for 2D-DE analysis. Increased initial DNA concentration may also be used to enhance the detection of large restriction fragments that are the result of incomplete or partial digestion with the first restriction enzyme.

4. By PFGE analysis, determine a set of useful rare-cutting restriction enzymes to use for the physical map. These should include a restriction enzyme that divides the region to be mapped into a small number of large fragments.

5. Use the same second restriction enzyme consistently with various first enzymes. This will allow easier reprobing of filters and reduce problems that are caused by genetic polymorphism. We have used *Eco* RI, *Hind* III, *Bgl* II, *Rsa* I, *Pst* I (NEBL) satisfactorily as second restriction enzymes.

6. Rare-cutting cleavage sites may be located in the fragments identified in conventional DNA electrophoresis. Although rare-cutting restriction enzyme sites are by definition infrequent, the large region of DNA surveyed by 2D-DE increases the likelihood of discovery of such sites. Appropriate double digest control lanes will help eliminate confusion that can result from the absence or alteration of restriction fragments normally detected in conventional electrophoresis.

7. If problems are encountered in achieving complete second enzyme digestion, first check to make sure that DNA-grade LMP agarose was used for the PFGE gel (we have had problems when we used non-LMP agarose) and that the vol of the gel slice was included in the calculations of second enzyme digest vol. If problems still persist, rinse the excised PFGE lane in 100 mL of sterile water for 1 h prior to incubation in 1X restriction enzyme buffer; double the BSA concentration for the second enzyme digestion to 200 µg/mL; change restriction enzyme suppliers. We find that addition of spermidine trihydrochloride in high salt restriction buffers decreases partial digestion, but results in problems (e.g., apparent DNA degradation) when included in low salt buffers.

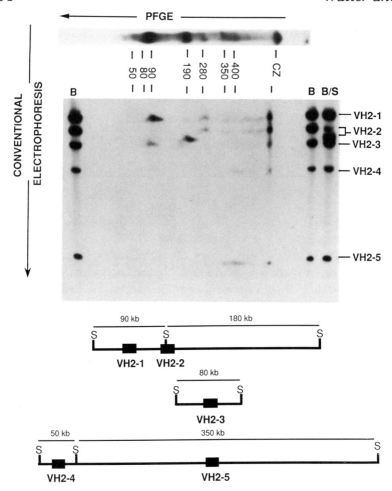

Fig. 3. Hybridization of a VH2 probe to an *Sfi* I/*Bgl* II 2D-DE gel, showing the effect of partial digestion of DNA with the first, rare-cutting, restriction enzyme. DNA was partially digested with *Sfi* I in the first enzymatic digestion step, separated by PFGE, and then completely digested with *Bgl*II in the second enzyme digestion step. CZ - compression zone. Control lanes were loaded to the right and left of the 2D-DE lane for comparison of DNA fragment migration. B - control lane of *Bgl*II-digested DNA, B/S - control lane of *Sfi* I/*Bgl*II double-digested DNA. VH2 gene segments are identified to the right. Numbers at the top and to the left of the 2D-DE gel indicate the sizes, in kbp, of DNA fragments in PFGE and conventional electrophoresis, respectively. A lane of DNA, partially digested with *Sfi* I, and PFGE separated, was aligned at the top of 2D-DE for comparison. Hybridization of the VH2 probe (VH2EB1.2) to the 2D-DE blot revealed that while the VH gene segments VH2-1 and VH2-2 lie on 180

8. 2D-DE-separated DNA migrates slightly slower than the DNA in the control lanes since the same amount of DNA in the 2D-DE as is in the control lanes is spread over a large area in the 2D-DE region. The lower concentration of DNA within the 2D-DE portion of the gel than in the control lanes causes slightly different migration speeds through the agarose in the 2D-DE and conventional portions of the gel.

9. Determine the sizes of the rare-cutting restriction enzyme fragments from PFGE gels, not from 2D-DE gels. Slight alterations in the migration of fragments in the second direction of 2D-DE, as a result of small fluctuations in electrical field strengths in a conventional gel apparatus, could result in inaccurate fragment size estimates.

10. Avoid analysis of more than 30 different fragments on the same 2D-DE gel. Larger fragments in the second direction will be difficult to identify consistently between gels. If necessary, do two 2D-DE gels, one to separate smaller fragments in the second direction (e.g., 5 kbp) and a second 2D-DE gel to separate larger (>10 kbp) fragments. Also note that analysis of large numbers of fragments (>30) requires longer exposure times.

11. For the VH region, we initially assigned VH gene segments to several large fragments using the rare-cutting restriction enzyme *Not* I as the first enzyme in the 2D-DE technique *(7)*. The restriction enzymes *Sfi* I and *Bss*H II were determined by PFGE analysis to each divide the VH region into well-separated fragments, between 50 and 600 kbp in size. We then determined the *Sfi* I and *Bss*H II fragments of the VH gene segments (now subdivided by *Not* I fragments). *Not* I/*Sfi* I, *Not* I/*Bss*H II, and *Sfi* I/*Bss*H II double digests were then carried out, PFGE separated, and analyzed by 2D-DE. Examination of this double digest information, together with the previous assignment of the VH gene segments to specific *Not* I, *Sfi* I, and *Bss*H II fragments, allowed the generation of a physical map of 1500 kbp of the VH region including the locations of 62 different VH gene segments *(7)*.

---

kbp and 90 kbp *Sfi* I fragments, respectively, they also detect a common 270–280 kbp partial *Sfi* I fragment. Interestingly, the *Sfi* I site that was only partially cleaved lies within the VH2-2 12 kbp *Bgl*II fragment. Therefore, the 180 kbp and 90 kbp *Sfi* I fragments are adjacent, as indicated in the schematic diagram at the bottom of the figure. Similarly, VH2-4 and VH2-5, lying on 50 kbp and 350 kbp *Sfi* I fragments respectively, detect a common partial *Sfi* I fragment of 400 kbp and are therefore also on adjacent *Sfi* I fragments.

12. Partial digestion in the first restriction enzyme dimension is useful to confirm the assignment of gene or repeat segments, normally detected in conventional electrophoresis, to common large restriction fragments. If the gene or repeat segments map to the same-sized large fragment in 2D-DE analysis of complete second enzyme digests, and also to the same partial digest band in incomplete second enzyme digestion, the fragments normally detected in conventional electrophoresis are on the same large restriction fragment. In addition, partial digestion information can allow the assignment of gene or repeat segments to adjacent large fragments, if different fragments detected in complete second enzyme digestion share a common partial restriction fragment (*see* Fig. 3 on p. 294).

13. Gene or repeat segments can map to two different fragments simultaneously, in 2D-DE analysis, if partial digestion occurred in the first enzyme digestion, or if the single gene or repeat segment is instead two or more segments of identical second enzyme restriction fragment size. 2D-DE analysis of deliberate partial first enzyme digestion may allow determination of whether or not the first alternative is true. If two or more fragments are consistently detected in 2D-DE analysis with several different first enzymes, the single gene, or repeat segment, is instead two or more segments of identical second enzyme restriction fragment size.

Acknowledgments

We thank Adam Chen and Louise Sefton for helpful discussion. The VH5 probe 2-V was a gift of F.Alt, Columbia University. The VH2 family probe VH2EB1.2 was derived from VCE-1, the latter a gift of T. Honjo, Kyoto University. This work was funded by the National Science and Engineering Research of Canada and by a Medical Research Council studentship to M. A. W.

References

1. Berman, J. E., Mellis, S. J., Pollock, R., Smith, C. L., Suh, H., Heinke, B., Kowal, C., Surti, U., Chess, L., Cantor, C. L., and Alt, F. W. (1988) Content and organization of the human Ig VH locus: Definition of three new VH families and linkage to the Ig CH locus. *EMBO J.* **7**, 727–738.

2. Schroeder H. W., Hillson, J. L., and Perlmutter, R. M. (1987) Early restriction of human antibody repertoire. *Science* **238**, 791–793.

3. Walter, M. A. and Cox, D. W. (1989) A method for two dimensional DNA electrophoresis (2D-DE): Application to the immunoglobulin heavy chain variable region. *Genomics* **5**, 157–159.

4. Church, G. M. and Gilbert, W. (1984) Genomic sequencing. *Proc. Natl. Acad. Sci. USA* **81,** 1991–1995.

5. Boyum, A. (1968) Isolation of lymphocytes, granulocytes and macrophages. *Scand. J. Invest.* **5** *(Suppl.)*, 9–15.

6. Gardiner, K., Laas, W., and Patterson, D. (1986) Fractionation of large mammalian DNA restriction fragments using vertical pulsed-field gradient gel electrophoresis. *Som. Cell Mol. Genet.* **12,** 185–195.

7. Walter, M. A., Surti, U., Hofker, M. H., and Cox, D. W. (1990) The physical organization of the human immunoglobulin heavy chain gene complex. *EMBO J.* **9,** 3303–3313.

CHAPTER 20

Pulsed-Field and Two-Dimensional Gel Electrophoresis of Long Arrays of Tandemly Repeated DNA

Analysis of Human Centromeric Alpha Satellite

Peter E. Warburton, Rachel Wevrick, Melanie M. Mahtani, and Huntington F. Willard

1. Introduction

Complex genomes are characterized by large amounts of tandemly repeated DNA that can comprise up to several percent of the genome in some organisms *(1,2)*. The analysis of the organization of this type of DNA presents certain challenges owing to its repetitive nature, genomic distribution, and large array size. The availability of the large-scale resolution of pulsed-field gel electrophoresis (PFGE) *(3,4)* has allowed an increased understanding of the genomic organization of long arrays of tandemly repeated DNA, including their overall size and internal polymorphic variation. Such analyses are useful for long-range physical mapping of the large blocks of repetitive DNA characteristic of complex genomes and allow genetic information to be obtained for these loci. Although described here for human centromeric alpha satellite DNA, these techniques are also applicable to other repetitive and multicopy DNA families.

From: *Methods in Molecular Biology, Vol. 12: Pulsed-Field Gel Electrophoresis*
Edited by: M. Burmeister and L. Ulanovsky
Copyright © 1992 The Humana Press Inc., Totowa, NJ

A two-dimensional gel electrophoretic technique that allows redigestion and resolution of PFGE fragments with a second restriction enzyme can be useful for the analysis of multicopy DNA families, within which probes can hybridize to large or multiple genomic locations *(5,6)*. Large restriction fragments are resolved by Contour-Clamped Homogeneous Electric Field (CHEF) Electrophoresis *(7)* (in the first dimension) and subsequently further digested into smaller fragments while still immobilized in the original CHEF gel agarose. The resulting digestion fragments are then run on conventional gel electrophoresis in the orthogonal direction (second dimension) to the first CHEF electrophoresis, effectively taking inventory of the family members contained in each large CHEF fragment. Alternatively, the secondary digestion products can also be run on a second CHEF gel that allows placement of rare restriction sites within the initial CHEF fragments. The resulting two-dimensional gels reveal information about the internal organization of these arrays not readily observable by either single-dimension technique alone *(6)* and can be used to generate long-range maps of multicopy DNA (ref. *5*; Mahtani and Willard, unpublished).

2. Organization of Arrays of Tandemly Repeated Satellite DNA

Arrays of tandemly repeated DNA are often composed of up to several thousand individual repeat units arranged in a head-to-tail fashion. Repeat units from different DNA families can range from only several nucleotides to several thousand bp in length, forming arrays as large as several million bp *(8)*. Often, members of a repetitive DNA family are found in several different arrays on multiple chromosomes throughout the genome *(9–13)*. Individual repeat units within an array can be greater than 99% homologous to one another *(14)*. However, repeat units from different chromosomes are generally sufficiently diverged to allow cloned repeat units to be used as specific hybridization probes for a particular array at high stringency *(11)*.

One such family is alpha satellite DNA, which is found at the centromere of each human chromosome *(15,16)*. Each centromere is characterized by a distinct linear arrangement of diverged 171 bp monomeric repeats, forming chromosome-specific higher-order repeat units that are in turn tandemly repeated to form each centromeric array *(17)*. These higher-order repeat units can consist of tens of monomers (ranging, therefore, up to several kbp in length) and form tandem arrays spanning millions of bp *(18–20)*. Long-range mapping by the techniques described in this chapter have allowed the sizing and polymorphic analysis of the chromosome-specific subset of alpha satellite from at least 11 different chromosomes (*see* Table 1).

These techniques have also aided in the analysis of other tandemly repeated DNA families that are unrelated to each other in primary sequence, and are found at various locations throughout genomes, often in large heterochromatic blocks. Individual cloned members from several of these abundant DNA families have been isolated and used as probes specific for certain regions and arrays. β-satellite, which is based on a 68 bp monomer, forms arrays up to several hundred kb. Specific subsets have been localized to chromosome 9, as well as to the short arms of the acrocentric chromosomes *(21,42)*. Cloned copies of human satellites 2 and 3, which are specific for the heterochromatic regions on chromosomes 16qh and 9qh, have been shown to hybridize to large PFG fragments *(22)*. Several repeat families from the human Y chromosome *(23,24)*, as well as a human midi-satellite described on chromosome 1 *(25)*, have also yielded to such analysis. In general, as different families of tandemly repeated DNA are isolated and examined, long-range analysis using the techniques described in this chapter will likely be an important aspect in the understanding of their genomic organization.

Chromosome-specific alpha satellite DNA probes have been isolated from well over half the human chromosomes *(17)*. The high degree of homology between repeat units from different chromosomes requires careful consideration of the stringency conditions used for chromosome-specific hybridization. Conditions can be obtained that allow varying degrees of hybridization mismatch, depending on temperature, salt and formamide concentrations at hybridization, and salt concentration and temperature of subsequent filter washes. In general, chromosome-specific hybridization with cloned alpha satellite probes utilizes high stringency conditions that theoretically require close to 100% identity (*see* Section 4.1.2. and Note 10) *(11,26)*. As stringency is reduced, other subsets of alpha satellite DNA from other chromosomes can be detected, until at low stringency, essentially all alpha satellite DNA can be detected with most hybridization probes *(11,16)*.

Table 1
Pulsed-Field Gel Enzymes for Alpha Satellite Arrays

Chromo- some	Locus	PFGE Enzyme	Array size (kb) Range	n*	ref.
1	D1Z5	*BamH* I, *Bgl* II, *Pvu* II	440–1510	8	18
3	D3Z1	*BamH* I, *Bgl* II *Xba* I	2400–2900	6	33,[a]
7	D7Z1	*BamH* I, *Bgl* II, *Pvu* II, *Xba* I	1530–3810	10	18
7	D7Z2	*BamH* I, *Bgl* II, *Pvu* II, *Xba* I, *BstE*	100–550	23	18,[a]
10	D10Z1	*BamH* I, *Bgl* II, *Xba* I	1390–2515	8	18
11	D11Z1	*Bgl* II, *Pvu* II	1960–4760	8	18
12	D12Z1	*Bgl* I	2250–4300	23	41
13/21	D13Z1, D21Z1	*Hind* III, *Sph* I, *Sst* I	>800–2000	6	[a]
16	D16Z2	*BamH* I, *Bgl* II, *Xba* I	430–1805	8	32
17	D17Z1	*BamH* I, *Bgl* II, *Apa* I, *Kpn* I, *BstE* II	1165–3710	11	6, 18
X	DXZ1	*Bgl* I, *Kpn* I, *Bgl* II, *Sca* I	1380–3730	37	20,[a]
Y	DYZ3	*BamH* I, *Bgl* II, *Pvu* II	200–1500	50	19,24,[a]

*Number of arrays estimated
[a]Unpublished data

2.1. Choice of Restriction Enzyme
for PFGE Analysis of Tandem Arrays

Analysis of these large arrays of alpha satellite DNA by PFGE requires digestion with an appropriately chosen restriction enzyme. Some restriction enzyme sites are found one or more times in each higher-order repeat unit on a certain chromosome. Upon digestion of genomic DNA with such enzymes, most or all of the tandem array is released into multiple characteristic repeat unit size fragments (Fig. 1A), that can then be resolved by conventional agarose gel electrophoresis (17). Because of the small fragments produced by such higher-order repeat enzymes, these enzymes are not suitable for PFGE. On the other hand, based on genomic restriction maps or sequenced copies of higher-order repeat units, other restriction enzymes are predicted to not cut in individual higher-order repeat units of certain arrays, and are therefore predicted to not cut within the overall tandem array (18,19). Restriction enzymes that cut single-copy genomic DNA into average size fragments of 15 kb or less (common cutting

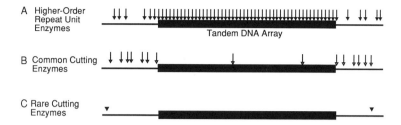

Fig. 1. Schematic of different types of restriction enzymes for digestion of tandem DNA arrays. **A:** Higher-order repeat enzymes cut at least once per repeat unit, giving multiple small fragments. **B:** Common cutting enzymes that do not cut repeat units, and therefore release tandem arrays in large fragments. Generally the recognition sites have 4–6 bp and no CpG. **C:** Rare-cutting enzymes that cut genomic DNA infrequently. These enzymes also release tandem arrays in large fragments, but include large amounts of flanking DNA. Generally their recognition sites have 6–8 bp and contain a CpG.

enzymes, e.g., *Bam*H I, *Xba* I; *see* Fig. 1B) provide a means to estimate the size of tandem DNA arrays by PFGE. Such enzymes are predicted to cut within a few kb of the edge of the array in the flanking genomic DNA, giving PFGE fragments that can be added together in order to estimate array size *(8)*. Restriction enzymes that cut the human genome into fragments on the order of 50 kb and greater (rare cutting enzymes, e.g., *Mlu* I and *Sfi* I; *see* Fig. 1C) would also release entire arrays in large fragments (if they do not cut the higher-order repeat unit). However, these enzymes are generally not useful for size estimates of arrays due to the large amount of nonrepetitive flanking DNA that might be included in the fragments (Fig. 1) *(19,27)*.

The differences between these two categories of restriction enzymes in the analysis of alpha satellite DNA is illustrated in Fig. 2. CHEF electrophoresis was performed on human male genomic DNA that had been digested with both common and rare-cutting enzymes. A Southern blot of the gel was hybridized to a cloned repeat unit pBamX 7 *(28)* that is specific for the X chromosome alpha satellite array at high stringency. The size estimates for this alpha satellite array obtained with the common cutting enzymes *Kpn* I, *Bgl* I, and *Bst*E II are in close agreement, at approx 3000 kb. (In this case, a polymorphic *Bst*E II site is located within the array; the size estimate from this enzyme is obtained by summing the two resulting CHEF bands.)

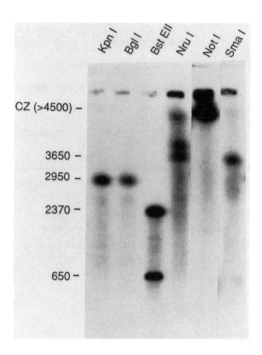

Fig. 2. CHEF analysis of X chromosome alpha satellite DNA with common and rare-cutting enzymes. High-mol-wt human male genomic DNA was digested with common and rare-cutting restriction enzymes. *Kpn* I, *Bgl* I, and *Bst*E II all release fragments totaling about the same size, which give an estimate of the array size. On the other hand, *Nru* I, *Sma* I, and *Not* I release larger fragments that presumably include flanking genomic DNA. The *Nru* I digest shown here is only partially complete. Filter hybridized to pBamX7 *(28)* at 53°C, with a final filter wash at 68°C in 0.1X SSC/0.1% SDS. Gel conditions: 0.8% agarose, 48 h, 150 V, 120–200 s ramped pulse time.

Further examples of the use of such enzymes in estimation of alpha satellite array sizes from the X and other chromosomes have been described *(6, 18–20)*. On the other hand, digestion with the rare-cutting enzymes *Nru* I, *Not* I, and *Sma* I gives a broad range of band sizes that decrease with the predicted frequency of the recognition site in the genome; indeed, the *Not* I fragment is too large to be resolved on this gel.

Table 1 shows common cutting enzymes that have proven useful in the analysis of 12 different alpha satellite DNA arrays. The enzymes *Bgl* II and *Bam*H I are useful for most (but not all) arrays. In general, restriction enzymes with a six base recognition sequence, where four out of six nucleotides are G or C (but not CpG), are most often found to be useful, which may reflect the A/T rich nature of alpha satellite DNA. Table 1 also shows the current range of sizes for 12 alpha satellite arrays, to be used as a guideline when choosing resolution parameters.

2.2. High Frequency Array Length and Restriction Site Polymorphisms

Mechanisms such as unequal crossing over *(29)* and molecular drive *(30)*, acting on tandem DNA arrays, can lead to the spread and fixation of polymorphic repeat variants within arrays and, further, to the contraction and expansion of overall array lengths within a population *(6,8)*. Certain arrays of tandemly repeated DNA are characterized by a distribution of total sizes and of polymorphic variation through a population, such as is seen in human minisatellite VNTRs *(31)*. With the advent of PFGE technology, an analysis of the variation in total size of larger tandem arrays has become possible *(18,32,33)*.

Although certain restriction enzymes are predicted to not cut within higher-order repeat units of some alpha satellite DNA arrays, sequence variation between repeat units can lead to the presence of polymorphic restriction sites within an array. Some arrays are characterized by a high percentage of repeat units containing polymorphic sites that can be detected on conventional gel electrophoresis *(34)*. Other polymorphic restriction sites are found in only a few higher-order repeat units and are thus detected infrequently in the arrays. Digestion with such enzymes cut the arrays into several large fragments that can be summed to estimate the size of the array (*see* Fig. 2). The presence and position of these polymorphic restriction sites are highly variable between different arrays. Indeed, in our experience, nearly all homologous centromeres show a different PFGE banding pattern with such enzymes. These high frequency polymorphisms, coupled with the distribution of overall array sizes in the population, can be used to genetically distinguish homologous centromeric arrays *(18,20,24,27)*.

The inheritance pattern of such PFGE polymorphisms for the alpha satellite array from chromosome 7 is shown in Fig. 3. High-mol-wt DNA from a three-generation family was digested with *Bgl* II, resolved on CHEF electrophoresis, Southern blotted, and hybridized to the chromosome 7-specific probe pMGB7 *(35)*. The Mendelian inheritance of the bands can be seen in this family; for example, band C can be followed from the paternal grandmother, through the father, to four of his children.

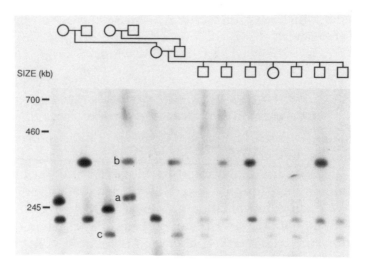

Fig. 3. Mendelian inheritance of polymorphic variation on pulsed-field gel electrophoresis. DNAs from a large three-generation family were digested with *Bgl* II, size fractionated on a CHEF gel, Southern blotted, and probed with pMGB7 *(35)*, that detects the centromeric alpha satellite locus D7Z2 on chromosome 7. Several bands are present in each lane, and others have run off the bottom of the gel. By analyzing the inheritance pattern of each fragment, the genotype of each person in this pedigree can be determined. For example, the band labeled "a" present in the paternal grandfather is not inherited by the father. The band labeled "b" is inherited by the father, who passes it on to three of his children. The other four children inherit the fragment from the paternal grandmother (labeled "c"). If there is any doubt as to genotype, a second enzyme can be used and the analysis repeated. Hybridization at 53°C, final filter wash at 65°C in 0.1X SSC/0.1% SDS gel conditions: 1% agarose, 24 h, 200 V, 10–80 s ramped pulse time.

2.3. Two-Dimensional Gel Analysis

A major difficulty in the detailed analysis of repetitive and multi-copy DNA families is the fact that DNA probes often hybridize to entire arrays and/or multiple locations in the genome, because of the homogeneous nature of their sequence. The two-dimensional technique, shown schematically in Fig. 4A, allows the separate analysis of individual PFGE fragments, despite hybridization to other fragments. Thus, localized information on the organization of tandem repeats within distinct regions of the array can be obtained. Knowledge of

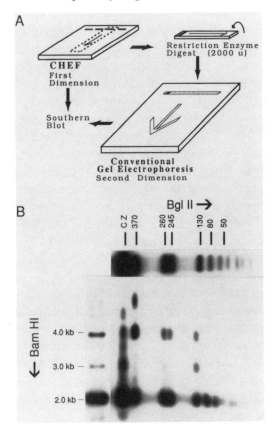

Fig. 4. Two-dimensional gel electrophoresis. **A:** Schematic representation of two-dimensional gel technique. CHEF electrophoresis is run in the first dimension. The lane of interest is excised and the DNA slice is redigested by bathing the gel slice in a solution containing restriction enzyme. The resulting agarose is embedded in conventional gel, and run in an orthogonal direction to the original CHEF gel. **B:** Two-dimensional analysis of chromosome X alpha satellite. The first dimension consists of a *Bgl* II digest (across the top of the gel). The second dimension consists of redigestion with *Bam*H I. A Southern blot of the gel is hybridized to pBamX7 as in Fig. 2. Gel conditions-First dimension: 1% agarose, 15 h, 200V, 20 s constant pulse time. Second dimension: 1.1% agarose, 60 V, 14 h.

polymorphic sites within the array (from conventional gels) can be used in conjunction with PFG maps to allow identification of multiple fragments that all hybridize to the same probe *(5,6)*.

Figure 4B shows an example of a two-dimensional gel hybridized to the X chromosome-specific alpha satellite repeat pBamX7 (28). High-mol-wt DNA from a mouse/human somatic cell hybrid that contains a single X chromosome was digested with *Bgl* II, which cuts this particular X chromosome alpha satellite array into 11 large fragments, and resolved on CHEF electrophoresis (shown across the top of Fig. 4B). The CHEF lane was excised from the gel, redigested with *Bam*H I, which cuts the X chromosome array into higher-order repeat units, and run on conventional gel electrophoresis. This gel clearly shows that different variants of alpha satellite repeat units, here illustrated by 3 kb and 4 kb repeat units, are found in distinct localized domains within the array. For example, the 245 kb *Bgl* II fragment contains only 2 and 4 kb bands, whereas the 130 kb *Bgl* II fragment contains 2, 3, and 4 kb bands. Similar observations have been made for chromosome 17 alpha satellite DNA (6).

Two-dimensional gel analysis can reveal genomic organization by several different strategies. PFGE fragments from digestion with the same enzyme in the first dimension can be digested with different enzymes or combinations of enzymes in the second dimension, that could allow short-range restriction maps of each separate PFGE fragment to be constructed. On the other hand, different restriction enzymes or combinations of enzymes can be used for digestion in the first dimension, and the same polymorphic enzyme used in the second dimension. By comparison of polymorphic repeat units or restriction fragments in the second dimension, large CHEF fragments in the first dimension can be joined together on a map (5,6). Further, by running both the first and second dimensions on CHEF gels, restriction sites in the second dimension can be placed directly in the first dimension fragments.

Figure 5 shows such a two-dimensional CHEF gel, in which DNA from a mouse/human somatic cell hybrid that contains a single human chromosome 17 was digested with *Bgl* II in the first dimension and redigested with *Bam*H I in the second dimension. The resulting Southern blot was hybridized to the chromosome 17-specific alpha satellite probe p17H8 (36). *Bam*H I sites can be placed within the 950, 375, and 275 kb *Bgl* II fragments, because these three fragments are seen to be cut by *Bam*H I in the second dimension. Similar conclusions were previously drawn (for the same chromosome 17 containing hybrid) from two-dimensional gels employing a different strategy (6). Both a *Bgl* II digest and a *Bgl* II/*Bam*H I double digest were run in the

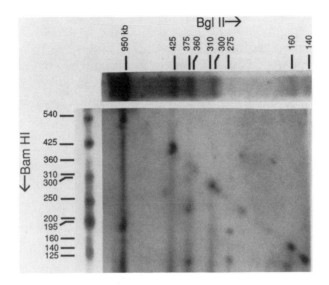

Fig. 5. Two-dimensional CHEF electrophoresis of chromosome 17 alpha satellite DNA. High mol wt DNA from a mouse/human somatic cell hybrid containing a single chromosome 17 (LT23-4C, *see* ref. *6*) was digested with *Bgl* II and subjected to CHEF electrophoresis. The lane was excised, redigested with *Bam*H I, and subjected to CHEF electrophoresis in the second dimension. Subsequent Southern blot filter was hybridized to p17H8 *(36)* at 53°C, with a final filter wash at 65°C in 0.5*M* NaCl.

first dimension, and each was subsequently digested in the second dimension with *Eco* RI, which reveals polymorphic repeat types. By following the different types of *Eco* RI defined repeat units contained in the large PFGE fragments from the single *Bgl* II digest to the bands seen in the *Bgl* II/*Bam*H I double digest, placement of the *Bam*H I sites within the *Bgl* II fragments was also possible *(6)*. Thus, different but complementary approaches using two-dimensional gels can be used in conjunction to confirm and extend results, illustrating the flexibility of this technique.

3. Materials

3.1. Pulsed-Field Gel Electrophoresis

1. Materials and methods for agarose block preparation for high-mol-wt mammalian DNA are required (*see* Chapter 8) *(37,38)*.
2. CHEF electrophoresis apparatus (CHEF-DR II, Bio-Rad- ~33 cm between electrodes) and associated materials and methods (*see*

Chapters 2 and 3), including gel pouring trays, combs, 0.5X TBE (45 mM Tris, 45 mM Boric acid, 1 mM EDTA, pH. 7.6), LE Agarose (Seakem) are required *(7) (see* Note 2).

3. Solutions and materials for Southern blotting and hybridization of filters at high and low stringency conditions *(11,39,40,* and *see* Section 4.1.).

3.2. Additional Materials for Two-Dimensional Gel Electrophoresis

4. Large horizontal conventional gel electrophoresis chamber (i.e., International Biotechnologies, Inc. New Haven, CT, Model HRH-~40 cm between electrodes), Tris Acetate buffer (1X: 40 mM Tris, 13 mM sodium acetate trihydrate, 2 mM EDTA, pH 8.0).

5. Restriction enzyme buffer ingredients, including 1 M spermidine; 50 mg/mL bovine serum albumin (BSA) (DNAse free); 0.5 M DTT; 5 M NaCl; 1 M KCl; 1 M MgCl$_2$; 1 M Tris-Cl (pH 7.4, 8.6); 0.5 M EDTA, and restriction enzymes *(see* Note 3).

6. Small (6.5" × 8") sealable freezer bags, 150-mm petri dishes, rotating table, shaking heated water bath, parafilm, tools to cut and manipulate agarose gels *(see* Note 3).

4. Methods

4.1. Pulsed-Field Gel Electrophoresis and Hybridization Stringency

1. Preparation of high mol wt DNA in agarose blocks for restriction enzyme digestion is described in Chapter 8, this vol. *(37,38).* In general, PFGE and associated protocols, such as preparation of high-mol-wt mammalian DNA and Southern blotting, are standard and present no special considerations in the technique for resolution of tandem arrays. The main consideration is the choice of restriction enzymes useful for tandem arrays *(see* Section 4.2. and refs. *8, 18,* and *19).*

2. High stringency conditions are required for chromosome-specific hybridization of alpha satellite probes. Low stringency will allow cross-hybridization to most or all other alpha satellite subsets (depending on the probe used). High stringency conditions include hybridization in 3X SSC, 50% formamide *(40)* at approx 53°C, and final filter washing in 0.1X SSC/0.1% SDS at 65–70°C. (The actual temperature best suited for a particular experiment must be determined empirically, *see* Note 10.) Low stringency conditions include hybridization in above solution at 42°C, and final filter washing in 0.5 M NaCl at 65°C.

4.2. Guidelines for Selection of Restriction Enzyme

General guidelines for selection of a restriction enzyme for analysis of a particular alpha satellite array are as follows. By extension, similiar considerations apply to other families of tandemly repeated DNA. It is assumed that a cloned copy of a higher-order repeat unit suitable for use as a chromosome-specific probe is available.

1. Obtain a restriction map of the higher-order repeat unit found on the chromosome of interest, e.g., p17H8 *(36)*, pBamX7 *(28)*. A restriction map from genomic DNA should also be obtained by digestion of genomic DNA, Southern blotting, and hybridization to the cloned probe in order to assure that the cloned probe is representative of the entire array.

2. If available, examine the sequence of the cloned repeat unit for restriction enzyme sites and for positions that have one bp mismatch to restriction enzyme sites. Computer programs that locate restriction sites such as MAPSORT, available in the University of Wisconsin Package, are extremely valuable. It is important to point out that the sequence of a single cloned copy may not be representative of the other repeat units in the same array.

3. Select a set of common cutting enzymes (*see* Fig. 1) that do not appear on the restriction map of the higher-order repeat unit. Enzymes whose recognition site are repeatedly found within one bp mismatch in the higher-order repeat unit generally cut more frequently in the array, and may be less useful for sizing arrays but suitable for polymorphism analysis.

4. In general, the enzymes chosen must be evaluated by digestion of high-mol-wt genomic DNA and several different ranges of size resolution on pulsed-field gels. We have not found a way to predict beyond these general guidelines which enzymes will be useful for individual arrays. Table 1 provides a list of enzymes found to be useful for 12 alpha satellite arrrays.

4.3. Two-Dimensional Gel Electrophoresis

1. Choose an appropriate restriction enzyme for the array of interest; it should cut the array into several easily resolvable CHEF fragments (*see* Section 4.2.). This may require several CHEF gels to determine the correct enzyme and CHEF resolution conditions. Potentially double digests may be suitable for two-dimensional analysis *(6)*.

2. Digest an agarose block containing approx 10 µg genomic DNA with an appropriately chosen restriction enzyme(s). Pour a very thin CHEF gel (75 mL agarose in 0.5X TBE for 12.5 × 14 cm Bio-Rad CHEF gel tray). Load the gel carefully with one half of the digested block/lane, which assures identical digestion of experimental and marker lanes (*see* next step). Run the CHEF gel as described (*see* Note 1).

3. After the CHEF electrophoresis is complete, cut the gel to separate duplicate lanes. Stain one of the lanes in ethidium bromide in order to check digestion and electrophoresis, and Southern blot (*see* Note 1). This lane will act as the marker for the first dimension. Immerse the other unstained lane in sterile distilled H_2O (or 1 mM EDTA) for 0.5 h in a petri dish sealed with parafilm on shaking table, to remove electrophoresis buffer (*see* Notes 3 and 4).

4. Mix the solutions required for 100 mL of restriction enzyme buffer as recommended by the manufacturer, supplemented with 1 mM DTT, 200 µg BSA, 2 mM spermidine, and 1 mM EDTA. Bring up to volume minus volume of gel slice (approx 7.5 mL/lane). Place the gel slice in a petri dish, pour buffer over the slice, and equilibrate for 2 h (*see* Notes 5 and 6).

5. Equilibrate a second time in fresh restriction buffer for two additional hours. This time, however, bring the buffer up to total volume, and omit EDTA.

6. Place the equilibrated gel slice in the bottom of a freezer bag. Add 10 mL of fresh restriction enzyme buffer, and 2000 (yes, 2×10^3) U restriction enzyme (*see* Note 7). Seal the bag, and submerge in a shaking water bath at the recommended temperature overnight. A digestion of the original sample of DNA with the second dimension enzyme(s) should also be performed at this step, to be loaded with the gel slice as a reference.

7. Melt approx 400 mL LE (Seakem) agarose in TAE (for a 20 × 24 cm gel), at a concentration slightly higher (approx 0.1%) than the original CHEF gel. Remove the CHEF gel slice from freezer bag. For two-dimensional CHEF, the second CHEF gel is run in 0.5X TBE.

8. Using an electrophoresis gel comb as a guide, place the CHEF gel slice on its edge against comb. Pour molten agarose (cooled to approx 55°C) around the gel slice and into tray. Make sure the CHEF slice remains flush against the comb (*see* Note 8).

9. Equilibrate the imbedded gel slice in electrophoresis buffer (1X TAE) in the gel box for approx 1 h. Run this gel at low field strength (approx 40 V, *see* Section 3.2.4.), stain and Southern blot normally.

5. Notes

1. When using common cutting enzymes on CHEF gels, the ethidium bromide staining of DNA will reflect the overall small size fragments (<50kb) on the CHEF gel. However, the tandem array DNA will primarily be in the upper portion of the gel, where relatively little genomic DNA can be seen by ethidium bromide.

2. It is imperative that the first dimension pulsed-field gel of two-dimensional gels run straight, which is why CHEF is suggested. If an apparatus is used that gives bell-shaped electrophoretic fields, the center most lanes can be used. The CHEF gel must be poured very thin, in order to get bands in the second dimension that are resolvable. When preparing and digesting agarose blocks, consider whether the block can be loaded in such a thin gel.

3. Nucleases are a potential serious hazard in the digestion of the CHEF lane for the second dimension. Always wear gloves when handling gels. Tools used to cut and handle gels should be autoclaved, and/or rinsed and flamed with ethanol. A stiff piece of plastic is useful for handling gel slices. Petri dishes should be clean and sterile. Autoclave all solutions for restriction enzyme buffers (except BSA, DTT, and spermidine). Also autoclave all glassware used to prepare restriction enzyme buffers.

4. In general, if multiple two-dimensional gels are being run from the same initial CHEF gel, the first dimension marker lanes can be grouped on one half of the gel and stained and Southern blotted together. Similarily, if the same enzyme is being used in the second dimension (i.e., with a different first dimension enzyme), these lanes can be grouped on the CHEF gel, and equilibrated and digested together.

5. A high concentration of BSA, approximately twice the recommended amount, was found to be important for complete digestion. This may be attributable to an affinity of the agarose for restriction enzyme in the digestion. Therefore, the long equilibration procedure might serve to saturate the agarose with protein, and allow the concentration of restriction enzyme to remain high enough to be effective. Be sure to purchase sufficient quantities of BSA for the large amounts of restriction buffer used.

6. The first equilibration allows for the volume of the gel slice. The second does not because the concentration of restriction buffer should be close to correct already. The first equilibration also contains small amounts of EDTA, found to be important to reduce degradation,

perhaps by removing small amount of residual trace metals from CHEF electrophoresis. Equilibrations are performed in parafilm-sealed petri dishes rotating gently enough to create a small wave of buffer over gel slice in dish.

7. When using restriction enzymes with low stability e.g., *Bam*H I, add 1000 U of the restriction enzyme, place in water bath for several hours, and then add the remaining 1000 U to the bag. This is not necessary when using stable enzymes e.g., *Eco*RI. A small amount of positive pressure in the freezer bag was useful to assure complete submersion of the agarose slice. When using expensive enzymes, e.g., *Sau*3A I, be sure to consider less expensive isoschizomers, e.g., *Mbo* I *(6)*.

8. After digestion of the gel slice is complete, remove it carefully from freezer bag. The edges of the slice can be trimmed, if necessary. Place the slice on its edge (*see* Fig. 4A), flat against an electrophoresis comb in conventional gel pouring tray. Carefully pour molten 55°C agarose around the slice, and use enough agarose to just cover it. Assure that the slice remains flat against the comb. The resulting gel will be somewhat thicker than a normal gel. After the agarose hardens, the gel slice and wells can be trimmed with a razor blade, if necessary.

9. For two-dimensional CHEF gel electrophoresis, the gel slice is embedded in the second CHEF gel similarly to the conventional gel. CHEF resolution conditions in both dimensions should be chosen so that bands remain on the gel in the second dimension, and a useful diagonal (*see* Fig. 5) is produced.

10. Hybridization of alpha satellite probes that have a G + C content of approximately 36%, in 3X SSC/50% formamide, should permit a 13–20% mismatch at 42°C and a 3–5% mismatch at 53°C *(11)*. Washing of Southern blot filters at 65°C/0.5M NaCl provides no increased stringency; however, filter washing at 65°C/ 0.1X SSC should increase stringency since the calculated T_m of alpha satellite at these conditions should theoretically disallow any mismatch *(26)*. Different alpha satellite probes require essentially an empirical determination of chromosome-specific hybridization conditions, which are, in general, hybridization between 51°C and 53°C, and filter washing in 0.1X SSC/ 0.1% SDS between 65°C and 70°C.

Acknowledgments

The authors wish to thank G. M. Greig for unpublished data. This work was supported by NIH grant HG00107.

References

1. Miklos, G. L. G. (1985) Localized highly repetitive DNA sequences in vertebrate and invertebrate genomes, in *Molecular Evolutionary Genetics* (Macintyre, J. R., ed.), Plenum, New York, pp. 241–321.
2. Singer, M. (1982) Highly repeated sequences in mammalian genomes. *Int. Rev. Cytol.* **76**, 67–112.
3. Schwartz, D. C. and Cantor, C. R. (1984) Separation of yeast chromosome sized DNA fragments by pulsed field gradient gel electrophoresis. *Cell* **37**, 67–75.
4. Smith, C. L., Warburton, P. E., Gaal, A., and Cantor, C. R. (1986) Analysis of genome organization and rearrangements by pulsed field gradient gel electrophoresis, in *Genetic Engineering*, vol. 8 (Setlow, J. K. and Hollaender, A., eds.), Plenum, New York, pp. 45–70.
5. Walter, M. A., Surti, U., Hofker, M. H., and Cox, D. W. (1990) The physical organization of the human immunoglobin heavy chain gene complex. *EMBO J.* **9**, 3303–3313.
6. Warburton, P. E. and Willard, H. F. (1990) Genomic analysis of sequence variation in tandemly repeated DNA: Evidence for localized homogeneous sequence domains within arrays of alpha satellite DNA. *J. Mol. Biol.* **216**, 3–16.
7. Chu, G., Vollrath, D., and Davis, R. W. (1986) Separation of large DNA molecules by contour-clamped homogeneous electric fields. *Science* **234**, 1582–1585.
8. Willard, H. F. (1989) The genomics of long tandem arrays of satellite DNA in the human genome. *Genome* **31**, 737–744.
9. Cooke, H. (1976) Repeated sequence specific to human males. *Nature* **262**, 182–186.
10. Beauchamp, R. S., Mitchell, A., Buckland, R., and Bostock, C. (1979) Specific arrangements of human satellite III DNA sequences on human chromosomes. *Chromosoma* **71**, 153–166.
11. Willard, H. F. (1985) Chromosome-specific organization of human alpha satellite DNA. *Am. J. Hum. Genet.* **37**, 524–532.
12. Alexandrov, I. A., Mitkevich, S. P., and Yurov, Y. B. (1988) The phylogeny of human chromosome specific alpha satellites. *Chromosoma (Berl.)* **96**, 443–453.
13. Cross, S., Lindsey, J., Fantes, J., McKay, S., McGill, N., and Cooke, H. (1990) The structure of a subterminal repeated sequence present on many human chromosomes. *Nucleic Acids Res.* **18**, 6649–6657.
14. Durfy, S. J. and Willard, H. F. (1989) Patterns of intra- and interarray sequence variation in alpha satellite from the human X chromosome: Evidence for short-range homogenization of tandemly repeated DNA sequences. *Genomics* **5**, 810–821.
15. Willard, H. F. (1990) Centromeres of mammalian chromosomes. *Trends Genet.* **6**, 410–416.

16. Manuelidis, L. (1978) Chromosomal location of complex and simple repeated human DNAs. *Chromosoma* **66,** 23–32.
17. Willard, H. F. and Waye, J. S. (1987) Hierarchical order in chromosome-specific human alpha satellite DNA. *Trends Genet.* **3,** 192–198.
18. Wevrick, R. and Willard, H. F. (1989) Long-range organization of tandem arrays of alpha satellite DNA at the centromeres of human chromosomes: High frequency array-length polymorphism and meiotic stability. *Proc. Natl. Acad. Sci., USA* **86,** 9394–9398.
19. Tyler-Smith, C. and Brown, W. R. A. (1987) Structure of the major block of alphoid satellite DNA on the human Y chromosome. *J. Mol. Biol.* **195,** 457–470.
20. Mahtani, M. M. and Willard, H. F. (1990) Pulsed-field gel analysis of alpha satellite DNA at the human X chromosome centromere: High frequency polymorphisms and array size estimate. *Genomics* **7,** 607–613.
21. Waye, J. S. and Willard, H. F. (1989) Human β satellite DNA: Genomic organization and sequence definition of a class of highly repetitive tandem DNA. *Proc. Natl. Acad. Sci., USA* **86,** 6250–6254.
22. Moyzis, R. K., Albright, K. L., Bartholdi, M. F., Cram, L. S., Deaven, L. L., Hildebrand, C. E., Joste, N. E., Longmire, J. L., Meyne, J., and Schwarzacher-Robinson, T. (1987) Human chromosome-specific repetitive DNA sequences: Novel markers for genetic analysis. *Chromosoma (Berl.)* **95,** 375–386.
23. Tyler-Smith, C., Taylor, L., and Muller, U. (1988) Structure of a hypervariable tandemly repeated repeated DNA sequence on the short arm of the human Y chromosome. *J. Mol. Biol.* **203,** 837–848.
24. Oakey, R. and Tyler-Smith, C. (1990) Y chromosome DNA haplotyping suggests that most European and Asian men are descended from one of two males. *Genomics* **7,** 325–330.
25. Nakamura, Y., Julier, C., Wolff, R., Holm, T., O'Connell, P., Leppert, M., and White, R. (1987) Characterization of a human midisatellite sequence. *Nucleic Acids Res.* **15,** 2537–2547.
26. Waye, J. S., Mitchell, A. R., and Willard, H. F. (1988) Organization and genomic distribution of "82H" alpha satellite DNA: Evidence for a low-copy or single-copy alphoid domain located on human chromosome 14. *Hum. Genet.* **78,** 27–32.
27. Jabs, E. W., Goble, C. A., and Cutting, G. R. (1989) Macromolecular organization of human centromeric regions reveals high-frequency, polymorphic macro DNA repeats. *Proc. Natl. Acad. Sci., USA* **86,** 202–206.
28. Waye, J. S. and Willard, H. F. (1985) Chromosome-specific alpha satellite DNA: Nucleotide sequence analysis of the 2.0 kilobasepair repeat from the human X chromosome. *Nucleic Acids Res* **13,** 2731–2743.
29. Smith, G. P. (1976) Evolution of repeated sequences by unequal crossover. *Science* **191,** 528–535.
30. Dover, G. A. (1986) Molecular drive in multigene families: how biological novelties arise, spread and are assimilated. *Trends Genet.* **2,** 159–165.

31. Jeffreys, A. J., Neumann, R., and Wilson, V. (1990) Repeat unit sequence variation in minisatellites: A novel source of DNA polymorphism for studying variation and mutation by single molecule analysis. *Cell* **60**, 473–485.
32. Greig, G. M., England, S. B., Bedford, H. M., and Willard, H. F. (1989) Chromosome-specific alpha satellite DNA from the centromere of chromosome 16. *Am. J. Hum. Genet.* **45**, 862–872.
33. Waye, J. S. and Willard, H. F. (1989) Chromosome specificity of satellite DNAs: Short- and long-range organization of a diverged dimeric subset of human alpha satellite from chromosome 3. *Chromosoma (Berl.)* **97**, 475–480.
34. Willard, H. F., Waye, J. S., Skolnick, M. H., Schwartz, C. E., Powers, V. E., and England, S. B. (1986) Detection of restriction fragment polymorphisms at the centromeres of human chromosomes by using chromosome-specific alpha satellite DNA probes: Implications for development of centromere-based genetic linkage maps. *Proc. Natl. Acad. Sci., USA* **83**, 5611–5615.
35. Waye, J. S., England, S. B., and Willard, H. F. (1987) Genomic organization of alpha satellite DNA on human chromosome 7: Evidence for two distinct alphoid domains on a single chromosome. *Mol. Cell. Biol.* **7**, 349–356.
36. Waye, J. S. and Willard, H. F. (1986) Structure, organization, and sequence of alpha satellite DNA from human chromosome 17: Evidence for evolution by unequal crossing-over and an ancestral pentamer repeat shared with the human X chromosome. *Mol. Cell Biol.* **6**, 3156–3165.
37. vanOmmen, G. V. B. and Verkerk, J. M. H. (1986) Restriction analysis of chromosomal DNA in a size range up to two million basepairs by pulsed field gradient electrophoresis, in *Human Genetic Disease: A Practical Approach* (Davies, K. E., ed.), IRL, Oxford, pp. 113–133.
38. Wevrick, R., Earnshaw, W. C., Howard-Peebles, P. N., and Willard, H. F. (1990) Partial deletion of alpha satellite DNA associated with reduced amounts of CENP-B in a mitotically stable human chromosome rearrangement. *Mol. Cell. Biol.* **10**, 6374–6380.
39. Schmeckpeper, B. J., Willard, H. F., and Smith, K. D. (1981) Isolation and characterization of cloned human DNA fragments carrying reiterated sequences common to both autosomes and the X chromosome. *Nucleic Acids Res.* **9**, 1853–1872.
40. Willard, H. F., Smith, K. D., and Sutherland, J. (1983) Isolation and characterization of a major tandem repeat family from the human X chromosome. *Nucleic Acids Res.* **11**, 2017–2033.
41. Greig, G. M., Parikh, S., George, J., Powers, V. E., and Willard, H. F. (1991) Molecular cytogenetics of alpha satellite DNA from chromosome 12: Fluorescence in situ hybridization and description of DNA and array length polymorphisms. *Cytogenet. Cell Genet.* **56**, 144–148.
42. Greig, G. M. and Willard, H. F. (1992) β-Satellite DNA: Characterization and localization of two subfamilies from the distal and proximal short arms of the human acrocentric chromosomes. *Genomics* **3**, 573–575.

CHAPTER 21

Construction of Lambda Libraries from Large PFGE Fragments

Catrin Pritchard and Margit Burmeister

1. Introduction

Pulsed-field gel electrophoresis (PFGE) has the capacity to fractionate large fragments of DNA up to thousands of kilobases in size. This aspect of the technique has been exploited for constructing long-range restriction maps of chromosomes from many different species including humans (*see* Chapters 14, 15, and 18). Besides its use for analytical purposes, PFGE has also been used as a preparative tool. Intact DNA obtained from preparative PFGE gels has been used for cloning into yeast artificial chromosome (YAC) vectors (*see* Chapter 16) and for constructing jumping libraries *(1)*. In addition, DNA eluted from PFGE gels has been used for generating libraries with a smaller insert size *(2–7)*. In this latter procedure, DNA from a somatic cell hybrid is digested with a rare-cutting restriction enzyme, separated by PFGE, and the DNA from a particular PFGE fragment is eluted, digested, and cloned into a plasmid or phage vector. The resulting library is then screened with a species-specific probe to identify DNA segments from the donor chromosome of the hybrid. This use of preparative PFGE has had widespread application in the cloning of DNA close to several important disease genes, namely cystic fibrosis *(4,6)*, Duchenne muscular dystrophy *(2)*, choroideremia *(7)*, polycystic kidney disease *(3)*, and Huntington disease *(5)*. The approach is

From: *Methods in Molecular Biology, Vol. 12: Pulsed-Field Gel Electrophoresis*
Edited by: M. Burmeister and L. Ulanovsky
Copyright © 1992 The Humana Press Inc., Totowa, NJ

attractive in that it is relatively quick to perform, technically straight-forward, allows end clones to be obtained for jumping into a neigh-boring region, and provides markers for saturating a region defined by a PFGE fragment. The strategy can also be adapted for cloning DNA from YACs into smaller capacity vectors.

The level of enrichment that can be achieved for a particular PFGE fragment is a major consideration in deciding whether to embark on preparative PFGE. The source of the starting cell line DNA is impor-tant as this will greatly influence the number of markers that can be cloned from the fragment. Good enrichment can be achieved by start-ing with a whole chromosome somatic cell hybrid (2–4, 7), provided that the fragment of interest lies outside of the region where most restriction fragments generated by the same enzyme migrate. How-ever, it is often worth searching for a hybrid that is more enriched for the region containing the PFGE fragment such as a radiation hybrid (5) or a chromosome-mediated gene transfer hybrid (6). These hybrids are now available for many regions of the human genome. Alternatively, future technological advancements may make it possible to obtain enrichment in other ways. For example, two-dimensional PFGE (*see also* Chapters 14, 19, and 20) may allow PFGE fragments to be cloned from total mammalian DNA rather than hybrid DNA. The method of coincidence cloning (8), in which DNA common to two different overlapping PFGE fragments can be selectively cloned, may also allow further enrichment in addition to overcoming the need to screen with species-specific repeats.

Cloning DNA from preparative PFGE gels can be divided into five procedures. In this chapter we describe the five procedures in detail: the best conditions for electrophoresis, identification of the fraction containing the PFGE fragment, elution of the DNA, construction of the lambda library, and screening of the lambda library. An outline of the procedures is illustrated in the Fig. 1.

2. Materials

1. DNA-agarose blocks made from cell hybrid each containing approx $4 \times 10^5 – 1 \times 10^6$ cells. (These should be made according to the method described in Chapter 8.)
2. Yeast and/or lambda concatamer DNA size markers (*see* Chapter 8 for construction).
3. PFGE buffer (TBE): 45 mM Tris-HCl, pH 8.0; 45 mM Boric acid; 1 mM EDTA, pH 8.0. Prepare a 5X stock by dissolving 108 g of Tris-

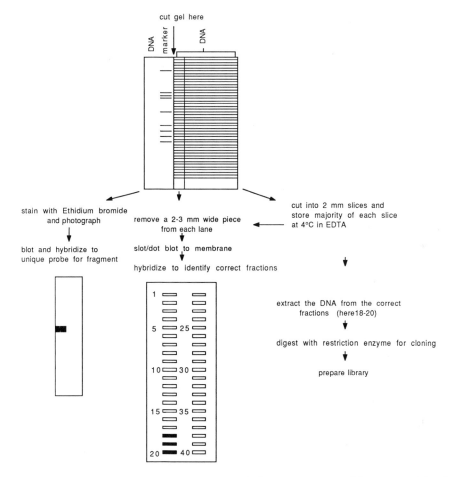

Fig. 1. Outline of the procedures needed to construct a library enriched for DNA from a specific PFGE fragment.

HCl, 55 g of boric acid, and 7.4 g of Na$_2$EDTA in 1 L of water. Store at room temperature.

4. Low melting point (LMP) agarose (BRL Ultrapure, Gaithersburg, MD).

5. Field-inversion gel electrophoresis (FIGE) or contour-clamped homogeneous electric field (CHEF) apparatus.

6. 0.5*M* EDTA, pH 8.0.

7. Proteinase K (Boehringer-Mannheim, Indianapolis, IN). Dissolved in water at a concentration of 10 mg/mL and stored at –20°C.

8. 5*M* NaOH. This should be stored at room temperature in a plastic container. Caution: Use gloves when preparing this solution as it causes burns.

9. Grid for guiding cuts in gel; this is simply a piece of paper or transparency with horizontal lines marked every 2 mm.
10. Slot-blot or dot-blot apparatus.
11. Hybond N-Plus membrane (Amersham, Arlington Heights, IL) or your favorite nylon membrane.
12. DNA elution solution: 100 mM NaCl, 10 mM EDTA, pH 8.0.
13. Agarase (Sigma, St. Louis, MO) dissolved in water at a concentration of 50 U/mL and stored at –20°C.
14. TE: 10 mM Tris-HCl, pH 8.0, 1 mM EDTA, pH 8.0.
15. Phenol equilibrated with several changes of 0.5M Tris-HCl, pH 8.0, and finally with 0.1M Tris-HCl, pH 8.0. Caution: Phenol causes severe burns and should be handled with gloves.
16. Chloroform.
17. 3M sodium acetate, pH 5.2.
18. 100% and 70% ethanol.
19. *Sau*3A I restriction enzyme. Dilutions of the enzyme should be made with 50 mM KCl, 10 mM Tris-HCl, pH 7.5, 0.1 mM EDTA, 1 mM dithiothreitol(DTT), 200 mg/mL, 50% glycerol; and stored at –20°C.
20. TAE buffer: 40 mM Tris-acetate, 2 mM EDTA. Prepare a 50X stock by dissolving 242 g of Tris-base in 800 mL of water. Add 57.1 mL of glacial acetic acid and 100 mL of 0.5M EDTA, pH 8.0. Adjust vol to 1 L with water and store at room temperature.
21. DNA size markers for analysis of partial digestion products. Convenient markers for this purpose are *Hind*III-digested wild-type lambda DNA in addition to uncut lambda DNA heated to 65°C, or the high-mol-wt markers available from BRL.
22. Calf intestinal alkaline phosphatase (CIP; Boehringer) at 1 U/µL
23. 10X CIP buffer: 10 mM ZnCl$_2$, 10 mM MgCl$_2$, 100 mM Tris-HCl, pH 8.3.
24. 20% sodium dodecyl sulphate (SDS; BRL Ultrapure).
25. Homemade lambda replacement vector arms cleaved with *Bgl*II or *Bam*HI at a concentration of 1 mg/mL and homemade packaging extract; or a lambda cloning kit, which can be purchased (e.g., from Stratagene, La Jolla, CA) consisting of *Bam*H I-cleaved arms of Lambda Zap II (at 1 mg/mL) and Gigapack Gold packaging extract.
26. T4 DNA ligase at 400 U/µL (New England Bio Labs., Beverly, MA; 1 U corresponds to 1/60 Weiss U).
27. 10X ligation buffer: 500 mM Tris-HCl, pH 7.8, 100 mM MgCl$_2$, 10 mM adenosine triphosphate, 200 mM DTT, 2 mg/mL bovine serum albumin, stored at –20°C.
28. SM medium: Mix 5.8 g of NaCl, 2 g of MgSO$_4$•7H$_2$O, 50 mL of 1M Tris-HCl, pH 7.5 and 5 mL of 2% gelatin. Adjust vol to 1 L and autoclave.

29. L Broth (LB): Dissolve 10 g of bactotryptone, 5 g of yeast extract, and 10 g of NaCl in 1 L of water. Adjust pH to 7.5 with sodium hydroxide and autoclave.
30. Magnesium cells: Inoculate a single colony of P2392 bacteria (or NM539 or other Spi-selecting bacteria) into LB containing 10 mM MgSO$_4$ and 2% maltose and grow overnight at 37°C with shaking.
31. NZCYM amine agar plates and top agarose: Dissolve 22 g of NZCYM mix (Gibco, Gaithersburg, MD) in 1 L of water. Add 15 g of bacto-agar for plates and 7 g of agarose for top agarose, then autoclave.

3. Methods
3.1. Preparative PFGE

1. Digest 20 of the DNA-agarose blocks with the appropriate restriction enzyme (*see* Note 1) by following the method described in Chapter 8 or 11. After the digestion, treat the blocks as also described in Chapter 8.
2. Pour a 1% LMP agarose gel made with 0.5X TBE. Use a comb with a continuous slot that will accommodate all of the blocks, as well as additional single lanes to run test samples and markers. Teflon or electrical tape can be used for combining slots.
3. Load all blocks except one into the large lane. The remaining block is used as a test sample and should be loaded in a separate slot (*see* Note 2). Load marker lanes on both sides of the large lane, separated from the large lane by a blank lane.
4. Run the gel in a CHEF or FIGE apparatus (*see* Note 3) under conditions that result in the best fractionation of your fragment (Chapters 1–7).
5. Stop the electrophoresis. Cut away the marker lanes and test lane. These should be placed in 10 mM EDTA containing 1 mg/mL ethidium bromide for 1 h, destained in 10 mM EDTA for 1 h, and then photographed (*see* Note 4). Take the rest of the gel and attach the grid to the underneath of the plate. With a sterile scalpel, cut 2-mm horizontal slices from the gel using the grid to guide you. It is best to do this when the gel is still attached to the plate as LMP agarose gels are very flimsy and tend to move away easily. Store the slices at 4°C in 15-mL tubes filled with 10 mM EDTA and 1 mg/mL proteinase K (*see* Note 5).

3.2. Identification of Correct PFGE Fraction

1. Cut off a 1/2 block from each gel slice and place in an Eppendorf tube for slot-blot analysis (*see* Note 6).
2. Assemble the slot-blot apparatus with a nylon membrane (*see* Note 7). Attach to a vacuum supply. Melt the samples at 95°C for 10 min

and then add 5M NaOH to a final concentration of 0.5M NaOH. Place at 65°C for 10 min.

3. Turn on the vacuum. Load each sample into the slots of the apparatus (*see* Note 8). After all the samples have been applied, dismantle the apparatus and place the membrane in 6X SSC for 10 min. Hybridize the filter to a probe for the fragment of interest, by using your standard hybridization protocol.

4. On the basis of the slot-blot results, identify the positive fraction(s). The signal is generally distributed over 2–3 positive fractions (*see* Note 9).

3.3. Elution of DNA from Correct Fraction

1. Take the positive fractions and remove the storage solution. Fill the tubes with elution solution, and rinse for 30 min with shaking. Repeat this procedure. Remove the liquid and add an equal vol of elution solution. Place the tubes at 65°C.

2. As soon as the agarose has melted, transfer the sample to Eppendorf tubes, placing 500 µL per tube (as many as 20 tubes may be required for this). Incubate at 37°C for 10 min. Add 50 U of agarase to each tube and leave overnight at 37°C.

3. Extract the DNA twice with an equal vol of phenol and then once with an equal vol of chloroform (*see* Note 10). Add 1/10 vol 3M sodium acetate and 2 vol of 100% ethanol and place at –20°C for at least 3 h. Spin in a 4°C Eppendorf centrifuge for 20 min. Remove the supernatant and add 1 mL of 70% ethanol to the pellet. Spin again for 5 min, remove the supernatant and dry the pellet in a speedvac. Dissolve the DNA in 50 µL of TE (*see* Note 11).

3.4. Construction of Lambda Library with the Eluted DNA

1. Partials are performed by the enzyme dilution method. On ice, add 15 µL of restriction enzyme buffer to the DNA and water to a final vole of 150 µL. Distribute the mix to five Eppendorf tubes, placing 30 µL per tube. Add varying concentrations of *Sau*3A I to each tube (a good range is between 0.0005 and 0.1 U; *see* Note 12). Incubate samples for 1 h at 37°C. Stop the reaction by adding EDTA to 20 mM.

2. Run 1/10 of each sample on a 0.5% agarose gel made with 1X TAE. Keep the remainder of the DNA at 4°C. Load the high-mol-wt DNA size markers in neighboring lanes. Perform electrophoresis slowly, at no more than 1 V/cm, for as long as it takes to obtain good size separation in the 10–50 kb size range (usually 24 h for a 10-cm minigel).

3. Stain the gel with ethidium bromide. If there is too little DNA to see, blot the DNA and hybridize with a probe for the DNA such as total hamster DNA if a hamster-human hybrid has been used as the DNA source.

4. For lambda libraries, DNA in the 10–20 kb size range is optimal for cloning. Decide which partial conditions result in the best distribution of fragments in this size range (i.e., the fraction with the highest signal or the highest ethidium bromide staining in the 20–40 kbp size range). Carefully phenol extract the DNA, add 1/10 vol of 3*M* sodium acetate and precipitate with 2.5 vol of 100% ethanol. After washing with 70% ethanol, dry the pellet and dissolve in 50 µL of TE.

5. Dephosphorylate the DNA to prevent self ligation of the *Sau*3A I fragments. Add 10 µL of CIP buffer, 39 µL of water, and 1 µL of CIP to the DNA. Incubate at 37°C for 1 h. Stop the reaction by adding EDTA to 50 m*M*, SDS to 1% and proteinase K to 50 µg/mL. Incubate at 55°C for 30 min then phenol extract the DNA twice, extract once with chloroform and then ethanol precipitate. Dissolve in 50 µL of TE (*see* Note 13).

6. Ligation is performed by using lambda arms at a concentration of not less than 30–50 µg/mL and phosphatased inserts of about 2–10 µg/mL. To the DNA sample add 2–5 µL of the lambda arms, 10 µL of 10X ligation buffer, 1 µL of ligase, and water to a final vol of 100 µL. Incubate at 15°C for 12–16 h (*see* Note 14).

7. Take 4 µL from the ligation mix and package into phage particles with packaging extract (*see* Note 15). Dilute the packaged phage with 500 µL of SM and add 20 µL of chloroform.

3.5. Identification of Cloned DNA Segments from the Fragment of Interest

1. Determine the titer of the library as follows: Take 1-µL, 10^{-1}-µL, and 10^{-2}-µL aliquots of the library stock and mix with 100 µL of magnesium cells. Incubate for 20 min at 37°C. Add 3 mL of NZYCM top agarose (melted and kept at 45°C) and pour each aliquot onto a 10-cm NZYCM agar plates. Incubate the plates for 8–12 h. Count the number of plaque forming units (PFU) on each plate and, from this, calculate the average number of PFU in the total library stock (*see* Note 16).

2. Plate out the library for screening with the species-specific probe (*see* Note 17) on large (13 cm diameter) NZYCM agar plates. Approximately 20,000 PFU can be plated per plate, by using 300 µL of magnesium cells per plate and 10 mL of NZYCM top agarose. Take three

filter lifts from each plate. Hybridize two with a species-specific probe for the fragment of interest, e.g., total human DNA or a probe for the human Alu repeat, if a human-hamster somatic cell hybrid has been used as the source of DNA. To ensure that the majority of clones contain inserts, hybridize one of the filters with a probe for the recipient parental species of the hybrid, e.g., total hamster DNA (*see* Note 18).

4. Notes

1. Successful enrichment by preparative PFGE depends on the following factors:
   a. DNA loading. There is a tendency to overload agarose blocks in order to obtain higher yields of DNA. DO NOT DO THIS as it results in upward smearing of fragments during electrophoresis; 20 blocks each containing $4 \times 10^5 - 1 \times 10^6$ cells yields plenty of DNA for constructing lambda libraries.
   b. Choice of restriction enzyme. Consider carefully which restriction enzyme results in the best enrichment for your fragment. You should test whether digesting the DNA with a second enzyme reduces the number of other fragments that comigrate with your fragment. In addition, try and avoid using an enzyme that results in partial cleavage of your fragment since this will reduce the yield of DNA. The basic idea is to cut out a fragment that does not comigrate with the majority of other fragments in the hybrid. This can generally be estimated by comparing the migration of your fragment with that of the total ethidium bromide staining. For example, human *Sfi*I fragments larger than 700 kb, or *Not*I fragments smaller than 600 kb are greatly enriched in most mammalian DNA samples insofar as they migrate outside of the region of brightest ethidium bromide fluorescence.
   c. PFGE fractionation conditions. Optimal conditions for FIGE or CHEF should be deduced in advance. These conditions should then be reproduced exactly for the preparative gels. It is important to use LMP agarose for the test experiments as DNA migrates slightly differently in this type of agarose.
2. The test sample is used for two purposes:
   a. As a check that the genomic DNA has been digested adequately and fractionated by the electrophoresis, and
   b. After electrophoresis, the DNA of this test lane can be blotted and hybridized to a probe for the PFGE fragment in order to confirm its position in the gel.

3. It is important to run CHEF or FIGE gels rather than OFAGE gels because OFAGE gels do not give straight migration of the DNA.
4. It is also possible to stain the preparative part of the gel *(4)*. This procedure is not recommended, however, since ethidium bromide staining and exposure to UV light is known to reduce DNA cloning efficiency. Staining and destaining is performed without shaking because LMP agarose gels are very fragile.
5. The slices can be stored for a long time in the EDTA solution in case you want to analyze other fragments from the same hybrid in the future.
6. It is usually not necessary to slot-blot all of the fractions if the approximate location of your fragment is known. This can be deduced either by referring to the marker lanes or by obtaining a result from blotting DNA in the test lane. If the approximate location is known then it is usually sufficient to slot-blot 20–30 fractions around this region. However, if other fragments from the same hybrid are going to be used in the future then it may be preferable to slot-blot all slices so that the slot-blot filter can be rehybridized with probes for these other fragments.
7. Previous reports have recommended warming the slot-blot apparatus to 37°C prior to and during application of the samples. We have found that this is not necessary because as the high concentration of NaOH in the sample degrades the agarose so that it cannot solidify.
8. Uneven cutting of slices will result in different sample volumes at this stage. It is important to load the same volume in each case because this will allow the fairest comparison between slices. Alternatively, to avoid bias due to uneven cutting or uneven distribution of DNA, slices can be melted at 65°C prior to removing an aliquot for slot-blot analysis. Melted slices can then be stored under a solution of 10 mM EDTA, 1 mg/mL proteinase K at 4°C.
9. The limiting mobility and load fractions, representing uncut DNA and DNA that has been trapped in the agarose blocks respectively, may also hybridize to the probe for your fraction. Be aware of this by noting the positions of these fractions when picking out your positives.
10. Shearing of the DNA can be minimized during phenol extraction by using a head-over-top rotating device or slow hand-tilting of the tubes. In general, only approx 100 kb can be retained intact by this elution method, generating DNA suitable for constructing phage, plasmid libraries, and for PCR. If larger DNA is required, e.g., for constructing cosmid, YAC, or jumping libraries, then alternative elution methods must be followed (*see* Chapter 16).

11. The yield is generally 2–4 μg of DNA from 20 agarose blocks. The agarose acts as a carrier for the DNA during the ethanol precipitation so there is no need to add other carriers such as yeast tRNA. Small amounts of agarose tend to remain in the DNA preparation, even after repeated phenol extractions. Do not worry about this as enzyme reactions are not significantly inhibited by it, provided the solution is dilute enough such that the agarose does not solidify. Alternatively, the agarose can be removed by adding TE to the pellet, leaving the DNA to dissolve overnight at 4°C and then centrifuging the sample briefly. The resulting supernatant is virtually agarose-free.

12. The correct partial conditions can also be established on a negative fraction and then reproduced exactly for the positive fraction. All of the negative fraction can be used for this, allowing visualization of the DNA by ethidium bromide staining rather than by hybridization. Occasionally it may be difficult to reproduce partial conditions on the positive fraction. If this is the case, then a few test partials should be performed on the positive fraction, using the results from the negative fraction as a guide for which range of enzyme concentration to use. As a general rule, try and restrict the test experiments to the positive fraction alone, because the other fractions may be needed for cloning at a later stage.

13. CIP can also be inactivated by adding nitrilotriacetic acid to 10 m*M*, and heating the sample for 20 min at 70°C. In this case there is no need to phenol extract the DNA.

14. The cloning strategy used will depend on the type of DNA segments one wants to obtain. For example, if small DNA segments are required, then the DNA can be digested to completion with a frequent-cutting enzyme such as *Bam*HI or *Eco*RI and then cloned into a plasmid or lambda insertion vector *(2,3,6,7)*. Although this approach is straightforward, it will result in the underrepresentation of many *Bam*HI or *Eco*RI fragments. The method of *Sau*3A I partial digestion does not introduce such a bias. However, it is technically more difficult because the yield of DNA from preparative PFGE is not high. Many different lambda replacement vectors are available for cloning *Bam*HI compatible ends. We have found that the most convenient vector to use is Lambda Dash (available from Stratagene) because it has multiple cloning sites and T7 and T3 promoters on either arm for making RNA probes.

The DNA generated from the partial digestion can also be used to rescue the ends of the large PFGE restriction fragment. Specially designed cloning vectors are required for this and will depend on the restriction enzyme that has been used to generate the large PFGE fragment. Lambda Dash I, II, or EMBL 6 (4) can be used for cloning *Not*I-*Sau*3A I, *Eag*I-*Sau*3A I, and *Sal* I-*Sau*3A I fragments. By mixing arms from two different vector preparations, *Sac*II, *Mlu*I, and *Bss*HIII compatible ends can be cloned *(4,9)*.

15. Gigapack Gold packaging extract (available from Stratagene) is probably the most efficient extract currently available. However, it is very expensive. Therefore, the success of the ligation should be tested before packaging with it. This can be done in two ways;
    a. By running a small aliquot (1/20 th) of the ligation on an 0.5% agarose gel and observing the formation of concatamerized DNA and
    b. By packaging a small amount of the ligation with a cheaper or homemade packaging extract before packaging with the Gigapack Gold.
16. In our experience, we have obtained libraries of 30,000–100,000 PFU from this method.
17. A species-specific probe must be available in order to screen the library. For human DNA clones in a rodent background, probes for the Alu and Line family of repeats can be used. Radiolabeled total human DNA can also be used to identify human sequences. However, be aware that some rodent sequences also hybridize strongly to this probe. An alternative method is to use Alu-Alu or Alu-Line PCR to isolate human sequences *(2,10)*. The DNA fragments obtained by this PCR approach will generally be small and will be biased in their distribution since two repeat sequences must be next to each other. A major disadvantage with using repeat probes is that some segments of DNA (approx 15% of phage in a human library) will be missed because they do not contain a highly repetitive sequence. It is therefore a good idea to use this cloning strategy in conjunction with other cloning methods, e.g., chromosome walking to generate completely overlapping clones for the PFGE fragment *(2,5)*.

    In our screens of phage libraries generated from PFGE fragments, we have occasionally identified lambda clones that contain the stuffer fragment from the lambda vector ligated to a small fragment of human DNA. Small fragments of human DNA will arise if overdigestion occurs during the partial digestion. These fragments can clone next to the stuffer fragment as it is not dephosphorylated at its 3' end. For DNA segments containing the stuffer fragment to clone in Spi-selecting bacteria, the stuffer fragment must carry a mutation in its Spi gene. The library can be screened with the stuffer fragment to identify such clones.

18. If you are starting with a whole chromosome hybrid, e.g., a hamster-human somatic cell hybrid, it may be worth determining the enrichment factor achieved by PFGE before spending time characterizing clones from the library. We have used two methods to determine this factor;

   a. Take a sample of gel-eluted DNA and cut it to completion with a restriction enzyme (for this, you can take an overdigested fraction from the test partial experiment and cut it to completion by adding more *Sau*3A I). Digest 10-µg, 1-µg, and 0.1-µg aliquots of the parental genomic DNAs (e.g., hamster and human DNA if you have used a hamster-human hybrid as your starting cell line) to completion with the same enzyme. Perform a genomic Southern blot with this DNA. Hybridize the filter to a unique probe for your fragment, strip the filter and then hybridize it with a total DNA probe from the recipient parental cell line (e.g., total hamster DNA in the case of a hamster-human hybrid). An enrichment of approx 100-fold is achieved when the hybridization signal with the total hamster DNA probe is approximately equivalent to that of 0.1 µg of hamster DNA, whereas the unique signal is equivalent to 10 µg of human DNA.

   b. The second method analyses the enrichment after the library has been constructed. Take a filter lift of the library and hybridize this with a unique probe for your fragment. The enrichment factor can be calculated from the following equation:

$$\text{enrichment factor} = \frac{\text{average no of positives} \times 6 \times 10^6}{\text{no of phages screened} \times \text{average insert size (in kbp)}}$$

(6×10^6 represents the size of the diploid human genome in kb; the average insert size for most replacement phages is 15–20 kbp). The enrichment factor is expected to be 20- to 200-fold.

   If the total amount of human DNA in the starting cell line is known, the fraction of human clones derived from the fragment of interest can be calculated as follows:

$$\text{fraction of clones from fragment} = \frac{\text{enrichment factor} \times \text{size of gel-eluted fragment}}{\text{total amount of human DNA in cell line}}$$

Acknowledgment

The experiments described in this protocol were performed in the laboratories of Richard M. Myers (University of California at San Francisco) and Hans Lehrach (European Molecular Biology Laboratory, Heidelberg; present address, Imperial Cancer Research Fund, London, UK). We thank them for their encouragement and support.

References

1. Collins, F. S., Drumm, M. L., Cole, J. L., Lockwood, W. K., Vande Woude, G. F., and Iannuzzi, M. C. (1987) Construction of a general human chromosome jumping library, with application to cystic fibrosis. *Science* **235**, 1046–1049.
2. Anand, R., Honeycombe, J., Whittaker, P. A., Elder, J. K., and Southern, E. M. (1988) Clones from an 840-kb fragment containing the 5' region of the DMD locus enriched by pulsed field gel electrophoresis. *Genomics* **3**, 177–186.
3. Harris, P. C., Barton, N. J., Higgs, D. R., Reeders, S. T., and Wilkie, A. O. (1990) A long-range restriction map between the alpha-globin complex and a marker closely linked to the polycystic kidney disease 1 (PKD1) locus. *Genomics* **7**, 195–206.
4. Michiels, F., Burmeister, M., and Lehrach, H. (1987) Derivation of clones close to met by preparative field inversion gel electrophoresis. *Science* **236**, 1305–1308.
5. Pritchard, C., Casher, D., Bull, L., Cox, D. R., and Myers, R. M. (1990) A cloned DNA segment from the telomeric region of human chromosome 4p is not detectably rearranged in Huntington disease patients. *Proc. Natl. Acad. Sci. USA* **87**, 7309–7313.
6. Ramsay, M., Wainwright, B. J., Farrall, M., Estivill, X., Sutherland, H., Ho, M. F., Davies, R., Halford, S., Tata, F., Wicking, C., Lench, N., Bauer, I., Ferec, C., Farndon, P., Kruyer, H., Stanier, P. Williamson, R. and Scambler, P. J. (1990) A new polymorphic locus, D7S411, isolated by cloning from preparative pulse-field gels is close to the mutation causing cystic fibrosis. *Genomics* **6**, 39–47.
7. van de Pol, T. J., Cremers, F. P., Brohet, R. M., Wieringa, B., and Ropers, H. H. (1990) Derivation of clones from the choroideremia locus by preparative field inversion gel electrophoresis. *Nucleic Acids Res.* **18**, 725–731.
8. Marchuk, D., Cole, J., Cantor, C., Weissman, S., and Collins, F. (1988) Coincidence cloning: A method for selective cloning of sequences shared between DNA samples. *Am. J. Hum. Genet.* **43S**, A194 (Abstract).
9. Frischauf, A. M., Murray, N., and Lehrach, H. (1987) Lambda phage vectors—EMBL series. *Methods Enzymol.* **153**, 103–115.
10. Nelson, D. L., Ledbetter, S. A., Corbo, L., Victoria, M. F., Ramirez, S. R., Webster, T. D., Ledbetter, D. H., and Caskey, C. T. (1989) Alu polymerase chain reaction: A method for rapid isolation of human-specific sequences from complex DNA sources. *Proc. Natl. Acad. Sci. USA* **86**, 6686–6690.

CHAPTER 22

In-Gel Detection of DNA

*Application to Study of Viral DNA Metabolism
by Use of Pulsed-Field Agarose Gel Electrophoresis*

Philip Serwer, Robert H. Watson, and Marjatta Son

1. Introduction

The introduction of pulsed-field agarose gel electrophoresis
(PFGE) has expanded the list of particles separable by use of gel electrophoresis to include: (1) linear DNAs as long as 3–6 Mbp, (2) DNA–
protein complexes and circular DNAs that become arrested during
invariant field agarose gel electrophoresis; and (3) micron-sized spheres
that also become arrested during invariant field agarose gel electrophoresis (reviewed in refs. *1–3*). During the replication, recombination and packaging of DNA by the various double-stranded DNA
bacteriophages, circular DNAs, protein–DNA complexes, and end-to-
end joined mature DNA multimers (concatemers) as long as 0.5–1.0
Mbp are formed (reviewed in refs. *4,5*). Thus, in addition to being
useful for genome mapping (*1,2,6,7* and *see also* Chapter 18), PFGE
also is useful for studying the DNA metabolism of both bacteriophages and, presumably, other viruses.

Study of the DNA metabolism of bacteriophage T7 provides
examples of the use of PFGE for the study of DNA metabolism in general. During infection of its host, *Escherichia coli*, T7 morphogenesis

From: *Methods in Molecular Biology, Vol. 12: Pulsed-Field Gel Electrophoresis*
Edited by: M. Burmeister and L. Ulanovsky
Copyright © 1992 The Humana Press Inc., Totowa, NJ

proceeds via two independent pathways that subsequently merge: (1) A DNA-free protein capsid is assembled from proteins that form an outer shell, inner scaffold and a pore at the point of eventual DNA entry. (2) DNA is replicated and forms end-to-end multimers (concatemers). Merging of these two pathways results in both the cutting of a mature, terminally repetitious, unique-ended genome from the concatemer and packaging of this genome in a capsid (reviewed in refs. *4,5*). When the concatemers in wild-type T7-infected cells were analyzed by use of rate zonal centrifugation, followed by gel electrophoresis, a background of concatemers heterogeneous in length was found superimposed on a set of discrete bands. The latter were formed by 1-mers, 2-mers, and so on. This experiment was first performed by use of invariant-field electrophoresis *(8)* and later, at improved resolution for longer than 2-mer length DNA, by use of PFGE (field inversion mode) *(9)*. The background, but not the bands, were observed when DNA packaging was blocked; this observation indicates that the band-forming DNA was produced during cutting of concatemers, not during their formation. During these experiments, attempts to detect circular DNA revealed only trace amounts of unbranched circular DNA (less than one genome in 10,000 was circular). Some branched DNA was found at the origin of electrophoresis. Other DNA that was thought, but not proven, to be branched was found during PFGE (field-inversion mode) between the origin and the zone of compression formed by linear DNA *(9)*. This same experiment revealed a complex of a T7 capsid and T7 DNA (DNA outside of the capsid). The T7 capsid–DNA complex was resolved from T7 DNA during the use of PFGE *(9)*. Although this complex was a degradation product of the mature bacteriophage T7, other capsid–DNA complexes are in the DNA packaging pathway (for an example in the case of the in vitro DNA packaging of the T7-related bacteriophage, T3, *see* ref. *10*). Thus, these procedures of PFGE should be capable of resolving DNA packaging intermediates that have either unpackaged or partially packaged DNA. Presumably, DNA–protein complexes that participate in other aspects of DNA metabolism will also be resolvable by use of PFGE.

For two reasons, staining of gels could not be used to perform the experiments of the previous paragraph *(8,9)*: (1) Overloading of the sucrose gradient would have occurred at the concentration of DNA necessary to provide enough DNA for analysis. (2) A background of stained host DNA would have confused the analysis. In addition, kinetic labeling of the DNA was needed for determination of the kinetics of passage of DNA through the various intermediate forms *(9)*. ^{3}H-thymidine is the label usually used for this type of study. In the case of refs. *8* and *9,* intracellular T7 DNA was labeled with ^{3}H-thymidine and the profile of T7 DNA after PFGE was determined by use of fluorography. During restriction endonuclease analysis of intracellular T7 DNA, fragments of DNA have been detected by use of either fluorography of DNA prelabeled with ^{3}H-thymidine *(8,9)* or autoradiography of DNA labeled by Southern blotting, followed by hybridization to ^{32}P-labeled DNA *(11)*.

As in the past, future use of PFGE for studying DNA metabolism will require the following, in addition to an adequate procedure of PFGE:

1. Detection of DNA radiolabeled with tritium before fractionation; and
2. Hybridization-based detection of unlabeled DNA fractionated by use of PFGE.

As shown in a subsequent section, at least one form of intracellular T7 DNA does not transfer efficiently during procedures of blotting. Thus, eliminating both transfer of the DNA and vacuum-drying of the gel should assist in accomplishing these goals. Such an in-gel procedure should also help to make quantification more accurate and, eventually, to automate the hybridization. Thus, in the present report are described in detail both a procedure for in-gel fluorography, and a procedure for in-gel DNA–DNA hybridization. Vacuum-drying of the gel is not used for either procedure.

2. Materials

2.1. In-Gel Fluorography

1. Salicylic acid. Crystalline salicylic acid forms a light powder that is best handled in a negative flow hood to prevent dispersal in the laboratory. Sneezing and allergy can be induced by salicylate. Thus, gloves and face mask are worn when working with the powder. On one occasion, some salicylate came into contact with the interior of a Plexiglas™ electrophoresis apparatus. Salicylate fluorescence made ethidium staining of DNA in agarose gels impossible for any gel subjected to electrophoresis in this apparatus and we were unsuccessful in cleaning the apparatus; the salicylate apparently eluted from the Plexiglas™ so slowly that it was impossible to clean. Thus, any container that contacts salicylate is not used for anything else.
2. Glycerol.

3. Plastic pans. To contain the gel, we have used plastic refrigerator containers purchased at a grocery store.
4. A 56°C oven. The oven (an incubator can also be used) optimally has a fan to speed the drying of gels.
5. Filter paper. To support gels during drying, they are placed on filter paper. Several companies provide filter paper for drying gels. We have used the filter paper backing sold for the Bio-Rad model 543 dryer.
6. X-ray film. We have used Kodak Diagnostic Film SB.
7. Plastic wrap. Plastic wrap has been used to separate gel from film during exposure of the film.
8. X-ray film cassettes. The size depends on the size of the gel to be exposed. Typically, we use 5" x 7" cassettes.
9. –70°C freezer.

2.2. In-Gel DNA–DNA Hybridization

1. Plastic pans for washing gels.
2. A 55°C oven and a 42°C oven, or one that can be operated at 42°C and at 55°C.
3. A platform with motor for shaking, in a water bath capable of 55–70°C.
4. A rocking platform that fits in the 42°C oven.
5. Flexible plastic sheets. To transfer gels, the procedure used for fluorography is also used here.
6. Denaturing solution: $0.5M$ NaOH, $1.5M$ NaCl.
7. SSPE buffer: $0.15M$ NaCl, $0.01M$ sodium phosphate, pH 7.4, $0.0012M$ EDTA *(12)*.
8. Hybridization buffer: 12.5% (w/v) dextran sulfate, 50% formamide, 0.1% sodium dodecyl sulfate, 5X SSPE buffer, 10 µg/mL sonicated and boiled salmon sperm DNA, 0.04% ficoll, 0.04% polyvinylpyrrolidone, 0.04% bovine serum albumen (Pentax, Fraction V). The last three components are those of Denhardt's solution *(12)*.
9. Sealable plastic sack and electrical sealer. We have used Micro-Seal bags from Dazey Corporation (Industrial Airport, KS 66031).
10. X-ray film. We have used Kodak Diagnostic Film SB.
11. X-ray film cassettes.

3. Methods

3.1. Fluorography

1. Add 99.43 g of salicylic acid/L of final solution to a beaker and add 700 mL of distilled water, followed by 72 mL of $10M$ NaOH. Stir the solution on a magnetic stirrer until clear. Adjust the pH

to 7 by use of either 10*M* NaOH or 10*M* HCl and a hand-held pH meter that is used only for this purpose. Add 100 mL of glycerol (*see* Note 1) and then increase the total vol to 1000 mL by adding distilled water.

2. After electrophoresis, remove an agarose gel from its electrophoresis apparatus and place it in a plastic pan containing sufficient solution of salicylate (from Step 1) to cover the gel. Typically, we use agarose gels that are 13 cm × 12 cm × 0.5 cm and 500 mL of the solution of salicylate. Soak the gel in the solution of salicylate at room temperature (25 ± 3°C) for 2 h.

3. After permeating the gel with salicylate, transfer the gel by use of the flexible plastic sheet to a double layer of filter paper that has previously been wet by soaking in distilled water.

4. Place the filter paper, with gel on top, on a rack in an oven held at 56°C. The gel dries in 15–20 h (i.e., overnight) (*see* Note 2).

5. The dried gel adheres to the top layer of filter paper. Prepare the dried gel for fluorography by trimming with scissors to fit in the X-ray film cassette.

6. Wrap the dried gel with plastic wrap and place in the X-ray film cassette, in opposition to the emulsion side of the X-ray film (*see* Notes 3 and 4).

7. Clamp the X-ray film cassette between two glass plates and place in a –70°C freezer. The use of –70°C during fluorography increases the efficiency of the response of the film *(13)*.

8. After exposure of the film for a time that depends on the amount of ^{3}H in the gel, remove the X-ray film from the cassette and develop it. The minimal exposure needed to detect a band formed by a DNA with 100 ^{3}H counts/min is about 4–5 d. Advantages of this procedure of fluorography are discussed in Note 5.

3.2. In-Gel DNA–DNA Hybridization

1. After electrophoresis, place the agarose gel (typically 13 cm × 6 cm × 0.5 cm) in a plastic pan and wash in three changes of distilled water for 2 h. Put the washed gel on a transparent plastic sheet and place in the oven to dry at 55°C. Drying occurs within 10–20 h (i.e., overnight) (*see* Notes 6 and 7).

2. Rehydrate the dried gel by soaking in distilled water for about 30 min and shaking on the shaking platform.

3. Soak the rehydrated gel in approx 500 mL denaturing solution for 2 h with shaking. Change the denaturing solution twice during this period.

4. Wash the alkalinized gel by soaking in distilled water (approx 500 mL) for 20 min with shaking; change the distilled water once during this time.

5. Neutralize the gel in 1X SSPE buffer (approx 500 mL) for 30 min with shaking. Subsequently, wash the gel with 0.1 SSPE for 1 h; change the buffer every 15 min.

6. After washing, seal the gel in the plastic sack, together with hybridization buffer (approx 1 mL/10 cm^2 of gel surface). Rock the gel gently on the rocking platform at 42°C for 15–20 h (overnight). Use only one gel per sack.

7. To hybridize a ^{32}P-labeled probe to DNA in the gel, add the probe to fresh hybridization buffer and seal this mixture (approx 1 mL/10 cm^2) in the sealable plastic sack, together with the gel. Hybridize the probe to DNA in the gel by rocking gently at 42°C for 24–48 h.

8. After hybridization, shake the gel at room temperature with 1X SSPE buffer that has 0.1% sodium dodecyl sulfate (approx 500 mL) for 1 h; change the buffer twice during this time.

9. Further shake the gel at 55°C (in the water bath) in 0.1X SSPE with 0.1% sodium dodecyl sulfate, for 2 h; change the buffer once.

10. To prepare the gel for drying, continue washing at 55°C for 1 h in 0.1X SSPE that has 0.1% sodium dodecyl sulfate and 12% w/v glycerol

11. Dry the gel at 42°C for 1–3 h.

12. Wrap the dried (flat) gel with plastic wrap and subject it to autoradiography by use of the X-ray film and cassettes.

Possible modifications and uses of this procedure of hybridization are discussed in Notes 8–11.

4. Notes

4.1. Fluorography

1. The 10% glycerol was added to the salicylate solution to prevent the extrusion of flakes of sodium salicylate during the air-drying of the agarose gel. Even in the absence of salicylate, adding 10% glycerol to an agarose gel changes the characteristics of the gel after drying. A gel dried from water without glycerol has a brittle, dry texture, and curls if more concentrated than 0.7–0.8%. A gel dried in the presence of 10% glycerol has a tough, rubbery, slightly sticky texture, and is flat for agarose concentrations as high as 2%. This observation inspired development of the procedure of in-gel DNA–DNA hybridization presented in Section 3.2.

2. The use of air-drying has the advantage of avoiding the cost, time, and space expended for the currently more often used procedures of vacuum-drying. We have dried as many as 20 gels simultaneously in one oven. Air-drying is also more fail-safe than vacuum-drying.
3. Dried gels with salicylate are hygroscopic. Wrapping should be done within 15 min of removal from the 56°C oven.
4. The only serious problem that we have had with this procedure of fluorography is the development of dark spots that are on the X-ray film, but that are not produced by tritium decay in the gel. Some of these spots are caused by sparks produced when pressure is applied to some plastic wraps. In our experience, Glad™ and Saran™ wraps typically do not produce sparks. We have not made a comprehensive survey of the different sources of plastic wrap. Because of the hazard of salicylate contamination of cassettes, we are reluctant to discontinue the wrapping of gels. A possible (though not proven) additional source of spots is the oozing of salicylate through the plastic wrap to contact the film; sodium salicylate, if deliberately brought into contact with film, will expose it. If this latter source of spots is suspected, a piece of transparent film can be placed between the wrapped gel and X-ray film, without detectable loss of the sharpness of bands.
5. The procedure of fluorography described here was originally presented in ref. *8* and was inspired by a previously-presented procedure of salicylate-based fluorography for polyacrylamide gels *(14)*. Its major advantages are simplicity and efficiency (in labor and cost per gel). Examples of the use of this procedure are also in ref. *8.*

4.2. In-Gel DNA–DNA Hybridization

6. As the concentration of both agarose and salt in a glycerol-free agarose gel increases, the curling that occurs during air-drying of the gel increases. Although addition of glycerol prevents curling, glycerol is not added for the first drying (Step 1 in Section 3.2.) because of a resultant increase in the nonspecific background retention of ^{32}P-labeled probe in the gel. The first washing of the gel (Step 1 in Section 3.2.) minimizes curling by removing salt. Without the salt, gels less concentrated than 0.7–0.8% do not curl, but more concentrated gels do. For gels that are at least as concentrated as 1.5%, the curling that occurs does not harm the DNA–DNA hybridization, if care is taken not to crack the curled gel by putting a weight on it. Dried gels can be stored at room temperature for at least 3 mo before hybridization.

7. After the first drying, the agarose gel is light enough to be blown away by air currents present in most laboratories. Therefore, care must be exercised when transporting the gel after the first drying.

8. To observe DNA without probing, the agarose gel can be stained with ethidium either before drying (i.e., while washing) or after. DNA in a gel stained before drying will remain stained after drying, during rehydration.

9. After drying of the gel, the DNA appears irreversibly trapped in the gel. Even the probe appears trapped after hybridization. Attempts to remove probes by posthybridization soaking of the gel in denaturation buffer have, thus far, failed. This difficulty in removing probes is a major limitation of in-gel DNA–DNA hybridization. That is, we have thus far been unable to reprobe.

10. To increase the specificity of hybridization, if needed, the temperature of washing can be raised. This is done by washing and drying the gel, as described in Methods and, then, performing: (a) additional washing at a temperature that can be as high as 70°C and (b) drying in the presence of glycerol. The drying in glycerol performed before the second wash toughens the gel so that a 0.4–1.5% agarose gel can withstand washing at the higher temperatures. Disintegration of agarose gels during washing was a greater problem before introduction of this procedure of toughening the gel *(15)*.

11. When we have compared the results of Southern transfer-hybridization *(16)* to those of in-gel hybridization, by use of a T7 DNA probe ^{32}P-labeled by use of nick translation *(12)*, we have usually found the two techniques to both have equivalent sensitivity (±50%) and produce bands with the same width (±20%) for 40 kbp double-stranded DNA. (Occasionally, we have had transfers that failed.) However, when we analyzed DNA present in lysates of bacteriophage T7-infected *E. coli*, we found that a major constituent of these lysates was revealed by in-gel DNA–DNA hybridization at levels that were much higher than the levels revealed by Southern transfer-hybridization. For example, in Fig. 1A is the pattern obtained when nonpermissive *E. coli* infected with T7 amber mutant in gene 19 (a gene whose product is necessary for DNA packaging, but not any other known process) were lysed at 14 min after infection and the DNA in the lysate was analyzed by field-inversion gel electrophoresis, followed by in-gel DNA–DNA hybridization. After probing by use of nick-translated ^{32}P-mature T7 DNA *(12)*, both a dark band near the origin of electrophroesis (arrow with an asterisk in Fig. 1A) and a band at the position of mature T7 DNA (1-mer in Fig. 1A) were revealed. However, when the same analysis

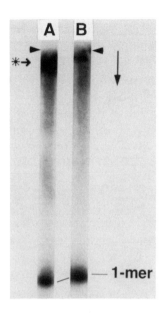

Fig. 1. DNA in a lysate of T7-infected *E. coli*. At 30°C, a culture of *E. coli* BB/1 was infected with T7 amber mutant in gene 19. At 14 min after infection, the culture was quenched and lysed *(8)*. The contents of the lysate were fractionated by use of field-inversion electrophoresis (apparatus from DNAStar) at 3 V/cm, 15°C, for 24 h through a 1.5% agarose gel cast in 0.01M sodium phosphate, pH 7.4, 0.001M EDTA. The time forward/time reverse ratio was 3 and the time forward was linearly ramped from 6 s to 18 s. After electrophoresis, the profile of T7 DNA in the gel was determined by hybridization with T7 DNA ^{32}P-labeled by use of nick translation. Hybridization was performed either (a) in-gel or (b) after Southern transfer. The arrowheads indicate the origins of electrophoresis; the vertical arrow indicates the direction of electrophoresis.

was performed after transfer to a nylon filter by use of denaturing solution (Fig. 1B), the origin-proximal band that had been observed in Fig. 1A was much weaker, even though the 1-mer was observed at a level comparable to that observed in Fig. 1A. Some continuously distributed DNA between these two bands was also observed in both Fig. 1A and B. The origin-proximal band was not present for lysates prepared before 8 min after infection and became more intense as time increased from 8 to 18 min after infection with T7. This origin-proximal DNA is, therefore, assumed to be the origin-proximal replicating DNA previously detected and characterized *(8)*. A detailed study of which forms of DNA transfer and which do not has not yet been made.

The reason for the inefficient transfer in Fig. 1B is not known, although failure of branched DNA to completely denature either before or during transfer seems likely to be a part of the explanation. In any case, to detect untransferred DNA after Southern transfer-hybridization, the gel used for transfer can be probed by in-gel hybridization. The basics of the present procedure of in-gel DNA–DNA hybridization were initially presented in ref. *17*.

Acknowledgments

We thank Cherie Shrewsbury and E. Anthony Meyer for technical assistance, and Linda C. Winchester for typing this manuscript. Support was received from the National Institutes of Health (GM-24365) and the Robert A. Welch Foundation (AQ-764).

References

1. Cantor, C. R., Smith, C. L., and Mathew, M. K. (1988) Pulsed-field gel electrophoresis of very large DNA molecules. *Ann. Rev. Biophys. Biophys. Chem.* **17**, 287–304.
2. Olson, M. V. (1989) Pulsed-field gel electrophoresis, in *Genetic Engineering, Principles and Methods*, vol. 11 (Setlow, J. K., ed.), Plenum, New York, p. 183.
3. Serwer, P. (1990) Sieving by agarose gels and its use during pulsed-field electrophoresis. *Biotechnol. and Genet. Eng. Rev.* **8**, 319–343.
4. Casjens, S. (1985) Nucleic acid packaging by viruses, in *Virus Structure and Assembly* (Casjens, S., ed.), Jones and Bartlett, Boston, p. 76.
5. Serwer, P. (1989) Double-stranded DNA packaged in bacteriophages: Conformation, energetics and packaging pathway, in *Chromosomes: Eukaryotic, Prokaryotic and Viral*, vol. 3 (Adolph, K. W., ed.), CRC Press, Boca Raton, FL, p. 203.
6. Gemmill, R. M., Coyle-Morris, J. F., McPeek, F. D., Jr., Ware-Uribe, L. F., and Hecht, F. (1987) Construction of long-range restriction maps in human DNA using pulsed field gel electrophoresis. *Gene Anal. Technol.* **4**, 119–131.
7. Weissman, S. M. (1987) Molecular genetic techniques for mapping the human genome. *Mol. Biol. Med.* **4**, 133–143.
8. Serwer, P., Watson, R. H., and Hayes, S. J. (1987) Multidimensional analysis of intracellular bacteriophage T7 DNA: Effects of amber mutations in genes 3 and 19. *J. Virol.* **61**, 3499–3509.
9. Serwer, P., Watson, R. H., and Son, M. (1990) Role of gene 6 exonuclease in the replication and packaging of bacteriophage T7 DNA. *J. Mol. Biol.* **215**, 287–299.
10. Shibata, H., Fujisawa, H., and Minigawa, T. (1987) Characterization of the bacteriophage T3 DNA packaging reaction in vitro in a defined system. *J. Mol. Biol.* **196**, 845–851.

11. Chung, Y.-B., Nardone, C., and Hinkle, D. C. (1990) Bacteriophage T7 DNA packaging: III. A "hairpin" end formed on T7 concatemers may be an intermediate in the processing reaction. *J. Mol. Biol.* **216,** 939–948.

12. Sambrook, J., Fritsch, E. F., and Maniatis, T. (1989) *Molecular Cloning: A Laboratory Manual,* vol. 2. Cold Spring Harbor Laboratory, Cold Spring Harbor, New York.

13. Laskey, R. A. and Mills, A. D. (1975) Quantitative film detection of ^{3}H and ^{14}C in polyacrylamide gels by fluorography. *Eur. J. Biochem.* **56,** 335–341.

14. Chamberlain, J. P. (1979) Fluorographic detection of radioactivity in polyacrylamide gels with the water-soluble fluor, sodium salicylate. *Anal. Biochem.* **98,** 132–135.

15. Mather, M. W. (1988) Base composition-independent hybridization in dried agarose gels: Screening and recovery for cloning of genomic DNA fragments. *Biotechniques* **6,** 444–447.

16. Southern, E. M. (1975) Detection of specific sequences among DNA fragments separated by gel electrophoresis. *J. Mol. Biol.* **98,** 503–517.

17. Son, M., Watson, R. H., and Serwer, P. (1990) In-gel hybridization of DNA separated by pulsed-field agarose gel electrophoresis. *Nucleic Acid Res.* **18,** 3098.

PART IV

THEORIES AND OBSERVATIONS

CHAPTER 23

Theories of Pulsed-Field Gel Electrophoresis

Stephen D. Levene

1. Introduction

Pulsed-field gel electrophoresis (PFGE) is one of the key techno-logical advances of the past ten years that has made the mapping of genomes of whole organisms possible. In conventional electrophore-sis, the mobility of DNA at almost any practical value of the field strength is essentially independent of mol wt above approx 30 kbp. Therefore, large (≥50 kbp) fragments of DNA required for mapping the genomes of entire organisms could not be separated prior to the introduc-tion of these new electrophoresis techniques.

PFGE, in its earliest incarnation introduced by Schwartz and Cantor (*1,2; see also* Chapter 4), used two sets of electrodes in a square configuration to generate highly inhomogeneous electric fields. The voltage was periodicially switched from one set to the other, swinging the field direction by an obtuse angle back and forth. In their original work, Schwartz and Cantor proposed that electric-field gradients were responsible for the dramatically improved separation of DNAs in the size range of 50 kbp to nearly 2 Mbp. In subsequent work by Chu, Vollrath, and Davis (*3; see also* Chapter 2), a hexagonal array of voltage-clamped electrodes was used to generate homogeneous alter-nating fields with a constant reorientation angle of 120°, an approach they denoted Contour-Clamped Homogeneous Electrophoresis

From: *Methods in Molecular Biology, Vol. 12: Pulsed-Field Gel Electrophoresis*
Edited by: M. Burmeister and L. Ulanovsky
Copyright © 1992 The Humana Press, Inc., Totowa, NJ

(CHEF). Separations of high-mol-wt DNA fragments qualitatively similar or superior to those of the Schwartz-Cantor approach were obtained using the CHEF system; thus, it became clear that inhomogeneous fields were not required in order to effect electrophoretic separations of large DNAs.

At about the same time, Carle, Frank, and Olson introduced another development (*4; see also* Chapter 1). They showed that periodically inverting the field in a conventional two-electrode gel electrophoresis experiment (but using a duty cycle greater than 50%, so that the field is on longer in one direction than in the opposite direction) leads to a striking improvement in the resolution of DNAs within a restricted size range. Moreover, the overall dependence of mobility on size was generally observed to be a nonmonotonic function of DNA length; the mobilities of some of the DNA fragments in these experiments were reduced by 1–2 orders of magnitude relative to their steady-field values. The expected decreasing mobility-to-mol wt relationship was retained for molecules smaller than this critical mol wt, whereas mobility increased with mol wt for DNAs larger than the critical value. Above a particular DNA mol wt larger than the value at the mobility minimum, DNA mobility was once again independent of mol wt. Carle, Frank, and Olson called this intermediate range of DNA lengths the "window" of improved resolution.

A quantitative understanding of the effects of rapid changes in field strength or direction on mobility remains largely incomplete, owing to the relative simplicity of the models used to describe gel electrophoresis. In this review, we discuss principles of a number of statistical-mechanical theories for the gel electrophoresis of DNA and their application to the analysis of PFGE experiments. We consider first a class of models based on the simple assumption that DNA molecules migrate end-on through the gel much like a snake in a thicket of bamboo. This motion is called reptation, a concept introduced into polymer physics by de Gennes (*5*) and Doi and Edwards (*6,7*). Reptation models succeed fairly well in describing the behavior of DNA undergoing electrophoresis in steady fields in dilute gels, such as agarose, and also predict effects of chain reorientation on mobility. Where conventional reptation theories fail dramatically is in their inability to account for the improvement in resolution obtained when the field is inverted. This failure has led to the development of another class of models that allow for other motions of the DNA chain in addition to end-on migration. All of the latter models treat the DNA chain in considerable detail. This invariably requires that the equations of motion of the chain be solved using computer simulation techniques. The details of the interaction between the chain and the gel are considered in various degrees. Some of the simulations are sophisticated extensions of tube models, whereas others abandon the reptation framework entirely and consider the interactions of the chain with individual obstacles in the gel. We conclude the review with a comparison of the theories to some available experimental data.

2. Electrophoresis Models

2.1. Tube-Reptation Models

We consider here those models in which the DNA chain is constrained to move only along its axis, as though confined to a tube in the gel. Originally introduced by de Gennes *(5)* and elaborated by Doi and Edwards *(6,7)* (in their terminology, the tube is generally referred to as the "primitive path"), the tube concept permits a simple description of the curvilinear path of the DNA chain constrained by contacts with the gel fibers, and avoids considering interactions of the chain with the gel in detail. Limited folding of the chain inside the tube is permitted, however, penetration of the wall of the tube by the chain is not. Thus, the tube is always a linear, as opposed to a branched, structure.

An expression for the electrophoretic mobility of the chain confined to the tube was derived by Lumpkin and Zimm *(8)* by considering the balance between the electrostatic force on the chain and the chain's frictional drag. Lerman and Frisch derived a similar expression somewhat earlier using a scaling argument *(9)*. The mobility, μ, is given by

$$\mu = \frac{Q}{\zeta} \langle h_x^2 / L^2 \rangle \tag{1}$$

where Q is the total charge on the DNA chain, ζ is the friction coefficient of the chain for motion along the tube, h_x is the component of the tube's end-to-end vector, h, along the field direction, L is the contour length of the tube, and the angle brackets denote an average over all conformations of the tube. This equation expresses the fact that the mobility of the chain will be greatest whenever the chain is maximally extended along the field direction, namely when h_x is equal to L. When $h_x = L$, all of the electrostatic force is directed downfield and the force has no component directed at the walls of the tube. Conversely, the mobility is zero when h_x is exactly zero. In this case, there can be no work done by the field to move the chain because all of the electrical force applied to the chain is directed at the walls of the tube. The derivation of this expression is given in the Appendix.

Equation 1 is not very useful for molecules of appreciable size when applied to DNA undergoing electrophoresis in agarose at any practical field strength. This is because the conformation of the tube is determined entirely by the stochastically distributed orientation of the leading segment of the chain, which is biased by the electric field. As the chain moves, the leading segment moves to form an extension of the tube while the other end of the chain vacates the trailing end of the tube. The leading segment tends to choose orientations biased in the downfield direction. The cumulative effects of this bias eventually cause the entire chain to become oriented. The field-induced bias increases the mobility of the chain and also leads to a dramatic decrease in resolution because h_x becomes more dependent on the field and less dependent on the mol wt of the DNA chain.

Lumpkin, Dejardin, and Zimm *(10)* explored this effect quantitatively and showed that the orienting effect of the field increases strongly with chain length, so that the mobility of high-mol-wt DNA is virtually independent of mol wt at surprisingly low fields, around 100 V/m (1 V/cm). Their expression for the mobility is a sum of two terms. The first term is DNA mol wt-dependent and accounts for the resolving power of gel electrophoresis. With increasing field strength, however, the effect of field-induced orientation increases the magnitude of the second mol wt-independent term relative to the first, hence the mobility becomes progressively less dependent on mol wt as the second term becomes dominant. Slater and Noolandi *(11,12)*, using a similar approach, derive a slightly different expression that leads to qualitatively similar saturation of the mobility with increasing field strength.

Because the mobility depends only on the value of $<h_x^2>$ in this model, Lumpkin, Dejardin, and Zimm were able to estimate the effect of field-induced reorientation for a simple case in which the reorientation angle was equal to 90°. We have generalized their formula to the case of a pulsed field with an arbitrary reorientation angle. The results are shown in Fig. 1. Consider a case in which the direction of the field, applied for long enough time that the tube has reached steady state, is rotated by a specified angle in the x,y-plane. We calculate

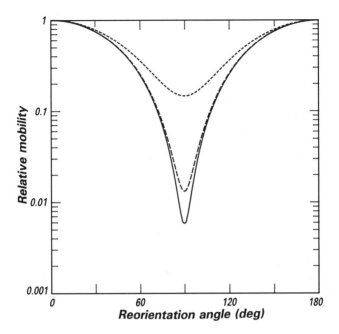

Fig 1. Dependence of the mobility in the alternate field direction on reorientation angle at three values of the reduced electric field strength, $E' = aqE/2k_BT$, calculated using Eq. 13 of ref. *10*. E' = 1.0 (solid curve), E' = 5.0 (simple dashed curve), E' = 10.0 (dashed-dotted curve). Mobilities are expressed relative to the steady-field value, corresponding to 0°. Numerical values of the parameters used were mean interfiber spacing,$<a> = 300$ nm, mean chain contour length per tube segment,$<l> = 900$ nm, DNA persistence length, $p = 50$ nm, number of tube segments, $N' = 340$. The assumed charge/segment, q, is taken to be 15% of the formal charge of two electrons/bp (D. Stigter, personal communication); the lowest value of E' shown here corresponds to an applied field E of 230 V/m (2.3 V/cm).

the initial projection of h along the new field direction and then substitute this expression for $<h_x^2>$ into Eq. 1 to obtain the initial mobility along the new field direction. The calculated mobility of a 50 kbp molecule in the new field direction is plotted in Fig. 1 as a function of reorientation angle for three values of the reduced field E'. There is a large field strength-dependent effect of reorientation angle on the mobility in this highly simplified model; at a reduced field $E' = 10$ (approximately equivalent to an applied field of 2000 V/m with the parameters assumed in Fig. 1), the initial mobility is more than 100-fold lower at 90° than at 0°, the case of a steady field. At $E' = 1$, the effect is smaller but still substantial, resulting in an approximate sevenfold decrease in the mobility with 90° switching. The greatest resolution is predicted by this model to occur for the cases in which the reorientation angle is exactly equal to 90°. Under these conditions, the field-independent term exactly vanishes. It should also be noted that the dependence of mobility on reorientation angle is symmetric about an angle of 90°; obtuse and acute reorientation angles are therefore predicted to have the same effect on $<h_x^2>$.

Extensive experimental data are available on the dependence of DNA mobility on reorientation angle *(13,14)*. Both predictions of tube-reptation models, the predicted optimum reorientation angle, and the symmetrical angular dependence of resolution, disagree with the experimental results, which indicate that the optimum reorientation angle lies near 120°. In most experimental configurations used, the increased resolution of high-mol-wt DNA is obtained with reorientation angles >90°, and little or no improvement in resolution over that of steady-field electrophoresis could be detected for values <90°. This broken symmetry is related to the effect of field inversion on mobility, which we now discuss in greater detail.

The expression for the field-dependent mobility in reptation models given by Lumpkin, Dejardin, and Zimm *(10)* is symmetric with respect to field inversion. Orientation of the chain is the main source of nonlinear behavior in this model, and arises from the biased choice of direction by the chain's leading segment. After a time long enough for the chain to migrate its own length, the trailing segment of the tube has, on average, the same degree of orientation in the field direction as the leading segment, thus, changing the sign of the field has no effect on the extent of orientation in simple reptation models, but merely affects the likelihood of which end of the tube becomes the leading segment. Tube-reptation theories, therefore, predict that the mobility of the chain in the reverse direction should be the same as that in the forward direction.

The inadequacy of these models became strikingly apparent with the field-inversion electrophoresis (FIGE) results of Carle, Frank, and Olson *(4)*. In these experiments, the field was reversed for periods one-half or one-third that of the field pulse in the direction of net migration. Typical results, such as those obtained by Kobayashi, et al. *(15)* for three different DNA mol wt, are shown in Fig. 2. The mobilities of each of the samples have strong minima near critical values of

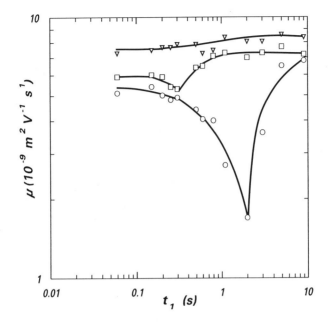

Fig. 2. Mobility of DNA in FIGE as a function of the period of the forward pulse, t_1. Replotted from Kobayashi, et al. (ref. *15*). The ratio of reverse-pulse period to t_1 was constant and equal to 1:3. The reduced field E', was equal to 670 V/m (6.7 V/cm, *see* caption to Fig. 1). Symbols correspond to DNA samples differing in mol wt: ($\triangledown$) 6.6 kbp, ($\square$) 23.1 kbp, ($\bigcirc$) 166 kbp. Curves are merely to guide the eye.

the forward pulse width, termed "antiresonance" conditions. Indeed, Carle, Frank, and Olson showed that for DNA of a given mol wt, the mobility can become negligible at the antiresonance. In this case, the mobilities of DNAs of lower *and* higher mol wt can exceed that of a given fragment, leading to a nonmonotonic dependence of mobility on mol wt in FIGE. Choice of the proper cycle times results in a "window" of resolution, in which molecules with a distribution of mol wt within this window are well-separated. Similar effects are observed with forward and reverse pulses of the same duration but unequal field strengths.

2.2. Simulation Models

The inadequacy of tube models is attributable to at least two significant assumptions. First, fluctuations in the contour length of the tube are assumed to be small and uncorrelated from those in h_x. Second, loops of DNA are assumed not to penetrate the walls of the tube, so that branched tube conformations are forbidden. In this section, we describe various theories that attempt to overcome one or both of these limitations. A common feature of these models is that they describe the detailed dynamics of the DNA chain explicitly, in contrast to the rather indirect description characteristic of tube models. In all of the cases considered here, the equations of motion of chains undergoing electrophoresis are solved numerically. These approaches treat the DNA chain as an array of discrete, charged segments. The chain may consist of charged spheres connected by neutral springs with varying degrees of extensibility, or it may be comprised of rigid, uniformly charged cylinders connected by flexible joints.

The earliest and conceptually simplest simulation model was developed by Deutsch *(16)* and Deutsch and Madden (*17; see also* Chapter 24). They solved the equations of motion of a two-dimensional chain undergoing electrophoresis in a lattice of point obstacles. In this treatment, the chain is composed of charged spheres connected by electrically neutral springs. The forces acting on each sphere include an electrostatic force, a random force due to Brownian motion, and a repulsive force between the chain and obstacles in the gel. In this model, the tension in the springs is continually adjusted during the simulation, so that the springs connecting the beads of the chain are almost inextensible. This condition increases the simulation's computational efficiency.

The behavior of the chain in this model, as shown in Fig. 3, is completely unlike that of the tube model, even in steady fields. In steady fields, the chain alternates between highly extended conformations, in which the chain is extended nearly to its contour length, and compact conformations, in which large lengths of chain fold back among themselves in the spaces between the obstacles. These large folds (kinks, in Deutsch's terminology) arise because the front end of the chain encounters obstacles as it moves, and hence moves slower than the trailing end of the chain. The high density of chain segments near the leading end frequently leads to the formation of loops of chain, which Deutsch calls a "bunching instability." Eventually, the chain unwinds around the obstacles. The last phase of this process usually leaves the chain hooked on one of the obstacles. The small loops disappear until the chain becomes almost completely extended, with the two arms of the chain on opposite sides of the hooked obstacle both pulled downfield. Because one of these arms is usually longer than the other, the longer arm wins, and the chain repeats the cycle of extended and

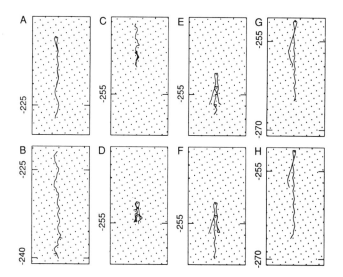

Fig. 3. High-field behavior of chain in steady fields according to Deutsch's computer simulation. Reproduced from ref. 16. The value of the reduced field was $E' = 21.3$. Relative times for each panel are : (A) 100, (B) 105, (C) 110, (D) 112, (E) 114, (F) 116, (G) 118, (H) 119. Reprinted with permission.

folded conformations. Although the extended chain conformations conform quite well to the notion of a chain confined to a tube in the gel, bunching of the chain and the formation of loops clearly violate the basic assumptions of the tube model.

At fields that are more typical of laboratory electrophoresis experiments, the motion of the chain differs somewhat from that described above, as shown in Fig. 4. Instead of the coiling process taking place at the leading end of the chain, loops tend to form near the middle of the chain. Relaxation of these conformations leads to an inverted-U conformation that has qualitatively the same behavior as a chain hooked on a single obstacle at high field. The inverted-U conformations have a significant role in the mechanism of field-inversion electrophoresis. These conformations have a characteristic lifetime, which when matched by the period of the reverse-field pulse, results in the chain undergoing little or no net motion. Deutsch *(18)* examined the effects of field inversion on the chains in his simulation, and found that if the reverse-pulse duration is short compared to the time required for formation of an inverted-U conformation, then the U conformation does not have time to reorient, and the chain merely

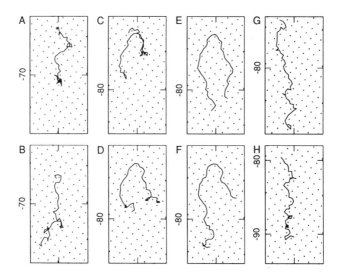

Fig. 4. Low-field behavior of a chain in steady fields according to Deutsch's simulation. Reproduced from ref. *16*. The value of the reduced field was *E'* = 0.30. Relative times for each panel are: (**A**) 260, (**B**) 270, (**C**) 280, (**D**) 290, (**E**) 330, (**F**) 340, (**G**) 350, (**H**) 360. Reprinted with permission.

experiences a rapidly averaged, and slightly lower, net field. If, however, the pulse time is long compared to the lifetime of the U conformations, then the chain will reach steady-state conditions during both phases of the cycle and migrate back and forth along the axis of migration. The time-averaged mobility observed with long pulses is, therefore, comparable to that obtained with steady fields. It is only when the pulse time is similar to the lifetime of the U conformations that the time-averaged mobility of the chain is significantly lower than in steady fields.

Zimm *(19,20)* and Duke *(21)* introduced theories that extend the tube model by removing the constraint of constant tube length. These models have some important advantages because they apply to chains moving in three dimensions, however, the models contain somewhat less detail than the Deutsch model. Neither the Zimm nor the Duke simulation uses an explicit model for the gel obstacles, so detailed information about the winding of chains around obstacles is not available. Nevertheless, these two models show behavior in field inversion electrophoresis that is strikingly similar to that of Deutsch's simulation.

In the Zimm model, a chain of rigid, freely jointed segments occupies a series of voids in a gel ("lakes") connected by narrow channels ("straits"). The lakes correspond to large spaces in the gel that typically contain two or more chain segments. The segments can flow through the straits between adjacent lakes. The number of lakes can fluctuate, thus the succession of lakes represents a tube of varying contour length. The flow of chain segments into (or out of) a lake is proportional to the local force acting on the chain segments. This force is a sum consisting of the usual terms, an electrostatic force and Brownian motion. There is, in addition, a tension that is related to the occupancy of the lakes. This tension is derived from formulas analogous to the force-extension relationships in rubber elasticity; there is an entropic force that opposes stretching and compression of the chain. The tension increases rapidly when the occupancy of a lake is low, passes through zero, and rapidly approaches a small, negative asymptotic value for high occupancy. When the difference in tension between two lakes is positive, a force favors a flow of segments into the lake with fewer segments.

The formation of new lakes, a process that is favored when the density of segments in a lake is high, is permitted anywhere along the tube of lakes and straits. This naturally occurs at the end lakes, but results in loops when overflows occur from lakes in the middle. In the absence of a field, or at low fields, there is an entropic barrier opposing formation of a new lake; a lake can store a large amount of chain before chain segments are forced out. Lake overflow, therefore, occurs slowly. At high fields, the electrostatic force can overcome this entropic barrier given enough time. The chain alternates between compact conformations, in which the average occupancy of the lakes is high, and extended conformations, in which the average density of segments in the lakes is low. Strong fields can pull segments of chain out of the two end lakes simultaneously, leading to inverted-U conformations similar to those observed by Deutsch (Zimm calls these lambda conformations, after the Greek letter Λ). As in Deutsch's model, the mobility of the lambda conformations is low when the period of field-switching is comparable to the lifetime of the lambda conformations. The high tension built up in the chain when the lambda conformations are formed leads to a rapid recoil of the arms of the lambda when the field is reversed. This rapid motion during the reverse phase of the cycle nearly cancels the motion that takes place during the forward part of the cycle, and accounts for the low mobilities that occur at the antiresonance.

The "repton" model presented by Duke *(21)* is very similar to the earlier version of Zimm's lakes-and-straits model *(19)*, but contains more assumptions. In this treatment, a repton corresponds to a segment of chain whose radius is approximately equal to the radius of a pore in the gel. The dynamics of this model are given by the rates at which reptons are allowed to hop to adjacent sites on a lattice. The transition probabilities for this process are determined by the electrostatic energy of a repton in the field. No tension exists between adjacent reptons. At the ends of the tube an entropic force is added to prevent the contraction of the chain inside the tube that would otherwise occur. Overflows of loops from the middle of the tube are not allowed.

As with Deutsch's and Zimm's models, the chain cycles between extended and compact conformations and spends a considerable amount of time in inverted-U conformations, which Duke calls hairpins. The close agreement of these three distinct treatments is remarkable given the differences in the detailed dynamics of the models. All three theories predict that U-shaped conformations play an important role in the dynamics of DNA in gel electrophoresis. The existence of U-shaped conformations and their important effects on the dynamics of DNA in gel electrophoresis, are underscored by observations of DNA molecules in gels by fluorescence microscopy *(22–24)*.

The transient formation and relaxation of U conformations are important in the context of orthogonal pulsed fields also. Olvera de la Cruz, Gersappe, and Shaffer *(25)* extended the earlier work of two of the authors on steady-field electrophoresis *(26)* to the case in which the field axis is rotated by 90°. Their two-dimensional model is very similar to Deutsch's, except that the beads are connected by extensible springs, hence there is no constraint on the contour length of the chain. The behavior of the chain is likewise similar to that in Deutsch's model in steady fields: the chain cycles through compact, extended and inverted-U conformations.

Complex behavior is observed when the orientation of the field is switched, as shown in Fig. 5. The detailed behavior of the chain depends on its instantaneous conformation at the time the field is switched. At high fields, field switching leads to very open inverted-U conformations, with arms pointing down in the new field direction if the chain is initially strongly stretched in the previous field direction. These conformations presumably result from the field pulling on the two chain ends. If the chain is initially compact, however, instabilities form along the chain, leading to the formation of loops. At lower fields,

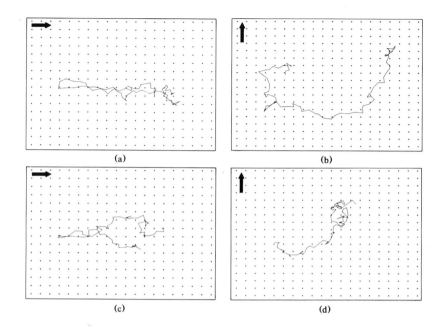

Fig. 5. Behavior of a chain in orthogonal pulsed fields. Reproduced from ref. 25. Arrows indicate direction of the electric field; **(A)** corresponds to $E' = 0.1$ and **(C)** corresponds to $E' = 0.15$. The chain consists of 80 segments. **(B)** and **(D)** show the conformation 400 time steps after the field direction was switched. Reprinted with permission.

open-U conformations form when the chain segment density is high near the ends so that any instabilities that form quickly become disentangled from the obstacles. If the chain is stretched and there are no dense regions, then the instabilities that form along the chain pull the ends through the obstacles.

The net result of orthogonal field switching prevents the collapse of the chain into compact conformations if the period of field switching is comparable to the time required for formation of the inverted-U conformations. Olvera de la Cruz, Gersappe, and Shaffer studied the mobility of chains in their pulsed-field simulation that spanned a five-fold range of mol wt, from 20–100 Kuhn lengths (a Kuhn length is the length of a flight in an equivalent random-flight chain, and is equal to approx 100 nm). The mobility of the lowest mol wt chain along one of the field directions was reduced by a factor of two relative to that in steady fields, as expected. With longer chains, however, the mobility reduction was greater; a factor of four for chains of 60 and 100 Kuhn lengths when the pulse time was equal to 800 time steps, and a factor of three for chains 40 Kuhn segments and larger when the pulse time was equal to 100 time steps. The results suggest that an appropriate choice of pulse time generates a mol wt-dependent window of resolution, as in the case of field inversion.

3. PFGE Theory:
Prospects and Future Directions

Operationally, there are at least five parameters that affect the resolution of DNA molecules in PFGE: field strength, pulse time, reorientation angle, gel concentration, and temperature *(13,14,27,28)*. Optimum values of these parameters appear to be interdependent, making the process of determining ideal PFGE conditions tedious and generally impractical. Unfortunately, even the most sophisticated models reviewed here are probably too simplified to be quantitatively useful in estimating the optimum parameters for a given pulsed-field separation. Clearly, there are additional important effects that need to be incorporated into a quantitatively useful theory.

The strength of the electric field has profound effects on PFGE separations. High fields degrade the resolution of very high-mol-wt DNA molecules, possibly by trapping these molecules at the origin of electrophoresis *(29; see also* Chapter 7). Separations involving giant molecules, such as *Neurospora crassa* chromosomes believed to be of the order of 5–10 Mbp *(30)*, require extremely low fields, of the order of 1 V/cm or less. The requirement for such low fields leads to extremely long electrophoresis times, often more than 1 wk. Understanding the effect of high fields in detail might lead to strategies that avoid degrading the resolution of large DNAs at high fields, and could reduce substantially the time required to separate chromosome-sized DNAs. The optimum pulse time for separations over a given range of mol wt is likely to be closely coupled to values of the field strength and the reorientation angle, also.

Only obtuse reorientation angles seem to be effective in PFGE. Generally, useful separations are obtained when the reorientation angle is between 90–180° *(13,14)*. The mol wt range resolved depends strongly on the reorientation angle. In PFGE experiments in which the reorientation angle is 120°, the mobility is generally a monotonically decreasing function of mol wt below some critical value of the mol wt, above which all DNAs have the same mobility. In FIGE, however, the window of resolution is relatively narrow and nonmonotonic. This fact suggests the possibility that two relaxation processes occur; one corresponding to perturbations of the chain that occur on field reversal, the other associated with overall reorientation of the chain. The characteristic times of these processes may be quite different and they may have relative amplitudes that depend strongly on the reorientation angle. Interestingly, Olvera de la Cruz, Gersappe, and Shaffer *(25)* predicted an effect of 90° pulsed fields on the mobility. This result contrasts with the fact that little effect is observed experimentally with 90° pulses *(13)*. The effect of reorientation angle in their model should be studied further.

A serious deficiency in all of the models reviewed here may be the failure to account adequately for the properties of the gel. The models considered here take the gel to be either a featureless tube, possibly allowing some variation in the size of spaces comprising the tube, or a set of point obstacles in two dimensions. Electron microscopic evidence, however, suggests that agarose gels are networks of agarose fibers *(31)*. These fibers are arranged randomly in space, leading to a complex structure for the gel with a broad distribution of void sizes. We are currently developing a full three-dimensional model of gel electrophoresis that uses a fairly realistic and detailed model for the gel (Levene and Hasenfeld, unpublished data). Although present theories are remarkably consistent in reproducing the effect of field inversion on mobility, and one study has begun to address the behavior of DNA molecules in orthogonal fields, there remain other important effects that should be studied. One important question is whether high fields lead to trapping of large DNAs. Trapping effects with large DNAs might be analogous to the arrest of large DNA circles observed at high fields in agarose *(29,32,33)*. A detailed theoretical treatment of this effect may require a fairly realistic model for the gel matrix.

We are unaware of any current efforts to understand in detail the dependence of PFGE behavior on temperature. Temperature effects on mobility in steady-field electrophoresis arise mainly from changes in the chain's friction coefficient owing to the temperature dependence of the viscosity of water *(34)*. It is possible, however, that temperature may exert some nontrivial effect in PFGE through effects on the local tensions in the chain. Examining the temperature dependence of the tension function in Zimm's lakes-and-straits model may, therefore, be a good starting point for investigating this effect.

It now appears that the prospects for a practical theory of PFGE are good. Exploiting parallel experimental and theoretical approaches to understanding pulsed-field electrophoresis may eventually be key in developing new PFGE technologies.

Acknowledgments

I would like to thank B. H. Zimm and D. Stigter for communicating their results prior to publication. I also wish to thank N. R. Cozzarelli for support during the writing of this review.

Appendix

Derivation of the Lumpkin–Zimm Equation

Lumpkin and Zimm *(8)* expressed the local component of the electrical force along the tube acting on a segment of the chain, F_{el}, as follows:

$$F_{el} = \frac{Q}{L} E \bullet \Delta s \tag{A1}$$

where Q is the chain's effective charge, E is the electric field, L is the contour length of the tube, and Δs is a unit vector tangent to a short segment of the tube. Let the direction of the electric field define the x-axis; the total force acting in this direction is given by

$$\frac{QE}{L} \Sigma i \bullet \Delta s = \frac{QEh_x}{L} \tag{A2}$$

where i is a unit vector in the x-direction and h_x is the component of the chain's end-to-end vector in the field direction, equal to the sum of all the x-components of the vectors Δs. The electrical force acting on the chain is balanced by the chain's frictional drag, $\zeta \dot{s}$, where ζ is the friction coefficient for motion along the tube and $\dot{s}$, is the chain's curvilinear velocity, its velocity along the tube axis. Solving for the curvilinear velocity we obtain

$$\dot{s} = \frac{QEh_x}{\zeta L} \tag{A3}$$

We now need to relate the chain's curvilinear velocity to the velocity of the chain center of mass normally measured in an electrophoresis experiment. The position of the chain center of mass relative to the origin of an arbitrary coordinate system, R_{cm}, is given by

$$R_{cm} = \frac{1}{M}\Sigma m_i r_i \tag{A4}$$

where M is the total mass of the chain and m_i and r_i are the mass and position of the ith segment of the chain, respectively. The mass associated with this segment is given by

$$m_i = M\Delta s_i/L \tag{A5}$$

and the velocity of the segment is just its local velocity along the tube

$$\dot{r}_i = \dot{s} t_i \tag{A6}$$

where t_i is a unit vector tangent to the tube at the position of segment i. The velocity of the chain center of mass is therefore given by

$$\dot{R}_{cm} = \frac{\dot{s}}{L}\Sigma t_i \Delta s_i = \frac{\dot{s}}{L} h \tag{A7}$$

where h is the end-to-end vector of the tube. We are interested only in the component of the center-of-mass velocity along the field direction, x_{cm}, averaged over all conformations

$$<\dot{x}_{cm}> = <\dot{s}h_x/L> = \frac{QE}{\zeta}<h_x^2/L^2> \tag{A8}$$

Dividing the center-of-mass velocity, $<\dot{x}_{cm}>$, by the field strength, E, gives the expression for the mobility given in Eq. 1.

References

1. Schwartz, D. C., Safran, W., Welch, J., Haas, J., Goldenberg, R. M., and Cantor, C. R. (1983) New techniques for purifying large DNAs and studying their properties and packaging. *Cold Spring Harbor Symposium* **47**, 189–195.
2. Schwartz, D. C. and Cantor, C. R. (1984) Separation of yeast chromosome-sized DNAs by pulsed field gradient electrophoresis. *Cell* **37**, 67–75.
3. Chu, G., Vollrath, D., and Davis, R. W. (1986) Separation of large DNA molecules by contour-clamped homogeneous electric fields. *Science* **232**, 1582–1585.
4. Carle, G. F., Frank, M., and Olson, M. V. (1986) Electrophoretic separations of large DNA molecules by periodic inversion of the electric field. *Science* **232**, 65–68.
5. de Gennes, P. -G. (1971) Reptation of a polymer chain in the presence of fixed obstacles. *J. Chem. Phys.* **55**, 572–579.
6. Doi, M. and Edwards, S. F. (1978) Dynamics of concentrated polymer solutions. *J. Chem. Soc. Farady Trans.* **274**, 1789–1801.

7. Doi, M. and Edwards, S. F. (1986) *The Theory of Polymer Dynamics*. Oxford University Press, New York, pp. 205,206.

8. Lumpkin, O. J. and Zimm, B. H. (1982) Mobility of DNA in gel electrophoresis. *Biopolymers* **21**, 2315,2316.

9. Lerman, L. S. and Frisch, H. L. (1982) Why does the mobility of DNA in gels vary with the length of the molecule? *Biopolymers* **21**, 995–997.

10. Lumpkin, O. J., Dejardin, P., and Zimm, B. H. (1985) Theory of gel electrophoresis of DNA. *Biopolymers* **24**, 1573–1593.

11. Slater, G. W. and Noolandi, J. (1985) Prediction of chain elongation in the reptation theory of gel electrophoresis. *Biopolymers* **24**, 2181–2184.

12. Slater, G. W. and Noolandi, J. (1986) On the reptation theory of gel electrophoresis. *Biopolymers* **25**, 431–454.

13. Southern, E. M., Anand, R., Brown, W. R. A., and Fletcher, D. S. (1987) A Model for the separation of large DNA molecules by crossed field electrophoresis. *Nucleic Acids Res.* **15**, 5925–5943.

14. Cantor, C. R., Gaal, A., and Smith, C. L. (1988) High-resolution and accurate size determination in pulsed-field gel electrophoresis of DNA. 3. Effect of electrical field shape. *Biochemistry* **27**, 9216–9221.

15. Kobayashi, T., Doi, M., Makino, Y., and Ogawa, M. (1990) Mobility minima in field-inversion gel electrophoresis. *Macromolecules* **23**, 4480–4481.

16. Deutsch, J. M. (1986) Theoretical studies of DNA during gel electrophoresis. *Science* **240**, 922–924.

17. Deutsch, J. M. and Madden, T. L. (1989) Theoretical studies of DNA during gel electrophoresis. *J. Chem. Phys.* **90**, 2476–2485.

18. Deutsch, J. M. (1989) Explanation of the anomalous mobility and birefringence measurements found in pulsed field electrophoresis. *J. Chem. Phys.* **90**, 7436–7441.

19. Zimm, B. H. (1988) Size fluctuations can explain anomalous mobility in field-inversion electrophoresis. *Phys. Rev. Lett.* **61**, 2965–2968.

20. Zimm, B. H. (1991) "Lakes-Straits" model of field-inversion electrophoresis of DNA. *J. Chem. Phys.* **94**, 2187–2206.

21. Duke, T. A. J. (1989) Tube model of field-inversion gel electrophoresis. *Phys. Rev. Lett.* **62**, 2877–2880.

22. Smith, S. B., Aldridge, P. K., and Callis, J. B. (1989) Observation of individual DNA molecules undergoing gel electrophoresis. *Science* **243**, 203–206.

23. Schwartz, D. C. and Koval, M. (1989) Conformational dynamics of individual DNA molecules during gel electrophoresis. *Nature* **338**, 520–522.

24. Gurrieri, S., Rizzarelli, E., Beach, D., and Bustamante, C. (1990) Imaging of kinked configurations of DNA molecules undergoing OFAGE using fluorescence microscopy. *Biochemistry* **29**, 3396–3401.

25. Olvera de la Cruz, M., Gersappe, D., and Shaffer, E. O. (1990) Dynamics of DNA during pulsed-field electrophoresis. *Phys. Rev. Lett.* **64**, 2324–2327.

26. Shaffer, E. O. II and Olvera de la Cruz, M. (1989) Dynamics of gel electrophoresis. *Macromolecules* **22**, 1351–1355.

27. Matthew, M. K., Smith, C. L., and Cantor, C. R. (1988) High-resolution and accurate size determination in pulsed-field gel electrophoresis of DNA. 1. DNA size standards and the effect of agarose and temperature. *Biochemistry* **27,** 9204–9210.

28. Matthew, M. K., Smith, C. L., and Cantor, C. R. (1988) High-resolution and accurate size determination in pulsed-field gel electrophoresis of DNA. 2. Effect of pulse time and electric field strength and implications for models of the separation process. *Biochemistry* **27,** 9210–9216.

29. Levene, S. D. and Zimm, B. H. (1987) Separations of open-circular DNA using pulsed-field electrophoresis. *Proc. Natl. Acad. Sci. USA* **84,** 4054–4057.

30. Orbach, M. J., Vollrath, D., Davis, R. W., and Yanofsky, C. (1988) An electrophoretic karyotype of *Neurospora crassa. Mol. Cell Biol.* **8,** 1469–1473.

31. Attwood, T. K., Nelmes, B. J., and Sellen, D. B. (1988) Electron microscopy of beaded agarose gels. *Biopolymers* **27,** 201–212.

32. Mickel, S., Arena, V., and Bauer, W. (1977) Physical properties and gel electrophoresis behavior of R12-derived plasmid DNAs. *Nucleic Acids Res.* **4,** 1465–1482.

33. Serwer, P. and Hayes, S. J. (1989) Atypical sieving of open-circular DNA during pulsed-field electrophoresis. *Biochemistry* **28,** 5827–5832.

34. Hervet, H. and Bean, C. P (1987) Electrophoretic mobility of lambda phage Hind III and Hae III DNA fragments in agarose gels: A detailed study. *Biopolymers* **26,** 727–742.

CHAPTER 24

The Dynamics of DNA Gel Electrophoresis

Joshua M. Deutsch

1. Introduction

Gel electrophoresis of DNA has improved greatly in the last decade, mostly owing to the invention of pulsed-field gel electrophoresis *(1)*. This has enabled the separation of DNA more than two orders of magnitude longer than what was previously possible *(2–4)*. This chapter will review what is understood about the microscopic behavior of DNA undergoing electrophoresis, which has reached a stage of sophistication allowing one to predict the mobility of a specified DNA molecule in a given system. The author's own research will be emphasized, much of the most elaborate calculations done in collaboration with Thomas L. Madden *(5–9)* and J. D. Reger *(10)*. It is hoped that this chapter will be comprehensible to most biologists, and to this end, it eschews a complete technical treatment of the subject that can be found in the cited literature.

The following premise has been adopted and has indeed proven quite successful. We will try to elucidate the microscopic behavior of DNA during electrophoresis at a level that takes into account the important internal degrees of freedom of the molecule. Once the problem is understood at this level, more practically important macroscopic quantities can be calculated, for example, the mobility of DNA. From

From: *Methods in Molecular Biology, Vol. 12: Pulsed-Field Gel Electrophoresis*
Edited by: M. Burmeister and L. Ulanovsky

there, more complicated experimental effects can be incorporated. For example, spatial inhomogeneities in the electric field are often important experimentally, but almost always at a length-scale much longer than individual DNA molecules. Thus, knowing the mobility of DNA as a function of a spatially uniform electric field, as would be modeled microscopically, allows one to compute the average position of DNA as a function of time in the inhomogeneous case.

Theoretical models used to describe gel electrophoresis have been forced to become more elaborate in order to account for experimental observations. They are now quite detailed and even stretch the limits of what can be feasibly simulated on a Cray YMP, although they still simplify the actual physical situation. However, even if one were to implement an extremely sophisticated model incorporating every physical detail, although it would reproduce experimental results, it would be difficult to ascertain which physical detail was primarily responsible for the phenomena being observed. Fortunately, most numerical models now studied, as will be described later in this chapter, appear to have captured most of the essential phenomena, and are accurate enough to explain and even predict many experimental findings.

The outline of this chapter is as follows: Section 2. describes the basic physical situation that will be analyzed and what simplifications are made. Section 3. reviews the "reptation" model in connection with gel electrophoresis. This model is quite simple, and as a result, is amenable to analytic calculations. It appears to describe electrophoresis well in a high-density matrix, such as polyacrylamide gel. At low densities, e.g., dilute agarose gels, the reptation model does not appear to apply. Much of the interesting phenomena seen in pulsed-field electrophoresis are attributable to types of motion that were excluded from this model.

Section 4. reviews three effects outside the scope of the reptation model. First, including the formation of kinks leads to a very different kind of motion then reptation, but gives results in agreement with experimental observation. Second, it is predicted that there can be a large frictional force between gel fibers and DNA because of local corrugations in its shape. This phenomenon is called "chain pinning." Third, it is expected that under certain circumstances, self-tightening could occur. These last two effects may be quite important when trying to understand DNA molecules greater than a megabase in length.

Section 5. reviews the simulation work of the author and collaborators on the microscopic dynamics of DNA undergoing electrophoresis, using a model that incorporates many internal degrees of freedom of the DNA. The results of these simulations show that DNA moves in an almost cyclical manner. In a constant field, the DNA molecule moves in the field direction, and alternates between extended and U-shaped configurations. These simulations have been used to predict many results in pulsed-field experiments, some of which are quite counterintuitive. A variety of experimental results are compared with the numerical predictions of this model.

Section 6. considers some of the problems associated with the separation of megabase DNA. For these very large sizes, the motion of DNA appears impeded by unknown effects, perhaps by increased frictional forces owing to chain pinning, and possibly by tight knots (Section 3.). Some suggestions as to how to overcome these problems are given.

2. The Physical Situation

We are interested in examining the dynamics of a DNA molecule in an agarose gel. DNA in this situation has a persistence length of about 300 bp, where the distance between bps is 0.34 nm. At scales below the persistence length, the DNA can be regarded as almost rigid. For scales larger than this, the DNA can move freely, taking the shape of a random coil in free solution.

The structure of agarose gels is still not completely understood. There is a large spread in the distribution of voids, that are typically 300 nm across. Direct observation of individual DNA molecules shows that the gel fibers are capable of supporting quite large DNA molecules without giving way.

Because DNA in salt solution is a polyelectrolyte, the application of a field will cause the small ions around the DNA to drift in one direction, and the DNA will drift in the other. There is a large amount of shielding of the DNA charge by the ions in solution. The amount of shielding will determine the magnitude of the net force on the DNA. This problem has been recently analyzed in detail *(11),* and it appears that the effective charge of the DNA is reduced greatly. In addition, it appears that the effective force is a strong function of both velocity and orientation of DNA. For a stationary DNA molecule, the effective charge appears to be from $0.06-0.1e$/bp.

Many approximations have been employed in simulating DNA electrophoresis. The dependence of the effective charge on velocity and orientation has been ignored. The gel has been taken to be perfectly rigid, and most results have been obtained in two dimensions. The repulsive interactions between different segments of the chain have been ignored. The neglect of velocity dependence of the effective charge may be the most severe approximation made, and further work is needed to incorporate a more realistic velocity dependence.

It is quite reassuring to find that despite these simplifications, the results of simulations and theoretical analysis coincide to a remarkable degree with experimental observation.

3. The Reptation Model

Even without an external field, the behavior of long polymers in a gel is rather complicated. The presence of the random gel has a large effect on some equilibrium statistical properties *(12)*. The topological constraints imposed by the gel have a drastic effect on the motion of a polymer, as was first recognized by de Gennes *(13)*, who introduced the reptation model in this context. de Gennes accounted for the restricting effect the gel has on a polymer by constraining the polymer to a tube. The polymer can move longitudinally but not transversely to the tube. Both experimental work and computer simulations *(14)* are in good agreement with de Gennes' predictions. The reptation model was first applied to gel electrophoresis by Lerman and Frisch *(15)*, and then more thoroughly investigated by Lumpkin and Zimm *(16)*, and Lumpkin, Dejardin, and Zimm *(17)*, who extended the zero field case to finite fields. However, in doing so, some additional assumptions were required, owing to the fact that the statistics of chain configurations were not understood for finite fields. They assumed that the internal links of the chain were frozen, and the direction of the head of the chain was chosen randomly, but biased by the electric field. As a consequence, the conformations of such chains have the statistics of biased random walks.

There is extensive literature making the same basic assumptions, but more accurately solving the dynamics using a computer (*see* ref. *18*). The most recent improvements on the reptation model have allowed links in the middle of the tube to move *(19–21)*, yielding results in much better agreement with pulsed-field experiments.

The theory of Lumpkin, Dejardin, and Zimm explained many important experimental observations in constant fields. The mobility decreases with mol wt and plateaus at a field-dependent, non-zero value. The shape of these mobility plots, both as a function of chain length and electric field, is qualitatively correct. Furthermore, the reptation model predicts an interesting phenomenon known as "band inversion," that, under some conditions, has been seen in constant field experiments *(18,22)*.

Unfortunately, data on pulsed-field gel electrophoresis was found to differ qualitatively from the predictions of simple reptation theories. For example, data on field inverse gel electrophoresis shows that the mobility as a function of chain length has a deep minimum *(23–25)* at odds with the predictions of reptation. With more realistic theoretical treatments that abandon the overly simple assumptions of reptation theory, this observation has been accounted for, as will be seen later in this chapter. Another situation where reptation theory fails to account for measurements is the orientation of DNA undergoing pulsed-field gel electrophoresis. For example, if a DNA chain is in a relaxed coil state and an electric field is suddenly turned on, the molecule will show a rather surprising orientational response. First it becomes highly aligned with the field, with the average orientation showing an overshoot, then this decreases and has an undershoot, before the orientation plateaus to a steady-state value *(26,27)*.

It should be noted that the previous contradictions with the reptation model only occur in agarose gels that have quite a low density, where the persistence length of the DNA is of the order of the pore size of the gel. In high-density gels, however, reptation theory appears to be an excellent description of DNA. The reptation model appears accurate enough to predict how the sequence of DNA effects mobility *(28)*.

In order to understand the experimental pulsed-field data, it is first necessary to discuss some important effects left out of the simplest versions of the reptation model. These are now briefly discussed.

4. Beyond the Reptation Model

4.1. The Motion of Kinks

The reptation model prohibits the chain doubling back on itself and moving transversely, outside the region of the tube. However, such configurations may occur in certain situations, and are called "kinks," "hairpin" configurations, or "herniations." The reptation model was first applied to an equilibrium situation with no applied field, and in this case it can be easily demonstrated that long kinks are exponentially suppressed, and can therefore be ignored. On the other hand, as will be seen later in this chapter, in the presence of an electric field, long kinks do indeed form. This is most easily illustrated in orthogonal pulsed-field geometry *(5)*, and is now described.

Referring to Fig. 1A, a chain is shown inside a gel and is initially oriented horizontally. The picture shows the simulation results for a chain length of 9000, using a method that will be discussed in Section 5. In Fig. 1B, the field is applied downward, causing the chain to move in this direction. Because of the presence of the gel, its motion downward is severely impeded. The best the chain can do is to form vertically oriented kinks. However, whether or not it will do this depends on three different factors. Kinks are stretched and doubled back on themselves, and consequently, have less entropy than the chain inside the original tube, so entropy tends to disfavor kink formation. This entropy reduction is linearly proportional to the length of the kink. Another effect hindering kink formation is an elastic curvature energy needed to bend the tip of the kink. Since the pore size in agarose is close to the persistence length of DNA, this effect is not too important. The final factor is the potential energy gain, which favors kink formation. It varies quadratically with kink length and is the dominant effect in most experiments, so that kinks form very quickly on the time scale of relaxation of the chain.

After kinks start to form, a long chain will contain many tiny kinks. Because chain length is locally conserved, the fraction of a molecule in kink conformations will depend on the initial amount of slack.

The evolution of the molecule as a function of time is shown in Fig. 1C–F. What is seen is a process of "cannibalism," where the larger kinks will devour the shorter ones, causing the latter to disappear in a finite time. At some point, a previously growing kink may have eaten its nearest neighbors, finding suddenly that its new neighbors are now bigger than it is, and so eventually it becomes consumed by them. This problem was treated analytically, and it was shown that the number of kinks decreases inversely with time. Because the total percentage of kinked chain remains constant, this implies that the average kink length grows linearly with time. The average velocity of kinks depends on the percentage of kinked chain. The velocity goes to zero as this percentage goes to zero. In the inset to Fig. 1, a kink, shown by the asterisk in Fig. 1D, has been magnified showing random bending, but no additional subkinks are present.

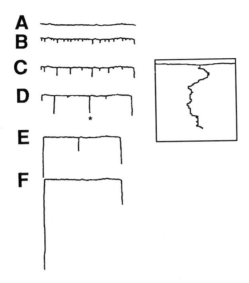

Fig. 1. **A:** A DNA molecule oriented horizontally, in the same direction as the electric field. **B:** The field is rotated 90° downward, the DNA moves downward but is hindered by the gel fibers in the way, and so forms vertical kinks. **C, D, E, F:** A sequence of snapshots showing the growth of larger kinks at the expense of the smaller ones. The inset shows the kink labeled by the asterisk in **D:** magnified several times.

This process will eventually terminate in one of two ways: either all the kinks will disappear, being eaten by the chain ends, or the entire chain will be consumed by one large kink. Which scenario occurs will depend on the initial configuration of the chain.

Recently, experiments were set up to investigate this situation using fluorescence microscopy *(29)*. The results are largely in accord with the theoretical analysis *(5)* summarized here. There are two other effects discussed below that may be particularly relevant to the motion of very long DNA molecules.

4.2. Chain Pinning

As DNA moves through a gel, we shall see that it often becomes "hooked" over gel fibers. In this case, there is an additional friction that can slow down the motion of the DNA as it slides over the top of the fibers. The additional frictional forces can be understood as being similar to static friction familiar from classical mechanics. A DNA molecule hooked over a gel fiber will experience several different forces. There are the forces generated by the electric field that pull the two sides of the chain vertically downward. Call the force to the left f_1 and the force to the right f_2. It might be possible for the chain to become badly wedged on a gel fiber because of local corrugations in the molecular structure of the DNA, and the agarose fibers. For this to happen, f_1/f_2 must be greater than some threshold value that depends on the geometry of the situation. Also, thermal fluctuations must be small enough to prevent the DNA from hopping out of the groove that it is stuck into. Thermal motion will be suppressed for large enough electric fields and long enough chains, as will be seen later in this chapter. However, because the interaction between agarose and DNA has not been investigated in detail, only a rough analysis of this pinning effect can be given. If there is no thermal noise, the DNA is assumed to be stuck. For it to move, a thermal fluctuation has to lift it out of the groove. A groove is expected to have a depth on the order of several angstroms. By equating the thermal energy and the electrostatic energy, it is straightforward to find the minimum length of DNA where thermal fluctuations cannot easily surmount the electrostatic potential (5). For example, take a hooked DNA in an electric field of 5 V/cm. This pinning effect begins to dominate for chains longer than 300 kb. Note that for longer chain lengths the force from the electric field increases, owing to the increased total charge, and therefore, this pinning effect becomes stronger.

4.3. Formation of Knots

In equilibrium, a long, linear, self-avoiding polymer will be found in any allowable state with equal probability, and will frequently contain internal knots. The longer the chain, the more numerous and complicated the knots become. One would qualitatively expect the number of knots to grow at least linearly with the length of the polymer. Because the dynamics of a chain undergoing gel electrophoresis do not allow the chain to double back frequently on itself, in such a situation it is less knotted than the same chain in a coiled equilibrium state.

Theoretical treatments of a single swollen chain ignore knots, and are in excellent agreement with experiments, from which it can be concluded that knots have only a minor effect in this case. In other situations, the effect of knots can be rather important. de Gennes (30) proposed the idea of a "tight knot," which is the microscopic analog of shoelaces that have become so taut, they are difficult to untie. A large chain tension is a prerequisite for the formation of a tight knot. An increased tension will increase the elastic bending energy of the chain, and therefore, frictional forces between chain segments in contact will become greater.

In the initial phases of electrophoresis, when the DNA is highly coiled and embedded in an agarose plug, it will contain many knots. In the last section it was seen that a long DNA molecule, even in a moderate field, could develop large chain tensions, and therefore, it seems a reasonable hypothesis that if a long DNA molecule, greater than a megabase, was knotted around an agarose fiber, it could often find itself pulled into a tight knot. Once formed, a tight knot is difficult to unravel. If the applied field is switched off or changes direction, a tight knot should remain, and will continue to hinder DNA migration.

5. Numerical Investigation of DNA Motion

Many researchers have now developed numerical schemes for modeling DNA electrophoresis. One approach has been direct numerical simulations *(6–10,31)*, the other has been improvements in tube models that allow for internal motion inside the tube *(19–21)*. Standard Monte-Carlo techniques that incorporate local motions of the chain, which have worked well in other systems, must be avoided *(32)*. This technique has been extended by the incorporation of long-range moves, and appears useful in probing megabase-length DNA *(10)*. An example of the results of such a simulation is shown in Fig. 1. Here, the persistence length l has been set equal to the pore size, and this chain is 9000 monomers in length. One monomer is equal to approx 150 bp, and $qEl/K_BT = 1$.

One of the most accurate treatments of DNA electrophoresis used "off lattice" models investigated by the author and T. L. Madden *(6–9)*. The simulation has made a number of predictions in good agreement with other experiments. The DNA molecule is modeled as a freely hinged chain connected by links of fixed length. At each pivot point is a charged bead that feels a force in the direction of the electric field. The gel is represented as a two dimensional network of obstacles that interact repulsively with the beads on the DNA. Thermal noise is modeled as a random force acting independently on each bead. An efficient algorithm was developed to update each time step.

The first off lattice simulations made by the author *(6)* used a regular array of obstacles. This was extended with T. L. Madden to give the obstacles random positions *(9)*. The results are in excellent agreement with direct observations of DNA. This technique can probe chains up to about 100 kb. To analyze longer chains requires the use of the long-range Monte-Carlo simulation *(10)* mentioned earlier. A sequence of chain configurations of length 3000 is shown in Fig. 2, where the electric field is constant, $qEl/K_BT= 1$, and pointing downward. The results are in good qualitative agreement with the more accurate simulation results discussed previously *(9)*. The motion has a roughly cyclical pattern that can be described qualitatively as follows:

1. The chain is hooked in an inverted U-shape around an obstacle, and is highly stretched (Fig. 2A).
2. The chain slides past the obstacle to the point where it has almost disengaged from it, but still remains almost fully stretched (Fig. 2B).
3. The chain, having disengaged from the obstacle, migrates downward and begins to contract mostly at the leading end (Fig. 2C and D). The leading end moves more slowly than the trailing portions.
4. A kink forms at the front and starts to grow (Fig. 2D). The inset box shows the kink magnified.
5. The kink continues to grow (Fig. 2E and F), and finally grows large enough to cause the chain to form another U-shape (Fig. 2G).

The whole cycle will now repeat starting with Stage 1. The different phases of this cyclical motion can be understood theoretically. The most interesting aspect of this motion is the instability that leads to kink formation. This is a result of the electric field, as kinks do not occur in equilibrium. First the leading edge of the chain bunches, and then turns into a kink. It has often been stated that this bunching is attributable to inhomogeneities in the gel. However, this cannot be true, as this simulation has been done with a regular lattice of obstacles, and has the same instability as when a random lattice is used. Thus, this instability is mostly attributable to the internal dynamics of the chain, although it is enhanced by gel inhomogeneity.

When the chain is hooked over an obstacle, it becomes highly extended as a result of the increased chain tension. In this case, chain tension caused by the electric field dominates over thermal forces that contribute to the entropy and favor a coiled state. It is straightforward to show that it is only when the chain has almost slid out of the obstacle that chain tension diminishes enough to become comparable with thermal forces. This is why DNA molecules remain extended in U-shape conformations. Experimental confirmation of these predictions was made by microscopic observation of individual DNA molecules *(33–35)*.

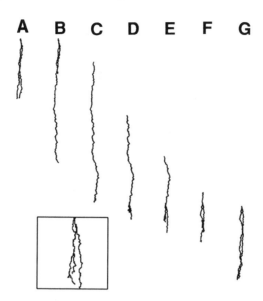

Fig. 2. A 3000-link chain moving through a regular lattice under the influence of a constant field using a long range lattice Monte-Carlo simulation, described in the text. The results look very similar to those for smaller chains, but using a more accurate off lattice simulation. Inset shows the kink in **D** magnified.

This long-range Monte-Carlo model and the more accurate off lattice model, have been applied to analyze a variety of experiments. In general, it appears these are in quite good agreement with them. Experiments *(26,27)* on DNA orientation after the application of a field to a coiled molecule, as mentioned earlier, are well-described by simulation. The orientation shows an overshoot and then an undershoot, eventually reaching a steady-state value. With the time step as the only adjustable parameter, the experimental and numerical curves agree very well. The reason for this behavior can be understood using the cyclical description above *(6,7)*.

The sharp mobility minimum as a function of pulse frequency seen in field inverse gel electrophoresis experiments has also been replicated by a number of numerical simulations *(6,7,10,19,20)*. The essential reason for this minimum can be clearly seen from the frequent occurrence of U-shaped configurations as in Fig. 2A. Amolecule in such a conformation has a higher mobility for an upward applied field, as compared with a downward field of the same magnitude. Thus, in this case, although a downward field is applied for two or three times as long as the backward (upward) pulse, it will not travel as far, at least for short enough pulsing. As a result, the average mobility of the DNA is greatly reduced. The results of the long-range Monte-Carlo simulation are displayed in Fig. 3, the ratio of forward to backward time being 2:1. The minimum velocity is approximately one-third of its asymptotic value. This is higher than what is obtained from the off lattice simulation *(7)*. The difference is probably attributable to some chain pinning that is present with the off lattice simulation, particularly for the simulation parameters chosen in this case. The minimum in the mobility seen experimentally is closer to the off lattice case.

Orthogonal field alternation gel electrophoresis (OFAGE) has been the subject of much detailed experimental analysis *(36–39)*. Numerical simulations have been carried out using the models described previously *(9,10)*. Again, for this situation, the mobility as a function of pulse frequency for both theory and experiment follow each other rather closely. For both curves, the mobility exhibits a single minimum of depth about 50% of the steady-state value, and the width of the minimum in both cases is also quite similar. The minimum in mobility can be rather simply understood at a qualitative level. The orientation of DNA is also in similarly good agreement *(38)*. After a rotation of the electric field, the orientation in the new field direction increases monotonically and shows no undershoot or overshoot in contrast with the field inversion case.

During PFGE there is a wide range of parameters where a large increase in the width of a band is seen transverse to the direction of motion. An explanation for this experimental anomaly has recently been proposed by Chu *(39)*. Because this spread increases with time, it appears to be attributable to chains migrating in directions quite different to the average direction of motion. They must be doing this for long durations of time for this spreading to be so apparent. Chu's explanation assumes that a molecule has a memory of its previous configuration, and if the period of a cycle is sufficiently short, its present conformation is strongly affected by what it was during the previous phase of the cycle. In the case of OFAGE, consider a molecule oriented horizontally along the field, after which the field is rotated 90°. This will cause the chain to move vertically downward by kink motion, as illustrated in Section 4.1. When the cycle period is sufficiently short, the field is reoriented back to the horizontal direction when the chain in the interim will have only partially reoriented and will still be pre-

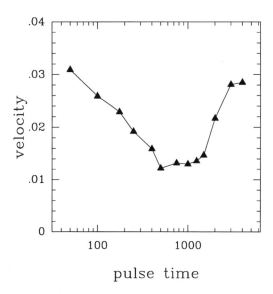

Fig. 3. Velocity vs pulse time for inverse field gel electrophoresis. The forward to backward time is 2:1, $qEl/K_BT = 1$, and there are 500 beads. These are the results for the long range Monte-Carlo simulation described in the text.

dominantly oriented horizontally. This suggests that the molecule does have a memory of its previous conformation. Chu devised a simple model of this effect which predicts that in many cases, the DNA remains trapped in one orientation indefinitely. Thus, if a DNA molecule is initially oriented along one field direction, it will migrate predominantly in that direction, not the time-averaged field direction. Because of the intrinsic randomness of the dynamics, the finite length of the chain should cause changes in chain orientation. One would expect that the longer the chain, the more pronounced this kind of hysteretic migration.

The off lattice simulation of OFAGE described here does point to an effect of orientational persistence. The direction of migration does not follow the field direction, but persists in one direction over a large timescale, longer than both the switching time of the field and the reorientation relaxation time of the chain. As suggested earlier, this effect was most pronounced for long chain lengths. For example, orientational persistence for chain lengths of 400 segments, which is approx 70 kb, was observed to last about 10 complete cycles. This should not lead to appreciable band spreading for such short chains, but may when extrapolated to longer chain lengths. This is in accord with experimental observation, as band spreading of this type becomes pronounced for chains at least several times this length.

6. Electrophoresis of Very Long DNA

Although PFGE techniques have succeeded in separating DNA of up to about 10 megabases, it would be highly desirable to increase this limit by improving the technique further, particularly given its importance to the Human Genome Project. In the megabase regime, unknown effects appear to hinder DNA separation and migration.

DNA molecules of order 10 megabases often remain trapped inside their agarose plugs, and consequently, do not enter the gel. More DNA can be made to enter if a lower electric field is initially employed. Once the DNA has succeeded in entering, the electric field can be increased by a factor of five without encountering major difficulties (E. Lai, private communication). Separation can be improved further by field "twitching," that is, adding short secondary pulses in a direction different from the primary ones *(40)*.

It is possible that the hindrances observed experimentally are a result of chain pinning and the formation of tight knots, as described in Sections 3.2. and 3.3. As will be argued, the improvements just described are at least in qualitative agreement with this hypothesis.

As discussed in Section 4.3., many more knots should be present before a field is applied, when the DNA is still inside its agarose plugs. As time progresses, it should become quite extended. In the process of orienting in the field, tight knots may occasionally form. These could be eliminated to some extent by using a low applied field, as tight knots can only form if the chain tension is sufficiently large. This explains qualitatively why a lower electric field is required in order to drive the DNA into the gel. Once this has happened, the chains are quite stretched and the density of knots will have decreased so that it is relatively unlikely that a chain will end up in a tight knot. Thus, at this stage the electric field can be increased considerably without the risk of trapping attributable to these knots, in agreement with experimental observation.

Unfortunately, using a low field to allow gel entry only works to some extent, and for DNA of lengths substantially greater than 10 megabases, this procedure becomes inadequate. For such long chains, knots could become so prevalent as to completely prevent the DNA from moving out of the plugs. This is related to an effect that occurs with circular DNA. Constant field electrophoresis appears ineffective for even quite short DNA, probably because it becomes caught on local barblike protrusions *(41,42)*. To prevent passage of DNA, these protrusions must be rigid enough not to give way. Thus, one way to eliminate the effect of such knots would be to employ some kind of "weak" gel. This is a gel network whose links are not completely permanent, and change slowly in time, both because of thermal motions, and in response to an applied force. Tightly knotted DNA wrapped around weak gel fibers will be freed when the fibers change connectivity. If agarose gel is employed at a high temperature still below the melting transition, it is expected to become weaker. There is evidence that running a gel under such conditions does improve electrophoretic performance for very long DNA *(43)*.

The utility of field twitching can be explained as a way to alleviate the effect of chain pinning. As discussed in Section 4.2., when a chain is hooked over an obstacle, thermal motion may be insufficient to allow the sliding of the chain because of local corrugations. By temporarily lifting the chain off the gel fiber, it can become unpinned, after which it will continue to move unimpeded. It is still not clear at the level of theory how to optimize the twitching field, and this problem deserves further attention.

7. Conclusion

This chapter presented an overview of the current theoretical understanding of DNA dynamics during gel electrophoresis, with a bias toward the work of the author. For chains below a megabase, the dynamics are rather well-understood. Using numerical and analytic methods, a variety of situations have been studied, particularly those where the data seems most counterintuitive. In all cases, the agreement with experimental results is rather good.

However, the behavior of megabase-length DNA has not yet been adequately understood. In order to test out the ideas presented in the last section, simulations incorporating these effects should be undertaken. Further improvements in direct imaging of DNA would also be of great benefit as computer simulations become more difficult for these long chain lengths. Theoretical understanding of the frictional forces between DNA and agarose fibers should be understood in more detail. The effects of local orientation and velocity on the effective charge of DNA *(11)* should be added to simulations, as should intrachain repulsion.

Acknowledgments

Much of this work was done in collaboration with T. L. Madden and J. D. Reger. This work is supported by NSF grant DMR 87-21673, and by the Nonlinear ORU at UCSC. Much of the computational work was done on the Cray X/MP and Y/MP at SDSC in San Diego, CA.

References

1. Schwartz, D. C. and Cantor, C. R. (1984) Separation of yeast chromosome-size DNAs by pulsed field gradient gel electrophoresis. *Cell* **37**, 67–75.
2. Chu, G., Vollrath, D., and Davis, R. W. (1986) Separation of large DNA molecules by contour-clamped homogeneous electric fields. *Science* **234**, 1582–1585.
3. Clark, S. M., Lai, E., Birren, B. W., and Hood, L. (1988) A novel instrument for separating large DNA molecules with pulsed homogeneous electric fields. *Science* **241**, 1203–1205.
4. Birren, B. W., Lai, E., Clark, S. M., Hood, L., and Simon, M. I. (1988) Optimized conditions for pulsed field gel electrophoretic separations of DNA. *Nucleic Acids Res.* **16**, 7563–7582.
5. Deutsch, J. M. (1987) The dynamics of pulsed field electrophoresis. *Phys. Rev. Lett.* **59**, 1255–1258.
6. Deutsch, J. M. (1988) Theoretical studies of DNA during gel electrophoresis. *Science* **240**, 922–924.
7. Deutsch, J. M. (1989) The explanation of anomalous mobility and birefringence measurements in pulsed field electrophoresis. *J. Chem. Phys.* **90**, 7436–7441.
8. Deutsch, J. M. and Madden, T. L. (1989) Theoretical studies of DNA during gel electrophoresis. *J. Chem. Phys.* **90**, 2476–2485.
9. Madden, T. L. and Deutsch, J. M. (1991) Theoretical study of DNA during orthogonal field alternating gel electrophoresis. *J. Chem. Phys.* **94**, 1584–1591.
10. Deutsch, J. M. and Reger, J. D. (1991) Simulation of highly stretched chains using long range Monte-Carlo. *J. Chem. Phys.* **95**, 2065–2071.
11. Stigter, D. (1991) Shielding effects of small ions in gel electrophoresis of DNA. *Biopolymers* **31**, 169–176.
12. Gersappe, D., Deutsch, J., and Olvera de la Cruz, M. (1991) Density fluctuations of self-avoiding walks in random systems. *Phys. Rev. Lett.* **66**, 731–734.
13. de Gennes, P. G. (1971) Reptation of a polymer chain in the presence of fixed obstacles. *J. Chem Phys.* **55**, 572–579.
14. Evans, K. E. and Edwards, S. F. (1981) Computer simulation of the dynamics of highly entangled polymers. *J. Chem. Soc. Faraday Trans. II* **77**, 1891–1912.

15. Lerman, L. S. and Frisch, H. L. (1982) Why does the electrophoretic mobility of DNA in gels vary with the length of the molecule? *Biopolymers* **21**, 995–997.
16. Lumpkin, O. J. and Zimm, B. H. (1982), Mobility of DNA in gel electrophoresis. *Biopolymers* **21**, 2315,2316.
17. Lumpkin, O. J., Dejardin, P., and Zimm, B. H. (1985) Theory of gel electrophoresis of DNA. *Biopolymers* **24**, 1575–1593.
18. Noolandi, J., Rousseau, J., Slater, G. W., Turmal, C., and Lalande, M. (1987) Self trapping and anomalous dispersion of DNA in electrophoresis. *Phys. Rev. Lett.* **58**, 2428–2431.
19. Zimm, B. H. (1988) Size fluctuations can explain anomalous mobility in reversing-field electrophoresis of DNA. *Phys. Rev. Lett.* **61**, 2965–2968.
20. Duke, T. A. J. (1989) Tube model of field-inversion electrophoresis. *Phys. Rev. Lett.* **62**, 2877–2880.
21. Noolandi, J., Slater, G. W., Lim, H. A., and Viovy, J. L. (1989) Generalized tube model of biased reptation for gel electrophoresis of DNA. *Science* **243**, 1456–1458.
22. Doi, M., Kobayashi, T., Makino, Y., Ogawa, M., Slater, G. W., and Noolandi, J. (1988) Band inversion in gel electrophoresis of DNA. *Phys. Rev. Lett.* **61**, 1893–1896; Lumpkin O. J., Levene S. D., and Zimm B. H. (1989) Exactly solvable reptation model. *Phys. Rev. A* **39**, 6557–6566.
23. Carle, G. F., Frank, M., and Olson, M. V. (1986) Electrophoretic separations of large DNA molecules by periodic inversion of the electric field. *Science* **232**, 65–68.
24. Crater, G. D., Gregg, M. R., and Holzwar, G. (1989) Mobility surfaces for field-inversion gel electrophoresis of linear DNA. *Electrophoresis* **10**, 310–315.
25. Heller, C. and Pohl, F. M. (1989) A systematic study of field inversion gel electrophoresis. *Nucleic Acids Res.* **17**, 5989–6003.
26. Holzwarth, G., McKee, C. B., Steiger, S., and Crater, G. (1987) Transient orientation of linear DNA molecules during pulsed-field gel electrophoresis. *Nucleic Acids Res.* **15**, 10031–10044; Akerman, B., Jonsson, M., Norden, B., and Lalande, M. (1989) Orientational dynamics of T2 DNA during agarose gel electrophoresis: Influence of gel concentration and electric field strength. *Biopolymers* **28**, 1541–1571.
27. Holzwarth, G., Platt, K. J., McKee, C. B., Whitcomb, R. W., and Crater, G. D. (1989) The acceleration of linear DNA during pulsed-field gd electrophoresis. *Biopolymers* **28**, 1043–1058.
28. Levene, S. D. and Zimm, B. H. (1989) Understanding the anomalous electrophoresis of bent DNA molecules - A reptation model. *Science* **245**, 396–399.
29. Gurrieri, S., Rizzarelli, E., Beach, D., and Bustamante, C. (1990) Imaging of kinked configurations of DNA molecules undergoing orthogonal field alternating gel electrophoresis by fluorescence microscopy. *Biochemistry* **29**, 3396–3401.
30. de Gennes, P. G. (1984) Tight Knots. *Macromolecules* **17**, 703–705.
31. Shaffer, E. O. and Olvera de la Cruz, M. (1989) Dynamics of gel electrophoresis. *Macromolecules* **22**, 1351–1355.

32. Olvera de la Cruz, M., Deutsch, J. M., and Edwards, S. F. (1986) Electrophoresis of a polymer in a strong field. *Phys. Rev. A* **33**, 2047–2055.

33. Smith, S. B., Aldridge, P. K., and Callis, J. B. (1989) Observation of individual DNA molecules undergoing gel electrophoresis. *Science* **243**, 203–206 [published erratum appears in *Science* (1989) **243**, 992].

34. Schwartz, D. C. and Koval, M. (1989) Conformational dynamics of individual DNA molecules during gel electrophoresis. *Nature* **338**, 520–522.

35. Houseal, T. W., Bustamante, C., Stump, R. F., and Maestre, M. F. (1989) Real-time imaging of single DNA molecules with fluorescence microscopy. *Biophys. J.* **56**, 507.

36. Mathew, M. K., Smith, C. L., and Cantor, C. R. (1988) High-resolution separation and accurate size determination in pulsed-field gel electrophoresis of DNA. 1. DNA size standards and the effect of agarose and temperature. *Biochemistry* **27**, 9204–9210.

37. Mathew, M. K., Smith, C. L., and Cantor, C. R. (1988) Effect of pulse time and electric field strength and implications for models of the separation process. *Biochemistry* **27**, 9210–9216.

38. Akerman, B. and Jonsson, M. (1990) Reorientational dynamics and mobility of DNA during pulsed field agarose gel electrophoresis. *J. Phys. Chem.* **94**, 3828–3838.

39. Chu, G. (1990) Pulsed-Field Gel Electrophoresis: Theory and Practice. *Methods: Companion to Methods in Enzymol.* **1**, 129–142.

40. Turmel, C., Brassard, E., Slater, G. W., and Noolandi, J. (1990) Molecular detrapping and band narrowing with high frequency modulation of pulsed field electrophoresis. *Nucleic Acids Res.* **18**, 569–575.

41. Levene, S. D. and Zimm, B. H. (1987) Separations of open-circular DNA using pulsed-field electrophoresis. *Proc. Natl. Acad. Sci. USA* **84**, 4054–4057.

42. Serwer, P. and Hayes, S. J. (1989) Atypical sieving of open circular DNA during pulsed field agarose gel electrophoresis. *Biochemistry* **28**, 5827–5832.

43. Snell, R. G. and Wilkins, R. J. (1986) Separation of chromosomal DNA molecules from C. albicans by pulsed field gel electrophoresis. *Nucleic Acids Res.* **14**, 4401–4407.

CHAPTER 25

Orientation of the Agarose Matrix by Pulsed Electric Fields

Nancy C. Stellwagen

1. Introduction

DNA molecules less than ~20 kilobase (kb) pairs in size are efficiently fractionated by unidirectional electrophoresis in agarose gels *(1)*. The variation of DNA mobility with gel concentration can be described by the Ogston mechanism of pore size distribution, if the observed mobilities are first extrapolated to zero electric field strength *(2)*. Therefore, sieving of the macromolecules by the matrix appears to be the dominant mechanism of mol wt separation in this size range.

DNA molecules larger than ~20 kb exhibit nearly constant electrophoretic mobilities in agarose gels *(3)*. However, mol wt separation can be achieved if pulsed electric fields are applied to the gel *(4)*. Separation is thought to occur because the time required for large DNA molecules to change their direction of migration in response to the changing direction of the electric field depends on DNA size. Various theories have been proposed to describe the center-of-mass velocity and the dynamics of reptating DNA molecules in pulsed electric fields of various configurations *(5–9)*. Some of these theories differ significantly in the mechanism proposed for the reorientation of the DNA molecules in the alternating electric field *(4)*.

The various theories proposed to explain the pulsed-field separation of very large DNA molecules assume that the agarose matrix pre-

From: *Methods in Molecular Biology, Vol. 12: Pulsed-Field Gel Electrophoresis*
Edited by: M. Burmeister and L. Ulanovsky
Copyright © 1992 The Humana Press, Inc., Totowa, NJ

sents an inert array of obstacles around or through which the DNA molecules must pass. However, electric birefringence studies *(10,11)* have shown that substantial orientation of the agarose matrix takes place in pulsed electric fields of the strength and duration used in pulsed-field gel electrophoresis. In addition, unusual orientation effects are observed when the electric field is reversed in polarity *(10,11)*. New results, to be described in this chapter, demonstrate that domains within the agarose matrix orient and disorient in phase with the alternating electric field.

This chapter describes different types of experiments that illustrate that the agarose matrix can be oriented by electric fields of the strength and duration commonly used in pulsed-field gel electrophoresis. The electrophoretic mobilities observed for DNA in unidirectional electric fields depend on whether the matrix is oriented by application of an electric field before electrophoresis is started *(12)*. The orientation of the matrix itself can be observed by the technique of transient electric birefringence *(10,11)*. Insofar as this technique is not well known, a brief description of the theory of electric birefringence is also given.

2. Methods

2.1. Electric Birefringence

2.1.1. Theory

The theory of electric birefringence has been described in several reviews *(13–15)*. Briefly, if a macromolecule (or a domain in a gel matrix) is asymmetric and polar, it will interact with an applied electric field and tend to orient in the direction of the applied field. Upon orientation, the refractive index of the solution (or gel) becomes different in directions parallel and perpendicular to the applied electric field. This difference in refractive index, called the birefringence, can be measured, and is a direct measure of orientation. At low electric field strengths the amplitude of the birefringence, Δn, is proportional to the square of the electric field strength, E, as shown in Eq. (1),

$$\Delta n = K n E^2 \qquad (1)$$

where n is the mean refractive index of the solution (or gel) and K is a proportionality constant, known as the Kerr constant, which is characteristic of each species being oriented in the electric field. Because

Δn is proportional to E^2, the sign of the birefringence is expected to be independent of the direction of the applied electric field. The decay of the birefringence after removal of the electric field is a result of the randomization of the orientation of the birefringent particles because of Brownian motion. For an assembly of axially symmetric, monodisperse rigid particles, the field-free decay of the birefringence is given by Eq. (2),

$$\Delta n / \Delta n_0 = e^{-t/\tau} \qquad (2)$$

where Δn is the birefringence at any time, t, after removal of the electric field, Δn_0 is the birefringence at the moment the field is removed, and τ is the rotational relaxation time characteristic of the oriented particle. The relaxation times can be related to particle dimensions by standard equations, if some assumptions can be made about particle shape *(13–15)*.

2.1.2. Methods

The apparatus and methods used for the electric birefringence measurements are described in *(16)*. The limiting time constant of the detecting system was 0.2 µs. The Kerr cell was a shortened 1.00-cm quartz spectrophotometer cell, chosen for its negligible strain birefringence. Parallel platinum plate electrodes, separated by 2.0- or 4.0-mm, were mounted on lexan supports of standard design *(13)*. Pulses ranging from 0.5–20 V in amplitude and 0.1–100 s in duration were generated by a home-built pulser (Craig Fastenow, University of Iowa, Department of Biomedical Engineering). The rise and decay times of the pulses were ≤10 µs. All measurements were carried out at 20°C; the calculated temperature rise during any one pulse or train of pulses was usually less than 0.1°C. Successive pulses or trains of pulses were separated by at least 10 min. Typical oscilloscope traces of the birefringence signals have been selected to illustrate the results.

Seakem LE agarose, purchased from FMC BioProducts, was used for all birefringence measurements. The gels illustrated here contained 1% agarose; similar results are observed in gels containing 0.6 and 1.4% agarose (J. Stellwagen, unpublished results). The agarose was dissolved in deionized water by boiling in a microwave oven. After cooling to ~50°C, the warm agarose solution was pipetted into the birefringence cell and the electrodes inserted. The warm agarose solidified almost immediately; measurements were begun about 1/2 h later. Results similar to the results described here were obtained if the agarose was dissolved in 1 mM Tris-HCl buffer instead of deionized water, or if the agarose was pipetted into the space between the electrodes and the remainder of the cell was filled with solvent *(10)*. Dilute (nongelling) agarose solutions were prepared in a similar manner *(17)*.

2.2. Oriented Gels

Oriented gels were prepared from Seakem LE agarose as described *(12)*. Control and oriented gels were prepared from the same agarose stock solution and cast in the same gel form, separated by a Plexiglas spacer, so they could be electrophoresed together. The oriented gels were prepared by applying an electric field to one side of the gel form as the agarose was cooling, using a pair of platinum wires taped to the bottom of the gel frame. Similar results were obtained if the agarose was allowed to solidify before the orienting field was applied *(12)*. After 2.5 h, the orienting electric field was turned off, the platinum wires were removed, DNA samples were applied to both the control and oriented portions of the gel, and electrophoresis was started. All electrophoresis experiments were carried out at room temperature (23°C), using an electric field strength of 2.6 V/cm. The gels illustrated here were cast and run in TAE buffer (0.04M Tris base, brought to pH 8.0 with glacial acetic acid, plus 1 mM EDTA), and contained 2.5 g/mL ethidium bromide (Sigma), added to the warm agarose just before pouring the gel. Similar results were observed in gels cast and run in other buffers and in gels without ethidium bromide.

The DNA samples used for the electrophoresis measurements were the 1-kb ladder (containing fragments ranging in size from 0.51 to 12.2 kb), the supercoil ladder (containing supercoiled DNA molecules ranging in size from 1.1 to 16.2 kb), and the 123-base pair (bp) ladder (containing multiples of the 123-bp monomer), all purchased from Bethesda Research Laboratories.

3. Electric Birefringence Experiments

3.1. Dilute Agarose Solutions

Dilute aqueous solutions containing 0.05–0.1% agarose exhibited positive birefringence when pulses just long enough to reach the steady state were used *(17)*, as shown in Fig. 1A. Because the linear dichroism of dilute agarose solutions is also positive *(18)*, positive birefringence

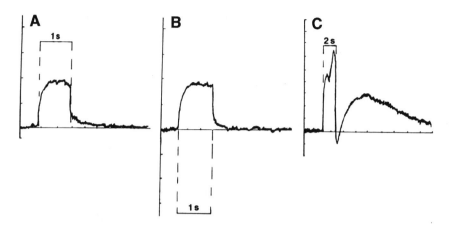

Fig. 1. Oscilloscope traces of the electric birefringence signals exhibited by dilute solutions of agarose in deionized water. **(A)** 0.05% agarose, 1-s orienting pulse, $E = 140$ V/cm; **(B)** same solution, after reversing the polarity of the applied pulse; **(C)** 0.1% agarose, 2-s orienting pulse, $E = 140$ V/cm. The applied pulses are indicated by vertical dashed lines; the noisier traces represent the observed birefringence signals. The vertical scale is arbitrary but constant for all traces; the horizontal scale is indicated by the length of the orienting pulse. The direction of positive birefringence is up.

indicates that the agarose molecules were aligned parallel to the electric field lines. The decay of the birefringence was characterized by relaxation times of ~20 ms, suggesting that individual, probably helical *(19,20)*, agarose molecules were being oriented by the electric field. Reversing the polarity of the electric field caused no change in the sign of the birefringence, as shown in Fig. 1B, in agreement with Eq. (1).

More complicated birefringence signals were observed if longer pulses were applied to the agarose solutions, as shown in Fig. 1C. The amplitude of the birefringence frequently exhibited a discontinuity during the pulse, either because of the orientation of a second birefringent species or because of the formation of more highly birefringent aggregates during the pulse *(17)*. After such complicated orienting signals, the decay of the birefringence frequently exhibited an initial rapid decrease, similar to that observed when short pulses were applied to the solution (Figs. 1A, B), and then went through a minimum and a maximum before decaying slowly to zero. The complex, slowly decaying birefringence signal indicates that various aggregates were present in the solution. The largest aggregates must have been positively birefringent, to explain the positive terminal decay of the birefringence. Other aggregates must have been negatively birefringent, to explain the minimum and maximum in the decay curve. The decay times of the largest aggregates correspond to the disorientation of micron-sized clusters of agarose molecules *(17)*. The presence of such large aggregates in dilute agarose solutions suggests that aggregates of this type may be intermediates in the process of gel formation.

3.2. Agarose Gels

3.2.1. Unidirectional Electric Fields

When pulsed unidirectional electric fields of 2.5-30 V/cm were applied to 1% agarose gels for periods of time ranging from 0.5–100 s, significant orientation of the matrix was observed, as shown in Fig. 2. If the applied pulse was ≤2.5 V/cm, the birefringence signal increased rapidly (in absolute value) to an apparent equilibrium value, as shown in Fig. 2A. After removal of the electric field, the birefringence decreased rapidly at first and then more slowly (Fig. 2A). At higher field strengths, e.g., 10 V/cm, the amplitude of the birefringence did not reach a steady state during the pulse, but continued to increase as long as the pulse was applied, as shown in Fig. 2B. Frequently, when a monotonically increasing birefringence signal was observed during the pulse, the signal continued to increase for 1–2 s after the pulse was removed before passing through a maximum and slowly decaying to zero (Fig. 2B). Higher voltages (e.g., 15 V/cm) often resulted in the appearance of a signal of opposite sign during the decay of the birefringence, as shown in Fig. 2C. Birefringence decay curves containing still more components were observed when longer pulses (e.g., 10 s) were applied to the gel, as shown in Fig. 2D. The decay of the birefringence in Fig. 2D suggests the presence of six different relaxation times of alternating polarity. Depending on the type of birefringence signal observed during and immediately after the pulse, the very slow terminal decay of the birefringence could be either negative (Figs. 2A, C) or positive (Figs. 2B, D). However, the time constants characterizing the decay of the various components of the birefringence were relatively independent of the sign of the birefringence, as observed previously *(10)*.

The sign of the birefringence observed during and after the pulse was usually positive, as shown in Figs. 2B and 2C. However, about 15%

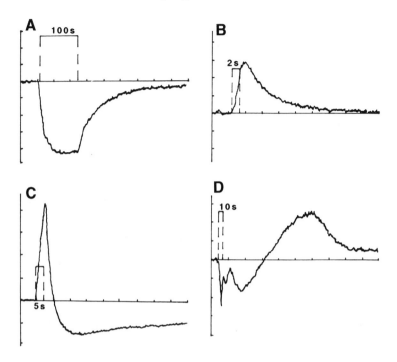

Fig. 2. Oscilloscope traces of the electric birefringence signals exhibited by LE agarose gels. (A) 1% gel, 100-s orienting pulse, E = 2.5 V/cm; (B) 1% gel, 2-s orienting pulse, E = 10 V/cm; (C) 1% gel, 5-s orienting pulse, E = 15 V/cm; (D) 0.6% gel, 10-s orienting pulse, E = 12.5 V/cm. The vertical scale is arbitrary; the horizontal scale is indicated by the pulse length. The direction of positive birefringence is up.

of the gels exhibited negative birefringence during (and after) the pulse, as shown in Figs. 2A and 2D. The reason for different independently prepared gels exhibiting birefringence signals of different signs is not known. It is possible that the agarose helices that form upon gelation *(19,20)* can be either right- or left-handed, leading to aggregates that have different structures in different independently prepared gels. Alternatively, the domains within the matrix that orient in the electric field may be only slightly anisometric, and their direction of orientation may depend on whether they are slightly prolate or oblate in shape. In a previous study *(10)*, about 60% of the gels exhibited negative birefringence, in contrast to the positive birefringence signals usually observed here. The reason for the preponderance of positively birefringence gels in this study and negatively birefringent gels in the previous study is not known. However, different agarose samples were used in the two studies, suggesting that intrinsic differences may exist between different lots of agarose.

The amplitudes of the birefringence signals illustrated in Fig. 2 are ~5 orders of magnitude larger than would have been expected from the Kerr law behavior of agarose gels oriented by short, high voltage electric field pulses *(16)*. Hence, very significant orientation of the agarose matrix takes place when long, relatively low voltage pulses are applied to the gel. The terminal birefringence relaxation times, which ranged from 9–50 s for the various gels illustrated in Fig. 2, are also ~5 orders of magnitude larger than observed in agarose gels oriented by short, high voltage pulses *(16)*. Relaxation times of this magnitude correspond to the reorientation of micron-sized domains within the agarose matrix *(10)*. Scattering domains of this size have also been observed in 1% agarose gels by quasielastic light scattering *(21)*.

3.2.2. Interval Pulses

Typical orientation patterns observed when pulses of the same polarity, separated by an equal interval with no pulsing, were applied to the gel, are illustrated in Fig. 3. If the decay of the birefringence after the removal of the first pulse exhibited a maximum (Fig. 3A), and the second pulse was applied while a significant portion of the original birefringence signal remained, an increased orientation of the gel was observed during the second pulse, as shown in Fig. 3B. However, if the applied pulses and the intervals between them were much longer, so that the birefringence had decayed essentially to zero before the second pulse was applied (Fig. 3C), the gel exhibited a periodic orientation in phase with the applied electric field, as shown in Fig. 3D.

The results in Fig. 3 suggest that the orientation of the matrix observed with interval pulsing depends on the applied voltage, the pulse duration, and the dead time between pulses. The first two parameters determine how quickly the oriented domains of the matrix lose their orientation when the electric field is removed; the interval between the pulses determines whether the orientation of the domains is additive or periodic.

3.2.3. Reversing Electric Fields

The sign of the birefringence of each individually prepared agarose gel remained constant upon repeated pulsing, independent of

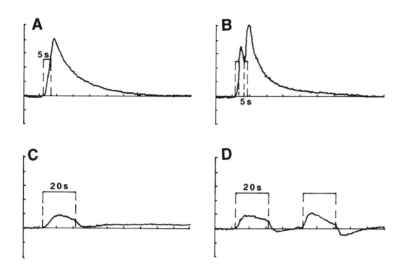

Fig. 3. Oscilloscope traces of typical electric birefringence signals observed for 1% gels using interval pulses. **(A)** 5-s orienting pulse, $E = 10$ V/cm; **(B)** two 5-s orienting pulses separated by 5 s, $E = 10$ V/cm; **(C)** 20-s orienting pulse, $E = 20$ V/cm; **(D)**, two 20-s orienting pulses separated by 20 s. The vertical scale is arbitrary but constant for all traces; the horizontal scale is indicated by the pulse length. The direction of positive birefringence is up.

the pulse length or the applied voltage. However, the sign of the birefringence of a given gel could be changed from positive to negative (or vice versa, depending on the sign of the original birefringence) by reversing the polarity of the applied electric field *(10,11)*. The resulting birefringence signal was approximately the mirror image of the original signal, as shown in Fig. 4.

Reversing the polarity of the electric field again to its original direction caused the sign of the birefringence to revert to its original value. In general, the sign of the birefringence followed the polarity of the applied electric field; repeated pulsing in one field direction did not change the sign of the birefringence, whereas repeatedly reversing the direction of the applied electric field caused the sign of the birefringence to reverse repeatedly. More complicated reversing field behavior was observed in samples that exhibited complicated birefringence decay patterns; however, these results will not be discussed here.

The reversal of the sign of the birefringence of agarose gels upon reversal of the direction of the applied electric field contradicts

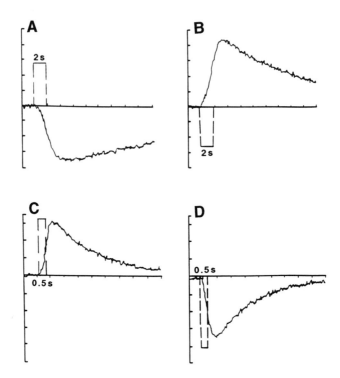

Fig. 4. Typical oscilloscope traces observed for 1% gels when the electric field is reversed in polarity. (**A, B**) 2-s orienting pulse, E = 10 V/cm; (**C, D**) 0.5-s orienting pulse, E = 50 V/cm. The vertical scale is arbitrary but constant for each pair (**A, B**) and (**C, D**); the horizontal scale is indicated by the pulse length. The direction of positive birefringence is up.

Eq. (1), which requires the sign of the birefringence to be independent of the polarity of the applied field. The reason for the failure of Eq. (1) is not known. The reversal of the sign of the birefringence upon field reversal must be related to gelation; dilute solutions of agarose exhibited normal (unchanged in sign) birefringence signals in reversing electric fields, as shown in Figs. 1A and 1B. It is possible that domains of the agarose matrix, once oriented in an electric field, tend to orient in the opposite direction upon field reversal, because of poorly understood polarization effects similar to those observed in micellar solutions (22). However, it is also possible that the reversal of the sign of the birefringence upon field reversal is related to other nonideal characteristics of the birefringence of agarose gels, such as the somewhat variable sign of the birefringence observed for different independently prepared gels and the constantly increasing amplitude of the birefringence observed with increasing pulse length in electric fields ≥2.5 V/cm (Fig. 2B). In addition, if pulses of constant length are applied to the gel, the amplitude of the birefringence does not obey Eq. (1) (J. Stellwagen, unpublished results). The relevance of these effects to the anomalous reversing field behavior exhibited by agarose gels is currently under investigation.

3.2.4. Bipolar Pulses and Pulse Trains

When bipolar pulsed electric fields were applied to the gel, the birefringence signal reversed sign in phase with, or slightly out of phase with, the bipolar pulse, as shown in Fig. 5. If the period of the bipolar pulse approximately matched the fast initial rise and decay of the birefringence, nearly symmetric positive and negative birefringence signals were observed, which oscillated in phase with the bipolar pulses, as shown in Figs. 5B-5E. If the two halves of the bipolar pulse were unequal instead of equal in duration, the birefringence signal oscillated in phase with the unequal bipolar pulses, as shown in Fig. 5F. However, in this case the birefringence signals observed during the two halves of the reversing pulse were not symmetric. The positive birefringence signal appeared to reach a constant positive value during the pulse of longer duration, whereas the negative birefringence signal was larger and continually increased in amplitude during the shorter pulse in the opposite direction. The terminal relaxation times observed after the end of the bipolar pulse train were similar to those observed after unidirectional pulses were applied to the gel, suggesting that domains of similar size were being oriented by both types of pulses.

Similar results were observed in higher electric fields, as shown in Fig. 6. If the birefringence signal continually increased in amplitude during the pulse before passing through a maximum and decaying slowly to zero, application of a bipolar pulse caused a constantly increasing negative signal to be observed during the second half of the bipolar pulse, as shown in Fig. 6b. If a train of 5-s/5-s bipolar pulses were applied to the gel with a 10-s delay between pulses, the amplitude of the birefringence in the positive and negative directions gradually equalized, as shown in Fig. 6C, with a negative residual birefringence signal observed between the pulses. If the bipolar pulses were applied

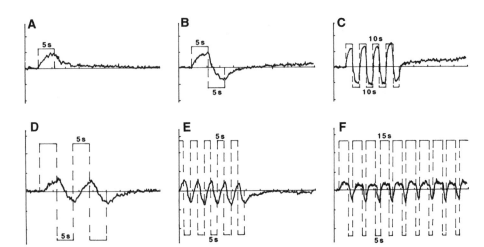

Fig. 5. Oscilloscope traces observed when bipolar pulses are applied to 1% agarose gels. (**A**) 5-s orienting pulse, E = 2.9 V/cm; (**B**) 5-s/5-s bipolar pulse, E = 2.6 V/cm; (**C**) train of 5-s/5-s bipolar pulses, E = 2.6 V/cm; (**D**) train of 5-s/5-s bipolar pulses, E = 5.0 V/cm; (**E**) longer train of 5-s/5-s bipolar pulses, E = 5.0 V/cm; (**F**) train of 15-s/5-s bipolar pulses, E = 5.0 V/cm. The vertical scale is arbitrary but constant for all traces; the horizontal scale is indicated by the pulse length. The direction of positive birefringence is up.

consecutively, the birefringence signal oscillated in phase with the bipolar pulses, as shown in Figs. 6D and 6E. However, the sign of the birefringence changed from predominantly positive in the initial bipolar pulse to predominantly negative after four successive bipolar pulses were applied to the gel, as shown in Fig. 6E. Similar results were observed with longer orienting pulses, as shown in Fig. 6F for 20-s/20-s bipolar pulses.

Taken together, the results in Figs. 5 and 6 suggest that domains within the agarose matrix orient and disorient with a periodicity equal to that of the applied alternating electric field. The extent of the orientation of the matrix during each half of the bipolar pulse depends on the amplitude and duration of each pulse. The periodic orientation and reorientation of domains within the agarose matrix in alternating electric fields may increase the microscopic fluidity of the matrix, making it easier for very large DNA molecules to migrate through the gel during pulsed-field electrophoresis *(10)*.

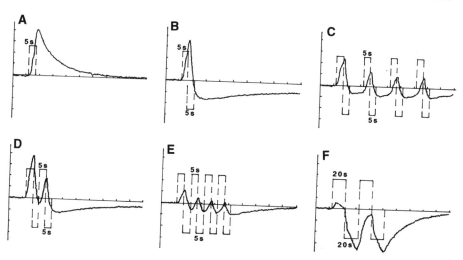

Fig. 6. Oscilloscope traces observed for 1% gels when bipolar pulses of 10 V/cm are applied to the gel. **(A)** 5-s orienting pulse; **(B)** 5-s/5-s bipolar pulse; **(C)** train of 5-s/5-s bipolar pulses, each separated by 10 s; **(D)** two consecutive 5-s/5-s bipolar pulses; **(E)** train of 5-s/5-s bipolar pulses; **(F)** two 20-s/20-s bipolar pulses. The vertical scale is arbitrary but constant for all traces; the horizontal scale is indicated by the pulse length. The direction of positive birefringence is up.

4. Electrophoresis in Oriented Gels

In order to demonstrate directly that the orientation of the agarose matrix could affect the mobilities of DNA molecules migrating in agarose gels, two-part oriented/unoriented gels were prepared as described in Section 2.2. The mobility of DNA molecules in the oriented gels was usually faster than observed in unoriented control gels run at the same time, as shown in Fig. 7, regardless of whether the orienting electric field was applied parallel or perpendicular to the eventual direction of electrophoresis. However, slower mobilities were observed in ~20% of the oriented gels, for reasons that are not known *(12)*. Identical mobilities were never observed in the control and oriented gels. The somewhat random difference in the sign of the mobility effect is similar to the somewhat random sign of the electric birefringence observed for different individually prepared gels, and is further evidence that the microstructure of the agarose matrix differs from gel to gel.

Stellwagen

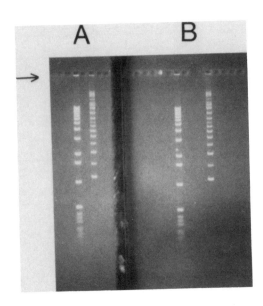

Fig. 7. Comparison of electrophoresis in oriented and unoriented gels. **(A)** control, unoriented gel; **(B)** gel oriented by an electric field of 4 V/cm applied parallel to the direction of electrophoresis. Left lanes in **A** and **B**: 1-kb ladder; right lanes, supercoil ladder. Note the increased distance of migration of the DNA fragments in **B**. The origin of electrophoresis is indicated by an arrow.

The absolute magnitude of the difference in mobility observed for DNA molecules in control and oriented gels varied from ~5–20%, depending on DNA mol wt, gel concentration, buffer, and voltage used to orient the gel *(12)*. Comparable mobility differences were observed when the orienting electric field was applied to molten agarose or when the agarose was allowed to solidify before the orienting voltage was applied. Therefore, both molten agarose and the solid agarose matrix can be oriented by an applied electric field. If an oriented gel was allowed to stand ~20 h before electrophoresis was started, the orientation of the matrix was lost and DNA molecules in both halves of the gel frame migrated with the same velocity. Therefore, very little energy is required to randomize the structure of an oriented agarose matrix.

In perpendicularly oriented gels, linear DNA molecules migrated in lanes skewed toward one side of the gel, as shown in Fig. 8A. As

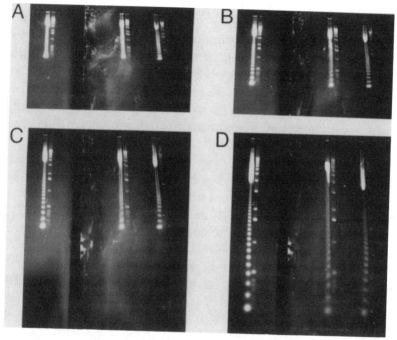

Fig. 8. Time course of electrophoresis in perpendicularly oriented gels. Left side of each panel, control, unoriented gel; right side of each panel, perpendicularly oriented gel. In each gel, the left lane of each pair contains the 123-bp ladder; the right lane of each pair is the 1-kb ladder. **(A)** after 2 h of electrophoresis; **(B)** after 3 h; **(C)** after 4 h; **(D)** after 20 h. Note the gradual straightening of the lanes on the right side of each panel with elapsed time. The magnification factor is constant for all photographs.

electrophoresis proceeded, the direction of migration gradually straightened, as shown in Figs. 8B-D, causing the lanes to become markedly nonlinear. The migration of linear DNA molecules in lanes skewed toward the side of the gel in perpendicularly oriented gels suggests that the orientation of the matrix is accompanied by the formation of pores or channels in the direction of the applied electric field. The gradual straightening of the lanes with time illustrated in Fig. 8 suggests that the original *perpendicular* orientation of the matrix was gradually lost during electrophoresis in the *parallel* direction. Therefore, the orientation of the agarose matrix can be altered during electrophoresis by electric fields applied to the gel.

5. Conclusions

The results presented here demonstrate that agarose gels can be oriented by electric fields of the amplitude and duration commonly used for pulsed-field gel electrophoresis. Both molten agarose solutions and solid agarose gels can be oriented by electric fields of a few V/cm. The orientation of the matrix can be detected by changes in the mobility and direction of migration of DNA molecules during electrophoresis. The orientation of the agarose matrix is lost if the gel is allowed to stand ~20 h after the orienting field is removed, suggesting that very little energy is required to rearrange the three-dimensional hydrogen-bonded structure *(19)* of the matrix. Because the matrix can be oriented by the electric field, mobilities observed for DNA molecules in agarose gels must be the combined result of matrix orientation and the migration of the DNA molecules toward the anode. Therefore, factors that affect the orientation of the matrix may affect the mobility of DNA in agarose gels, even if the DNA molecules themselves are unaffected.

The orientation of the agarose matrix was observed directly by electric birefringence experiments, using pulsed electric fields of the amplitude and duration commonly used for pulsed-field gel electrophoresis. The time constants characterizing the decay of the birefringence after removal of the electric field suggest that micron-sized domains within the agarose matrix are oriented by the electric field. The oriented domains vary in size and polarity depending on the length of the pulse and the magnitude of the applied voltage.

Reversing the polarity of the applied electric field causes an anomalous reversal in the sign of the birefringence exhibited by agarose gels *(10,11)*. As a result, domains within the matrix must be continually orienting and reorienting in phase with alternating bipolar electric fields. Since the size of the oscillating domains depends on pulse length and applied voltage, the permeability of the matrix must also depend on these variables. The orientation and reorientation of the matrix in alternating electric fields occurs in addition to the orientation and reorientation of DNA. Hence, theories describing the gel electrophoresis of DNA, especially in pulsed electric fields, need to incorporate the orientation of the matrix in order to describe the experiment completely.

Acknowledgments

The experiments summarized here were carried out by John Stellwagen, David Stellwagen, and Diana Holmes, to whom the author is greatly indebted. Financial support by Grant GM29690 from the National Institute of General Medical Sciences is also gratefully acknowledged.

References

1. Stellwagen, N. C. (1987) Electrophoresis of DNA in agarose and polyacrylamide gels. *Adv. Electrophoresis* 1, 179–228.
2. Stellwagen, N. C. (1985) Effect of the electric field on the apparent mobility of large DNA fragments in agarose gels. *Biopolymers* 24, 2243–2255.
3. McDonell, M. W., Simon, M. N., and Studier, F. W. (1977) Analysis of restriction fragments of T7 DNA and determination of molecular weights by electrophoresis in neutral and alkaline gels. *J. Mol. Biol.* 110, 119–146.
4. Cantor, C. R., Smith, C. L., and Mather, M. K. (1988) Pulsed-field gel electrophoresis of very large DNA molecules. *Ann. Rev. Biophys. Chem.* 17, 187–304.
5. Fesjian S., Frisch, H. L., and Jamil, T. (1986) Diffusion of DNA in a gel under an intermittent electric field. *Biopolymers* 25, 1179–1184.
6. Southern, E. M., Anand, R., Brown, W. R. A., and Fletcher, D. S. (1987) A model for the separation of large DNA molecules by crossed field gel electrophoresis. *Nucleic Acids. Res.* 15, 5925–5943.
7. Deutsch, J. M. (1987) Dynamics of pulsed-field electrophoresis. *Phys. Rev. Lett.* 59, 1255–1258.
8. Zimm, B. H. (1988) Size fluctuations can explain anomalous mobility in field-inversion electrophoresis of DNA. *Phys. Rev. Lett.* 61, 2965–2968.
9. Noolandi, J., Slater, G. W., Lim, H. A., and Viovy, J. L. (1989) Generalized tube model of biased reptation for gel electrophoresis of DNA. *Science* 243, 1456–1458.
10. Stellwagen, J. and Stellwagen, N. C. (1989) Orientation of the agarose gel matrix in pulsed electric fields. *Nucleic Acids Res.* 17, 1537–1548.
11. Sturm, J. and Weill, G. (1989) Direct observation of DNA chain orientation and relaxation by electric birefringence: Implications for the mechanism of separation during pulsed-field gel electrophoresis. *Phys. Rev. Lett.* 62, 1484–1487.
12. Holmes, D. L. and Stellwagen, N. C. (1989) Electrophoresis of DNA in oriented agarose gels. *J. Biomol. Struc. Dyn.* 7, 311–327.
13. Fredericq, E. and Houssier, C. (1959) *Electric Dichroism and Electric Birefringence*, Clarendon, Oxford.
14. Charney, E. (1988) Electric linear dichroism and birefringence of biological polyelectrolytes. *Qu. Rev. Biophys.* 21, 1–60.

15. García de la Torre, J. and Bloomfield, V. A. (1981) Hydrodynamic properties of complex, rigid, biological macromolecules: Theory and applications. *Qu. Rev. Biophys.* **14**, 81–139.

16. Stellwagen, N. C. and Stellwagen, J. (1989) Orientation of DNA and the agarose gel matrix in pulsed electric fields. *Electrophoresis* **10**, 332–344.

17. Stellwagen, N. C. and Stellwagen, D. (1990) Electric birefringence of dilute agarose solutions. *J. Biomol. Struct. Dyn.* **8**, 583–600.

18. Jonsson, M., Åkerman, B., and Nordén, B. (1988) Orientation of DNA during gel electrophoresis studied with linear dichroism spectroscopy. *Biopolymers* **27**, 381–414.

19. Arnott, S., Fulmer, A., Scott, W. E., Dea, I. C. M., Moorhouse, R., and Rees, D.A. (1974) The agarose double helix and its function in agarose gel structure. *J. Mol. Biol.* **90**, 269–284.

20. Foord, S. A. and Atkins, E. D. T. (1989) New x-ray diffraction results from agarose: Extended single helix structures and implications for gelation mechanism. *Biopolymers* **28**, 1345–1365.

21. Pines, E. and Prins, W. (1973) Structure-property relations of thermoreversible macromolecular hydrogels. *Macromolecules* **6**, 888–895.

22. Hoffman, H., Kramer, U., and Thurn, H. (1990) Anomalous behavior of micellar solutions in electric birefringence measurements. *J. Phys. Chem.* **94**, 2027–2033.

CHAPTER 26

Reptation Theory of Pulsed Electrophoresis and Trapping Electrophoresis

Jean Louis Viovy
and Anne Dominique Défontaines

1. Introduction

In the history of the still-young technique of pulsed electrophoresis experimental, theoretical, and numerical progress are increasingly intermingled. The first key to this field can be traced to theoretical letters of Lerman and Frisch and Lumpkin and Zimm *(1,2)*, who remarked that long flexible chains like chromosomal DNA cannot migrate in gels as random coils, and suggested a reptative-like motion in which chains thread their way among fibers like a snake among roots. They also predicted that large DNA chains orient in the field, and for the first time, made the connection between this orientation and the saturation of the mobility that causes the failure of conventional gel electrophoresis techniques. The idea of fighting this orientation by periodically changing the field appeared for the first time about two years later, in the founding experimental work of Schwartz and Cantor *(3)*, giving birth to the pulsed electrophoresis technique. Detailed developments of the reptation ideas sketched in 1982 were steadily pursued in 1985 and in the following years *(4–20)*, in particular by the groups of Lumkin, Zimm, and coworkers, and Slater, Noolandi, and coworkers. These first developments, however, remained

From: *Methods in Molecular Biology, Vol. 12: Pulsed-Field Gel Electrophoresis*
Edited by: M. Burmeister and L. Ulanovsky
Copyright © 1992 The Humana Press, Inc., Totowa, NJ

focused on migration in constant-field, and theorists got deeply involved in the understanding of pulsed electrophoresis only since 1987 *(21–39)*. This contrasts with the community of experimental biologists, in which frantic developments of pragmatic approaches readily followed Schwartz and Cantor's first papers *(38–74)* (*see also* other contributions to this vol.). A large number of variants of the general pulsed-field gel electrophoresis (PFGE) approach have been proposed. Their present nomenclature often obeys commercial rather than scientific requirements, and it is unusefully confusing. PFGE methods can be classified into three "families" recalled below in order of decreasing technical complication:

1. Crossed-fields gel electrophoresis (CFGE) *(38–60)*, in which at least two field orientations are used. This family can be divided again into subfamilies. An important distinction, from the point of view of theory and practical applications, is whether the fields are homogeneous or not. Orthogonal-field alternation gel electrophoresis (OFAGE) *(3,40* and *see* Chapter 4, this vol.) uses two fields inhomogeneous and parallel to the gel plane, leading to curved migration lanes. Transverse alternating field electrophoresis (TAFE) *(58* and *see* Chapter 1, this vol.) has the same type of inhomogeneous fields transverse to the gel, in order to restore linear migration. In clamped homogeneous electric field (CHEF) *(44,* and *see* Chapter 2, this vol.), linear migration is obtained by the use of two homogeneous electric fields at a fixed angle (generally around 120°). The programmable autonomously controlled electrode (PACE) *(48)* method is a variant of the latter, in which the field angle can be varied without changing the electrode geometry, thanks to an independent dynamic control of the tension on each electrode. Homogeneous crossed fields can also be applied to the gel by a mechanical rotation of the gel in a fixed field, rotating gel electrophoresis (RGE) *(38,53)*, or by a mechanical rotation of electrodes at a fixed potential, rotatory pulsed-field gel (RPFG) *(50)*. Finally, methods using more than two fields, such as pulsed homogeneous orthogonal-fields (PHOGE) *(60)*, in which a crenel-type migration is obtained by applying three different fields, have been proposed.

2. Field-inversion gel electrophoresis (FIGE) methods *(61–68* and *see* Chapter 1, this vol.), in which the field is alternately applied in the two directions along a single orientation. One can distinguish the "single field" FIGE, in which the field amplitude is the same in the two directions *(61* and *see* Chapter 1, this vol.), and the "dual

field" protocol (DFIGE) (*25* and *see* Chapter 7, this vol.), in which the forward and backward pulses have different amplitudes. Zero-integrated field electrophoresis (ZIFE) (*67* and *see* Chapter 7, this vol.) is a particular case of the latter, in which the product of pulse time by field amplitude just compensates between the forward and the backward pulses.

3. Intermittent-field gel electrophoresis (IFGE) methods (*29,69–71*) in which the field is just interrupted periodically.

4. Finally, it is worth pointing out that hybrid methods combining several of theses effects (for instance, introducing field interruptions in CFGE or FIGE, or field inversion in CFGE) are now very easy to attempt using computer-controlled apparatus, and are proposed for specialized applications (*46,55,67*).

Overall, it is probably fair to say that experiments had the most innovative role in the development of PFGE, and that theories have been more "explicative" than "predictive." There are notable exceptions to this, however, in which predictions of theory or computer simulations preceded experimental observations.

From the mechanistic point of view, the existence of "hernias" (a migration mechanism in which the chain migrates by developing one or several loops in the direction of the field in contrast with the "head on" biased reptation motion), and of "geometration" motions (an alternation of stretching and collapsing of the chain), was suggested by computer simulations on lattices of obstacles performed by Deutsch (*21–24*). These predictions present close similarities with the dynamics later observed in fluorescence videomicroscopy experiments (*75–77*).

More practically, the occurrence of a band-inversion in some constant-field cases (*6,13,16,17*) was predicted by the biased reptation model first and confirmed experimentally later on (*13,28*). The biased reptation model was also used by Slater et al. to study the principle of DFIGE (*25*), and to derive theoretical predictions for it. These authors predicted that DFIGE is more versatile than single-field FIGE, a fact now well acknowledged by experimentalists. For this latter technique, theory has been of less help: Several theoretical models have shown that the very striking band-inversion observed in FIGE (*61*) could indeed arise from an "antiresonance" between chain conformation fluctuations and the external field (*23,27,32,36*), but different mechanisms for this antiresonance have been proposed, and no theoretical agreement or quantitative modeling of data has yet been obtained.

The situation of theory is somewhat better for homogeneous crossed-field electrophoresis (CHEF, RGE, PACE, and so on): For these methods, the biased reptation model provided scaling laws and master curves that rather well describe the evolution of the mobility as a function of field amplitudes, field angles, pulse times, and molecular size *(25,26,33)*. Computer-assisted separation algorithms that associate these models with migration data bases to optimize pulsed-field conditions and avoid time-consuming trial-and-error approaches are now developed (*see*, e.g., the CHEF Gene Mapper® system by Bio-Rad). This indicates that PFGE is leaving childhood, though it is still far from maturity: In the early years of the method, it was sometimes assessed that separation was highly irreproducible from one laboratory to the other, or even from one experiment to the other, and rather pessimistic opinions about predictive approaches were not uncommon (*see*, e.g., *74*). The design of better controlled apparatus, the performance of systematic series of experiments, and the identification of relevant parameters by theoretical approaches have shown that pulsed electrophoresis is indeed a reproducible method obeying rational (although sometimes complicated) laws and amenable to quantification if enough knowledge and care is exerted.

On a more qualitative basis, current models issued from macromolecular theory, such as reptation *(4–14)*, bead-spring chain *(27)*, "straits and lakes" *(5)*, "bunching," "geometration" *(21–23)*, and so on, are convenient handwaving mental tools for imagining new geometries and protocols, and for guiding reflexions when something unexpected happens in new experiments.

The main aim of this chapter is to provide a simple description of these current theoretical tools and concepts, and to promote their use in daily laboratory life.

2. Biased Reptation

2.1. Mobility in Constant Field

2.1.1. Biased Reptation vs Molecular Sieving

In the most straightforward approach to gel electrophoresis, the gel essentially acts as a sieve with a distribution of pore sizes. Theories developed by Ogston *(78)*, Rodbard, Chrambach, and coworkers *(79–81)* assume that the ratio of the electrophoretic mobility in the gel μ to the mobility in free solvent μ_0 is f, the frac-tional volume available to the particle in the gel. The distribution of pore sizes is then derived by geometrical arguments, leading to a general logarithmic evolution of the mobility with size, better known as "Fergusson plot" *(82)*.

The Fergusson plot should be applicable to flexible macromolecules, such as DNA, provided the "particle radius" is replaced by the DNA hydrodynamic radius R_S. According to this model, however, particles with a radius of gyration much larger than the pores of the gel, such as chromosomal DNA, should not penetrate the gel at all. Actually, this is not observed, and large DNA migrate in gels at a velocity that saturates at a limiting finite value when the size of the chain increases *(83–85)*. The clue to this striking behavior was discovered in 1982, when Lumpkin and Zimm and Lerman and Frisch *(1,2)* independently remarked that thin flexible chains can thread their way in the gel (Fig. 1) by a reptation process similar to the one proposed by De Gennes for entangled synthetic polymers *(86–88)*.

In the reptation model, the sequence of pores followed by the DNA chain is called the "tube," whereas the section of chain contained in one pore (of typical size a) is called a "blob." If the chain is flexible on the scale of one pore, the tube can be approximated as a random sequence of N connected segments of size a. The square averaged end to end distance of the chain is $<R_g^2> = Na^2$. This approximation of random steps is quite valid for single-strand DNA in acrylamide or agarose, reasonable for duplex DNA in agarose (the persistence length is around 500 Å while the average pore size in 1% agarose is larger than 1000 Å), but it would not be so, for instance, for duplex DNA in acrylamide. Escape from the "tube" along the chain contour can occur only by loops or "hernias" (Fig. 1B), in which each pore of the gel is crossed by the chain twice instead of once. In the absence of electric field, De Gennes has shown that such events are exponentially unprobable owing to the entropy loss associated with hernias *(87,88)*. Then, the chain can move only by "sliding" along its tube. This reptation motion is activated by random thermal forces, i.e., it randomly occurs in one direction or the other. It is very slow, and it was shown that the time taken by a chain to leave its tube (reptation time) increases as the cube of the chain size (*see* Appendix). The entropic "elasticity" of flexible chains also implies that the tube length fluctuates by an amount of order $a(N)^{-1/2}$ around its average value aN. For long chains, these fluctuations can be neglected.

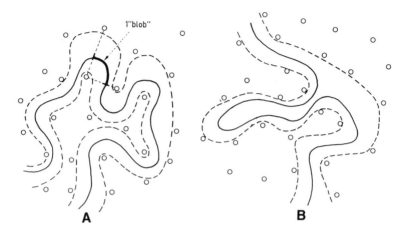

Fig. 1. (**A**)Schematic representation of the chain (full line) and its "tube" (dotted lines) in the gel (dots); (**B**)The same with a loop or "hernia."

In the usual buffer conditions of DNA electrophoresis (pH > 7), the chain is uniformly ionized and negatively charged. When an electric field along, say, z is applied, all segments are attracted to the anode. In particular, the field tends to "pull" hernias out of the tube, against the entropic barrier discussed above. Qualitatively, one expects that this "tube leakage" remains unprobable as long as the electric force that tends to "pull" one blob out of the tube is smaller than the entropic (thermal) force that tends to push it back, i.e., as long as

$$\varepsilon = Eqa/2kT \, c \qquad \qquad (1)$$

where q is the blob charge, T the absolute temperature, k the Boltzmann constant, and c a number of the order of unity. This dimensionless parameter ε proportional to the electric field, which reflects the balance between electric and thermal energy on the scale of a blob, is crucial in the theory of DNA electrophoresis. The actual value of the constant c is difficult to predict for at least two reasons: On the one hand, the finite persistence length of duplex DNA tends to resist the formation of loops, and to diminish tube leakage (i.e., increase c). On the other hand, the probability of tube leakage rapidly increases with the size of pores so that, in gels with a pore polydispersity, such as agarose, tube leakage will occur in the largest pores at smaller values of the field (i.e., c will be lowered). The applied field may also modify the tube length (equal to Na in the absence of field), and affect the reptation behavior accordingly. Duplex DNA, however, is rather rigid, and cannot adopt in a pore a very contorted conformation. In other words, the tube length is not much smaller than the curvilinear length of the DNA itself, so that the tube length cannot vary so much. These comments suggest that there should be a regime at sufficiently low fields in which tube leakage is negligible and tube length is not strongly perturbed. They provide the qualitative rationale for the biased reptation (BR) model, which relies on two basic assumptions: (1) no "hernias" or "tube leakage" and (2) fixed tube length Na.

2.1.2. Mobility in Constant Field

Once both assumptions in Section 2.1.1. are accepted, a rather simple analytical model can be developed, since it is no longer necessary to keep track of the positions of all chain segments, but only of the curvilinear center-of-mass of the chain in its tube. The components of the electric forces perpendicular to the local tube axis are canceled by the tube reaction, and the longitudinal components can be gathered in a single effective force, which induces the curvilinear biased reptation of the DNA chain along its tube *(4, 9–12)*. The mobility can be written as the product of two terms (*see* Appendix):

$$\mu \approx \mu_1 <\rho^2> \qquad (2)$$

$\mu_1 = q/\xi$ is the ratio of charge to friction, i.e., it is the limiting mobility that the chain would adopt in free solvent. $\rho = (R_z/Na)$ is the reduced orientation, i.e., the projection of the end-to-end vector of the chain onto the field direction, divided by the total tube length. Finally, the triangular brackets represent an average over all the conformations adopted by the chain. The second term in relation 2 reflects the loss of efficiency of electric forces, owing to the fact that they have to drag the chain along a tortuous path (the tube) instead of pulling it linearly as in a free solvent. The value of this orientation factor is of course crucial to electrophoresis theory.

2.1.3. Tube Orientation and Nonlinear Mobility

When a chain reptates, the head has to choose a new "gate" in the gel from one pore to the other. Because the head "blob" is charged, it is attracted by the anode, and the successive choices of pores, which in the long run define the conformation of the tube, is biased by the field. The tube is not completely aligned, however, because the electric forces are counteracted by random Brownian forces that tend to randomly kick the head into pores out of the field direction. The theory of this incomplete orientation raises some delicate questions, but the general idea is that it is again controlled by the balance between electric and Brownian forces, i.e., by the parameter ε (relation 1) and by the size of the chain (for longer chains, random forces cancel to some extent and the electric forces become more important). One can show (*see* the appendix for an outline) that $<\rho^2>$ is roughly equal to $1/3N$ for N smaller than a critical value $N^* \approx 3/\varepsilon^2$, and to $\varepsilon^2/9$ for N larger than N^*. Therefore, one can predict two regimes for the mobility in constant field:

*A linear low-field regime is obtained for $N < N^*$, with

$$\mu \sim E^0 N^{-1} \tag{3}$$

*A nonlinear intermediate-field regime is obtained for $N > N^*$, with

$$\mu \sim E^2 N^0 \tag{4}$$

*Finally, the model predicts that, for $\varepsilon > 1$, the mobility should saturate at μ_0 for all chains. This prediction has little meaning, however, since the tube hypothesis is expected to break down in such high field conditions as a result of tube leakage.

The simplified version of the biased reptation model presented above provides a rather convincing account of the saturation of the mobility for long chains. A more detailed treatment, in which averages are performed more carefully to account for the coupling between fluctuations in the tube projection and the reptation velocity, predicted a minimum of mobility around N^* *(6,13,16,17)*. This minimum was indeed observed experimentally in the conditions announced by the theory *(13,23)* (*see*, e.g., Fig. 2). The agreement between the biased reptation theory and experiments is not complete, however. The measured exponent of the mobility vs field, for instance, does not seem to exceed 1.5, whereas the theory predicts an exponent of 2 for long enough chains. Whether this is simply a crossover effect caused by the difficulty of performing constant-field electrophoresis with chains long enough to reach the asymptotic regime, or the evidence of a deeper discrepancy between the model and the reality, is still an unsolved question. Other problems will be described and discussed in the following sections.

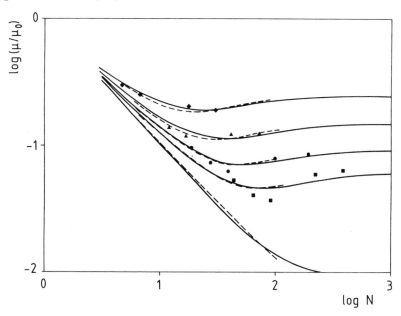

Fig. 2. Evolution of the mobility vs mol wt for different scaled electric fields (ε = 0.2, 0.435, 0.54, 0.69, and 0.9, from lower to upper curve). The full lines are analytical results from ref. *16*, the dotted lines are the computer simulations in ref. *13*. The points are fitted experimental data from ref. *13*.

2.2. Mobility in Crossed Fields

2.2.1. General Principle

In the previous section, we have shown that long charged chains in gels orient in the direction of the field when they reptate (*see*, e.g., Fig. 3A). For oriented chains, the effective electric force is proportional to the size, as is the friction, and separation is impossible. This orientation, however, is only progressively built up by reptation occurring after the onset of the field. The time necessary to reach the steady-state velocity is the time for complete disengagement of the chain from its old tube (orientation time t_{or}). Its precise value depends on the initial state, but at the first approximation t_{or} is proportional to $N E^{-2}$ for long chains (*9–14,31,33*). This is sufficient to qualitatively understand CFGE: When two fields E_A and E_B at an angle ϕ are alternately applied during a time τ to a mixture of chains, the smaller chains with $t_{or} \ll \tau$ rapidly reach the steady-state mobility at the beginning of each

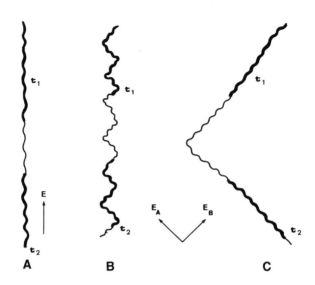

Fig. 3. Different regimes of migration in crossed-field PFG, in the "zig-zag" or "staircase" conformation. Bold lines represent chain conformations at different times t_1 and t_2 while the thin lines represent the tube. **(A)** Permanent field; **(B)** "short pulses" regime ($t < t_{or}$), showing the reduction of overall orientation; **(C)** "macroscopic" regime, ($t > t_{or}$), showing the path as a sequence of fully oriented sections.

pulse (Fig. 3C): They spend most of the duration of pulse A (resp. B) oriented along E_A (resp. E_B), and migrate at the steady-state velocity given by relation 2 (The same is valid for pulse B). The velocity along the diagonal z (effective field) is obtained by projecting the "zigzag" path onto z, i.e., it is reduced by a factor $cos(\phi/2)$ regarding the mobility in a constant field of amplitude E_A.

Longer chains with $t_{or} \gg \tau$ do not have the time to completely reorient in a single pulse, so that their tube is a sequence of several subsections oriented along E_A and E_B. On average, this tube is oriented along the diagonal (Fig. 2B), but its effective orientation ρ is weaker than it would be if the field was permanently oriented along z. Things happen as if such chains were feeling an effective field along z given by the vector sum of E_A and E_B. As shown in the Appendix, the velocity is reduced by a factor $cos^3(\phi/2)$ regarding the mobility in a constant field of amplitude E_A. Therefore, the model predicts that, for a given pulse time, the velocity decreases by a factor of order $cos(\phi/2)$ around a critical size $N^{or}(\tau)$ (the size of chains that just have the time to reorient in a time τ), which scales as τE^2. These predictions are in good qualitative agreement with experiments.

2.2.2. Comparison with Homogeneous Fields Data

This qualitative picture is not sufficient for practical separation applications. In particular, the resolution of CFGE will depend on the way the crossover occurs, and on the detailed evolution of the mobility when N is of the order of N^{pr} (τ). A complete calculation is not easy, because the velocity of the DNA in transient regimes continuously varies during time as a consequence of the progressive orientation. The reorientation time τ_{or}, for instance, does not only depend on field amplitude and chain size, but it also depends on the pulse time and field angle to some extent. For two homogeneous fields as applied in usual CHEF or RGE, the dynamic equation that describes chain reptation can be solved analytically using some approximations *(26,33)*, leading to universal curves such as those presented in Fig. 4. The region of high selectivity associated with $N^{pr}(\tau)$ can be shifted to larger DNA by increasing the pulse time. The mobility is also predicted to decrease linearly with molecular size on a large range. Although masked by the logarithmic representation chosen for Fig. 4, this behavior is apparent in Fig. 5, where the simplified analytical model *(33)* is compared with experimental data *(38)*. The agreement is not quantitative but encouraging, considering that all data corresponding to 17 sizes and 6 pulse times are fitted with only two fitting parameters, the critical size N^* and one reference pulse time $\tilde{\tau}$. Experimental data systematically present a steeper zone after the linear one, indicating that the resolution is higher than predicted. This is not so surprising, since we have been obliged to oversimplify the model in order to treat it analytically to the end. For instance, a complication arises from the fact that the extremity of a chain that plays the role of the "head" may become the "tail" when the field is changed: The reptation direction may be inverted, a situation that becomes dominant for obtuse angles. This "jigsaw" *(33)* or "ratchet" *(26,38)* effect is expected to depend strongly on fluctuations which are preaveraged in purely analytical calculations. We predicted that, for long enough chains and obtuse reorientation angles, this ratchet mechanism is able to induce a symmetry breakdown in the migration: Once chains have oriented along one of the fields (say E_A), they always present a smaller mobility along the other field. If the fields are obtuse enough and the chains long

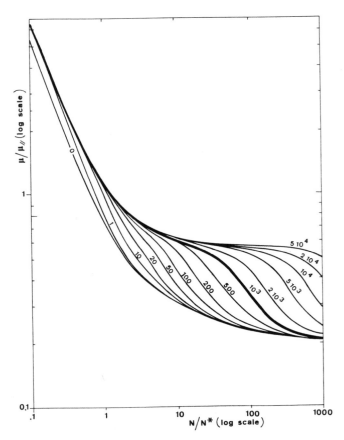

Fig. 4. Universal mobility vs mol wt curves for different reduced pulse times (field angle = 110°). Values of the reduced pulse time, τ/t_A, are indicated on the curves.

enough to ensure that the head and tail are exchanged at each field alternation, the tube sections built during pulse B are always destroyed at the beginning of pulse A, so that the chains macroscopically move along E_A, and not along the diagonal (Fig. 6). This effect was also observed in computer simulations *(24,26)*, but it has not been clearly demonstrated experimentally. Theory indicates *(33)* that it should occur only for chains with t_{or} larger than τ, i.e., in the vicinity of the compression band. It should also strongly depend on minor imbalances between the two alternated fields, and it may be at the origin of unexplained band distortions sometimes observed in experiments. This explanation, in particular, would be consistent with the observation that, when band distortions occur, the longer chains can suffer from

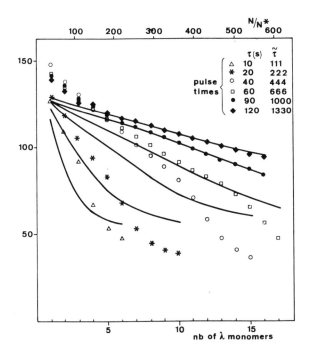

Fig. 5. Best-fit of the approximate analytical biased reptation model to rotating gel electrophoresis experiments *(38)*. The time scaling is the same for all curves, and N/N^* for one λ monomer was set to 36, following independent experiments *(12)*.

considerable band spreading and lateral deviation, whereas shorter ones generally behave perfectly well.

Slater et al. also performed a numerical simulation of the biased reptation model in CFGE conditions *(26)*: In this case, the steeper zone present in experimental plots appears in the theoretical predictions too, accompanied by a moderate band inversion effect just before the compression zone. All these effects are in close agreement with experiments, and the numerical solution of the theory, which does not involve as many approximations as the analytical one, becomes quantitative within typically 20% of experimental data.

2.2.3. Inhomogeneous Fields

In inhomogeneous field conditions, such as those obtained in OFAGE or TAFE, the fields vary little on the microscopic scale, so that theory should in principle be feasible if the fields are known at any point of the gel. In practice, however, quantitative predictions are

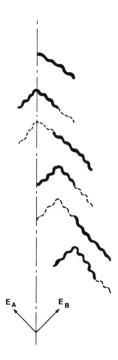

Fig. 6. Schematic representation of the "jigsaw" or "ratchet" migration mechanism for crossed-fields at obtuse angles. Figures from top to bottom alternatively represent the position of the chain at the end of a pulse pulling to the right, then at the end of the next pulse to the left, then at the end of the next pulse to the right, and so on. Dotted lines are guides for the eye, representing the position at earlier stages.

unrealistic. We have seen in the previous section that, with homogeneous and equal crossed fields, the migration can occur by two competing mechanisms, "zigzag" (or "staircase") and "ratchet." The balance between these two mechanisms, and also the net velocity predicted by each of them, depend on field angles and amplitudes, which continuously vary along the gel, as well as on temperature, gel concentration, and chain length. Qualitatively, the theory tells us that not only the migrated distance, but also the band pattern, will change each time any of these parameters, or the relative position of the gel and electrodes, is changed. This leaves little hope for ever quantifying accurately experiments in inhomogeneous fields. One of the initial advantages of OFAGE or TAFE, as compared to homogeneous field methods, was a better repartition of bands in the gel by compression of the low molecular range. This is now easy to achieve in any homogeneous field geometry, thanks to pulse ramping which is much easier to quantify than inhomogeneous fields. Therefore, we believe that the only advantage remaining in favor of inhomogeneous fields is the band-narrowing effect, which may allow a better sensitivity in the case of very dilute or small samples (in particular in TAFE).

2.2.4. Resolution of Topology

The biased reptation model also explains why asymmetrical crossed fields are able to resolve DNA by topology, using the method proposed by Chu *(46)*. In this method, an asymmetry is introduced in the "CHEF" protocol with homogeneous fields at 120°, by varying the amplitude and duration of pulses in the two directions while keeping the product of field strength by pulse time constant. If the mobility was linear with field strength, that would not alter the direction of migration, and indeed the smaller linear DNA in the sample studied by Chu et al. (1–2 kbp) migrate along the diagonal. The biased reptation model predicts that chains above a critical size N* have a mobility that increases with field strength, so that such chains will be biased in the direction of the strongest field, as observed in experiments (*see* Fig. 7). Supercoiled DNA, conversely, is not expected to reptate because of its conformation, and one expects a sieving mechanism of the Ogston type, as described at the beginning of Section 2.1. In this case, the mobility should remain linear in the field (i.e., bands should remain on the diagonal), but rather small since supercoiled DNA is bulky and its motion is strongly hindered. This is an elegant way to distinguish DNA with different size and different topology, which fortuitously migrate at the same velocity in conventional CHEF or constant field. Finally, it is worth pointing out that, in some conditions of field strengh, pulse duration, and agarose concentration, DNA with nonlinear topology (and in particular circular DNA, *55,56)* may undergo trapping processes, such as those described in Section 4. Then, the mobility decreases with field strength *(34)* so that, in the same type of field conditions as described by Chu *(46)*, circular DNA should be biased in the direction of the weaker field (*see* Fig. 7). Using the three different migration mechanisms of DNA in gels (Ogston sieving, reptation, and trapping), asymmetrical CHEF can be a very powerful tool for resolving topology.

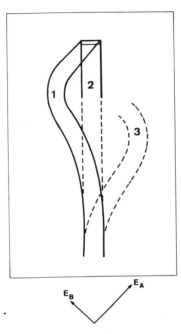

Fig. 7. Schematic picture of the resolution of the topology of chains by asymmetric-fields CHEF (following ref. *46*). Lane 1 represents DNA undergoing reptation; lane 2 represents DNA undergoing Ogston sieving; and lane 3 represents the type of migration we expect for DNA undergoing trapping *(see text)*.

2.3. Mobility in Field Inversion

The biased reptation approach to field inversion, given by Lalande et al. *(14,25)* and by us *(33)* closely follows the lines indicated in Section 2.1.1. If two opposed fields of amplitude E_a and E_b are alternatively applied during times τ_a and τ_b, respectively, chains with an orientation time much smaller than τ_a and τ_b migrate at a velocity reduced by a factor $[\tau_a-(E_b/E_a)^3 \tau_b]/(\tau_a+\tau_b)$ as compared to the velocity in a constant field E_a (this factor simply correponds to the time-average of the permanent-field mobilities corresponding to E_a and E_b). Conversely, chains with an orientation time much longer than the pulse times will not have time to adapt themselves to the reverse pulse. By a "jigsaw" mechanism very similar to the one happening for crossed fields at obtuse angles (*see* Section 2. 2. 2.), one can show that these long chains should retain the orientation corresponding to the forward pulse. Therefore, they have a mobility reduced by a different factor $[\tau_a - T_b E_b/E_a]/\tau_a = \tau_b)$ (*see* Appendix).

This approach accounts rather well for experimental separations of DNA in DFIGE *(25)* or ZIFE, when E_a and E_b are different enough (typically a factor of 2 or more) *(26)*. This model, however, predicts that the resolution vanishes for $E_a = E_b$, in contradiction with the well-established separation power of single-field FIGE *(61–65)*.

2.4. Intermittent Fields

The reptation model predicts that chains progressively lose their orientation by thermal activation when the field is switched off. This can be used to restore size-dependence since chains in random tubes have an initial mobility scaling as N^{-1}. Complete disorientation is expected to require a Brownian reptation time, which increases as N^3 (a fact now well verified experimentally, *89*), whereas the orientation time only increases as N^1. Therefore, the on/off ratio necessary to disorient large chains should indefinitely decrease with chain length, with catastrophic consequences on separation time for long chains. The reptation model indeed predicts that intermittent fields are even less efficient for separating large chains than conventional electrophoresis at very low fields *(33)*. It is worth pointing out, however, that orientational relaxation experiments also revealed a fast partial relaxation of the chains, which may help separation in intermittent fields. Experimentally, a reasonable efficiency is obtained for chains in the kbp range *(28,29,69–71)*, but, to our knowledge, no striking success has been reported on the separation of Mbp DNA using intermittent fields alone. Anticipating the following sections, it is also worth mentioning that intermittent fields may help to solve experimental problems not accounted for in the biased reptation theory, such as trapping (*see* Chapter 4, this vol.).

3. Beyond Biased Reptation

3.1. General Remarks

The biased reptation model rather well describes the mobility of long DNA in agarose and acrylamide, and under a wide variety of pulsed-field conditions. A notable exception is FIGE, for which biased reptation leads to qualitatively wrong predictions. Experiments investigating electrophoretic motions on a molecular scale and computer simulations were also developed in order to better understand the mechanisms of PFGE. These studies, presented in more detail in several sections of this volume, revealed new features of DNA dynamics in gels, and raised further criticism of the biased reptation model. In this section, we recall this criticism, the experiments it is based upon, and the theoretical developments it promoted. The section is organized from a "theorist's point of view," i.e., in terms of molecular mechanism rather than experimental technique or historical priority.

3.2. Field Inversion, Orientation Overshoot, and Internal Modes

3.2.1. Field Inversion

The germinal work of Carle, Frank, and Olson in FIGE reported a very strong band inversion effect, in which chromosomes VIII to II of *Saccharomyces Cerevisae* (600–800 kbp), for instance, migrated about ten times more slowly than chromosomes I ($\approx$250 kbp) and XII ($\approx$2000 kbp) *(61)*. Such a dramatic nonmonotonous behavior as a function of chain size is not always observed *(62,66)* in FIGE. In all cases, however, a minimum is observed when the mobility is plotted vs pulse times. This minimum, which is associated with an "antiresonance" of the field period with a molecular characteristic time of the chains, deepens and shifts to larger times (roughly linearly), when the DNA size is increased. It is not predicted by the biased reptation model.

3.2.2. Chain Orientation

The original biased reptation model (with constant tube length) proposes various predictions for the orientation of chains *(14)*. In permanent regime, the orientation is expected to scale as E^2 for $\varepsilon < 1$ and E^0 for $\varepsilon > 1$, and to be independent of N for sufficiently large chains. It also predicts that chains orient monotonously at the onset of the electric field, in a time scaling as $N E^{-2}$ for $\varepsilon < 1$ and $N E^{-1}$ for $\varepsilon > 1$ *(14)*. Conversely, they are expected to loose their orientation at the arrest of the field, in a reptation time scaling as N^3.

Fluorescence polarization *(99)*, linear dichroism *(100–103)*, and electric birefringence *(89,104–106)* confirmed the orientation of chains by the electric field. The time scales for reaching the steady state *(89,100–103)* and this for complete disorientation *(89)* are in qualitative agreement with scaling predictions of the reptation theory. These experiments, however, also revealed qualitative discrepancies with the biased reptation predictions. In constant field, the orientation seems to continuously increase as N^1 beyond the size at which a saturation is predicted by the theory. There is also a strong transient overshoot of the orientation at the onset of the field, and a fast relaxation process just after field interruption, which cannot be accounted for by the sole orientation of the tube in the field (other more subtle "undershoots" were reported, but we let them aside here to focus on the most spectacular observations). These effects were attributed to internal modes of the chains, i.e., to the failure of the assumption of constant tube length. The comparison of the main relaxation times of the orientation with mobility measurements on a molecular scale *(107,108)* and with field inversion results *(109)* support the idea that the "antiresonance" of the mobilty observed in FIGE and the orientation transient overshoot have a common molecular origin.

3.2.3. Generalization of Biased Reptation: Internal Modes

Taking internal modes into account immediately introduces terrible complications in the theory of biased reptation: Equations must take into account the positions of all chain sections and their interactions. Even without field, this problem could only receive an approximate solution, known as Doi's length fluctuation theory *(109)*. One of us proposed *(32)* a generalization of biased reptation based on this theory. Drastic approximations were necessary to achieve analytical solution, and the theory is only qualitative. The striking finding is that, when field-inversion is applied to a chain, the choices of the chain head can be considerably altered by thermally activated length fluctuations, if the time scale of these fluctuations is of the order of the pulse times. In this case, we have shown that the chains may have two locally stable conformations, the usual oriented state obtained in constant field, and a more collapsed state. The overall chain orientation, and the chain mobility as a consequence, are reduced by a factor that can reach 10. This negative resonance-type effect leads to a deep minimum in the mobility vs pulse time curve (Fig. 8), as experimentally observed. Our prediction for the evolution of this minimum with size, however, was smaller than the experimental one. In the light of further simulations and videomicroscopy experiements, we now believe that this discrepancy is attributable to the coupling of length fluctuations with the field, not considered in the model.

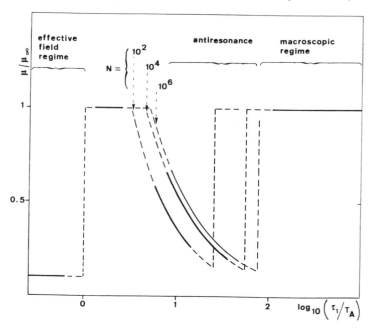

Fig. 8. Evolution of mobility vs pulse time in FIGE, as predicted by the biased reptation model with thermal length fluctuations *(32)*. The forward/ reverse ratio of times is 2, and the critical size *N** is fixed to 10. Full lines correspond to the quantitative predictions in fully established dynamic regimes, the dotted lines correspond to the crossover regions where the model provides no quantitative prediction.

A rather opposite and complementary route was taken by Lim et al. *(110)*, who modeled the DNA by a chain of non-Hookean springs (i.e., springs that cannot extend past a finite length corresponding to the fully extended length of the DNA). This model was solved analytically in the deterministic "preaveraged" approximation, i.e., without considering fluctuations. At the onset of the field, this chain of springs extends toward the field by its two ends, reaching a U-shaped very extended conformation. Then, it progressively slips around the bottom of the U like a rope on a pulley, and recover a more classical linear conformation. This analytical model describes well the experimentally observed orientation overshoot at the onset of the field, which is caused by this "pulley" effect. It does not, however, provide any insight into the mechanism of field-inversion, since it does not take into account fluctuations.

Lumpkin proposed a simpler model, consisting in only three beads with nonlinear drag connected by two springs *(35)*. This is a minimal type of chain with internal modes, and obviously very far from actual long DNA chains, but it can be solved with less approximations than the previous ones. The analogy of this model with the electrophoresis of real DNA is rather indirect, but it is interesting that, in field-inversion conditions, such a simple model also leads to a sharp dip in the mobility vs pulse times.

A very different theory was proposed by Zimm *(5)*. The underlying model, which represents the tube as a sequence of wide "lakes" and narrow "straits" joining them, puts the emphasis on the fluctuations in tube diameter, whereas all previous reptation models assumed the tube diameter to be constant, or at most affected by a rather narrow distribution of pore sizes *(92)*. Again, this model, which had to be solved numerically, presented a minimum in the mobility plotted as a function of pulse time. The evolution of the position of this minimum with chain length is in qualitative agreement with experiments. Its depth, however, seems too small.

3.2.4. Computer Simulations
of the Generalized Biased Reptation Model

Numerical simulations of the generalized biased reptation model *(27,110)* with internal modes have permitted progress beyond the limitations of analytical calculations. They did show that an extensible chain in a tube presents the following features:

1. Nonuniform extension, the head section being more compact than the tail.
2. An important overshoot of the extension and of the orientation at the onset of the field, owing to the U-shape effect already demonstrated by analytical calculations.
3. In agreement with experiment, distinct undershoots of the orientation when the field is reversed.
4. A minimum in the mobility as a function of pulse time in FIGE, which shifts to larger times when the size of the chain is increased.

The big disadvantage of these simulations as compared to analytical calculations, however, is that they have to be performed explicitly for a given chain size, and computation time did not allow the investigation of very long chains. In particular, it has not yet been possible to determine field-inversion mobilities for chains greater than the equivalent of about 50 kbp, far below the typical sizes at which strong band inversion is experimentally observed. For this reason, it is difficult to say if the weakness of the "mobility dip" observed in simulations is a consequence of the small size of the chains considered, or of a deeper inadequacy of the tube hypothesis.

Another way of simulating a chain in a tube with internal modes was recently proposed by Duke *(36,37)*: it is a discretized version of generalized reptation, called "repton model" *(111)*, in which the "blob" length can take values 0 or 1. This model is less time consuming than the previous one, so it can be applied to longer chains, but it remains limited to lower fields because of the discretization. Qualitatively, it leads to similar conclusions. Anticipating Section 3.3.1., it is interesting to notice that this repton model, which does not involve any tube leakage, presents the same type of "geometration" dynamics, which was associated by Deutsch with the presence of tube leakage in his own simulations.

3.3. Criticism of the Tube Assumption

In spite of its very attractive analytical simplicity, the tube hypothesis remains disputable in the presence of an external field (*see* discussion in Section 2.1.1.), and it was not universally accepted.

3.3.1. Simulations of a Chain Among Obstacles

As early as 1986, Olivera de la Cruz *(15)*, and then Deutsch and coworkers *(21–24)* and others *(18–20)* proposed a more direct approach, in which a charged flexible chain is simulated by Monte Carlo or Brownian Dynamics on a lattice of obstacles, without any tube assumption. These simulations permitted the investigation of mechanisms of motion more complicated than reptation for the first time. The most important is "tube leakage," i.e., formation of "hernias" (*see* Section 2.1.1.). These hernias seem to occur in several cases:

1. All along the chain, when the field is rotated by an angle around 90°, as in OFAGE *(24)*.
2. At the chain's head in constant field. In that case, the competition between hernias apparently reduces the head's mobility, and induces a "bunching" of the chain. After a stretching time, which may be rather long, one of the hernias wins against the others,

and stretch the chain in an aligned and reptative-like conformation again. Then, the motion of the chain is an alternation of strong collapsing and stretching events, called "geometration" by Deutsch *(22)*, and rather different from the more continuous motion predicted by the biased reptation model. As quoted in the previous section, however, similar motions resulted from Duke's *(36,37)* simulation of a tube model without leakage, so they probably require some internal modes (flexibility) of the DNA chain, but not necessarily hernias.

3. In field inversion, finally, some pulsing frequencies induce a resonance between bunching events that occur at the two extremities of the chain, which strongly reduces the mobility and lead again to a "dip" in the mobility vs pulse time plot *(22)*.

These simulations, more realistic than those of chains in a tube, are also more time-consuming, and they have not yet been able to provide accurate results on chains long enough to represent current pulsed electrophoresis situations (>100 kbp). Moreover, most of the simulations were performed in situations of high field (ε of order 1). It would be interesting to know how the tube leakage and bunching processes evolve when the field is progressively lowered.

Finally, progress toward still more realism in the simulations at the expense of computing time have started, by means of molecular dynamics in three-dimensional lattices of obstacles *(18,19)* and on random percolation clusters of obstacles *(20)*.

3.3.2. Discussion in the Light
of Fluorescence Videomicroscopy

Considerable progress in the qualitative understanding of DNA migration in agarose, both in conventional and PFGE conditions, was recently promoted by fluorescence videomicroscopy experiments *(75–77)*, by which individual molecules can be observed in real space and real time during their migration. These observations played in some sense the role of a "referee" in the field of theories and simulations. The first obvious observation is that, in average, the behavior of chains up to typically 200 kbp is in surprisingly good agreement with theoretical predictions and computer simulations:

1. The chains orient in the field, and orient more in higher fields, as predicted by the reptation model *(1,2,4–14)*.

2. At the onset of the field, they stretch in U-shaped conformations, as expected from simulations *(21–24)* and analytical treatment of generalized reptation *(110)*.
3. DNA seem to adopt two very different states of stretching, in agreement with theory *(32)* and simulations *(21–24)*.
4. They also present a denser "head" and alternances of stretching and collapsing phases between these two stretching states, as predicted by simulations of lattice chains *(21–24)* and of the generalized reptation model *(27,36,37)*.
5. The collapse of the head is often associated with the formation of hernias (i.e., loops), as in simulations in lattices of obstacles *(21–24)* and in contrast with tube models. These hernias lead to U-shaped conformations rather often in the course of migration.
6. A "resonance" in field inversion is also observed, in which the chain is "trapped" between two bunched extremities, in close resemblance with simulations *(22,37)*.
7. As far as CFGE is concerned, two rather different behaviors have been reported: At rather low fields (3 V/cm) *(76)*, DNA adopt for moderate angles "staircase" conformations equivalent to the "zigzag" model of reptation theory *(33)*. For obtuse angles, they also perform "hook inversion," a mechanism similar to the one called "jigsaw" *(33)* or "ratchet" *(26,38)* in earlier theoretical work. At high fields (14–20 V/cm) *(77)*, another mechanism prevails, dominated by the formation of hernias along the stretched chain as predicted by Deutsch *(21,23)*.

Overall, videomicroscopy provided much more confirmations than contradictions to qualitative theoretical predictions. About the key question of the tube hypothesis, the answer is not clear-cut. A very general point is that, as expected from entropic considerations, a linear tube seems a rather good description of the chain conformation in low fields (1–3 V/cm, depending on various factors as gel concentration, temperature, and buffer strength), and hernias become increasingly important when the field is increased. These hernias, however, do progress in the gel by a biased reptation mechanism too! So, in our sense, the tube as a representation of topological constraints remains correct at all field strengths (for chains much longer than the pore size). The hypothesis of a *linear tube* is wrong for high fields, however, it should be replaced by a model of hernias growing by reptation in a ramified tube. Unfortunately, this is much more difficult to model theoretically!

4. Trapping Electrophoresis

4.1. Principle

Pulsed electrophoresis was first applied to the separation of duplex DNA in agarose, and this remains by far the major application of the technique. The biased reptation model, and more generally any model for the dynamics of flexible charged chains, are also applicable to single-strand (SS) DNA in agarose or in acrylamide. In particular, a "compression band" is observed for SS DNA in acrylamide around 1 kb, which is expected to originate in the reptative orientation of chains, as was the case around 20 kb for duplex DNA in agarose. This compression of bands is as much of a nuisance here as there, and pulsed electrophoresis was recently applied to SS DNA with some success *(65,71,90)*. A particularly important application of SS DNA electrophoresis is sequencing. In this case, a further difficulty is encountered: Separating DNA molecules differing by one single basepair requires a relative accuracy that continuously increases with the size, because the "bare" relative difference in friction between two consecutive DNA roughly scales as N^{-1}.

A very interesting approach to that problem has recently been proposed by Ulanovsky et al . *(91)*, in which a bulky protein is attached at one end of the SS DNA fragment. This protein leads to a very strong size-dependence of the mobility in a narrow region of size, which has been attributed to "trapping" of the proteinated chain. Moreover, the authors showed that the selective region can be shifted by a periodic inversion of the field that "detraps" the chain. We share this qualitative picture, and describe in the next sections a theoretical model for this new method, based on the biased-reptation theory.

4.2. Proteinated Chain in Constant Field

4.2.1. Reptation

The radius of gyration of SS DNA of the order of 1 kb is larger than the average pore size of acrylamide gels usually used in sequencing, so that the reptation picture should be valid (this is confirmed by the linear dependence of the mobility on size; *91*). As in Part 2, we model the chain by a succession of N blobs, each one occupying one pore of the gel. Because the persistence length of SS DNA is in general much smaller than the pore size of the gel, the blob can take many contorted configurations inside the pore. In particular, it can be stretched by the electric field (*see* Section 2.1.). Evaluations of this effect for SS DNA *(34)*, however, suggest that stretching is weak in the conditions in which trapping electrophoresis is presently performed. Therefore, we use here the simplest version of the reptation theory, in which each DNA is represented by a chain of N blobs with charge q, friction ξ and fixed length a. At one end of the chain a big protein is fixed. This additional blob has no charge and a friction $\zeta' = \alpha\zeta$. It can be viewed as hard sphere or "balloon" of radius r. In the presence of a constant electric field this chain reptates and the major effect of the protein is to lower the mobility. The mobility can be calculated as in Section 2.1., taking into account the extra friction owing to the protein (*see* Appendix). If the friction of the protein is small as compared to the total friction of the chain (i.e., if $\alpha < N$), the velocity is not strongly perturbed. Comparison of quantitative results with experimental data on small chains (labeled and unlabeled) allows us to confirm the assumption of constant friction and to obtain the numerical value for α.

4.2.2. Model for Trapping and Detrapping

In actual gels, pores have a distribution of sizes. For an unproteinated chain it has been shown *(92)* that this distribution is averaged out as long as the largest pores remain much smaller than the coil size of the whole DNA, and it weakly affects migration. The situation is very different for the labeled chains. In particular, the protein cannot pass through pores with a size smaller than its radius. It is worth pointing out here that, in contrast with unlabeled chains in which both ends are symmetrical, the uncharged and bulky protein is expected to "trail" in the migration and play the role of the tail. Then, a chain may thread a pore with radius smaller than r, and be stopped only when its proteinated tail reaches this pore (hereafter called a "trap"). We assume that the friction coefficient of the protein is then infinite in the direction of trapping, and remains α in the reverse direction. Doing so, we also implicitly neglect the deformation of the gel, which would allow for a finite increase of α.

Once the chain is trapped, competitive effects come into play: The electric field tends to keep the chain trapped by pulling it downfield, whereas thermal fluctuations can detrap it. The general problem of detrapping of a flexible chain in a random gel, taking full account of fluctuations, is very difficult. A qualitative evaluation of the mean trapping time t_p, however, can be attempted by supposing that chains detrap mainly by following the "path" corresponding to the lowest energy barrier. The path we assumed in the present calculation is represented in Fig. 9 A–C: If the tail of the chain bearing the protein is from time to time "lifted" from the trap by thermal forces, it will tend to orient in the field, and this orientation facilitates the continuation of the detrapping process. This path, in which the trap plays the role of a "pulley," involves a much smaller activation barrier than, for instance, a simple "backwards reptation" as pictured in Fig. 9 D–F. Once the detrapping path is assumed, one can use the Kramers theory of activated processes to derive the average residence time in one trap, t_d. In the regime we expect to be relevant to present trapping experiments, t_d is predicted to scale as $\exp(\varepsilon N^{3/2})$, i.e., it increases exponentially with chain size and electric field.

4.2.3. Mobility in Constant Field

Labeled chains of a given size inside the gel may be divided in two fractions: a fraction ρ_o of untrapped chains, which migrate at a velocity given by Eq. 2 (modified to include the protein friction, as indicated in the appendix), and a fraction $(1-\rho_o)$ of trapped chains that do not migrate. On average, these chains remain trapped during a time t_d. Once detrapped, they migrate until they encounter a new trap, and remain trapped again. This problem can be solved completely for any initial condition by rate-equation theory, but it is also easy to understand without any calculation that, in permanent regime, the average velocity is proportional to the ratio of the time spent out of the traps, t_r to the time spent in traps t_d.

Two regimes can be predicted: For small enough chains, $t_d \ll t_r$ and the the mobility is unchanged by the presence of the traps. Chains longer than a critical size $N^{\mathrm{trap}} \approx \varepsilon^{-2/3}$ have a trapping time that increases exponentially with N and becomes much larger than t_r: As shown in the appendix, the mobility of these chains is multiplied by a factor of order $\approx \exp[-(\varepsilon \rho N^2)/2]$ regarding chains smaller than N^{trap}. This exponential dependence of the mobility in the trapped regime is responsible for the strong selectivity of the trapping electrophoresis method (*see* Fig. 10). Note that the critical size N^{trap} scales as a negative power of ε, so that the mobility drops faster at higher fields: The model predicts that the velocity obtained at two different fields intersect around N^{trap}, a very uncommon feature in electrophoresis also observed experimentally. Presently available data *(91)* are compatible with the value $-2/3$ we predict, but they are too scarce to be considered as a "confirmation" of the model.

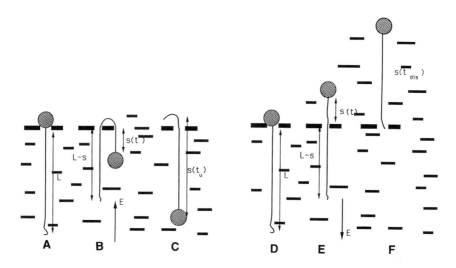

Fig. 9. Schematic representation of the trapping of protein-labeled DNA in a gel. **A:** Chain in the trap, with field pulling down; **B:** chain in the course of crossing the potential barrier (thermal detrapping), with field pulling down; **C:** chain detrapped; **D:** same as **A**; **E:** chain in the course of detrapping activated by a reverse field-pulse (field pulling up); **F:** chain detrapped by the reverse pulse.

In spite of the strong selectivity observed, the dramatic decrease of the mobility owing to trapping, and the weak E-dependence of the position of the crossover restrict the interest of constant-field trapping electrophoresis for sequencing DNA. They impose an important decrease of the field (i.e., a dramatic increase of the experimental time) to push the selective region into the kb range. This is clearly apparent in Fig. 11, where the mobility is displayed as a function of size and field. Trapping increases the slope in N, but it leads to no practical mobility at all in the region where good separation is most difficult to achieve in conventional methods (large chain, high fields). Fortunately as exposed by Ulanovsky et al. *(91)* and discussed in the next section, pulsed-field electrophoresis provides a more versatile tool.

4.3. Pulsed-Field Trapping Electrophoresis

4.3.1. Model for Field Inversion

The field-inversion method (FIGE) consists of applying the field in the usual direction during a time t_a (for "ahead"), then reverting it during $t_b < t_a$ (for "backwards"). For an untrapped chain the reverse

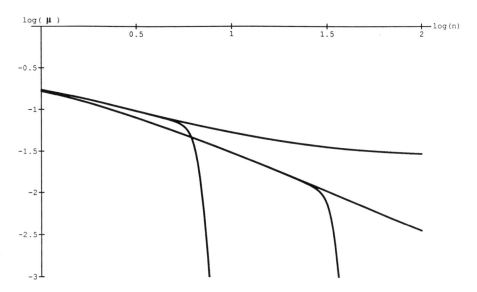

Fig. 10. Log(mobility) vs Log(*N*) for a proteinated DNA during constant field electrophoresis, with (lower curves) and without (upper curves) traps. The reduced field ε is .5 and .05, respectively. Other parameters are $p = 1000$ and $\alpha = 1$.

pulse doesn't change the dynamics because in average the first part of a forward pulse, of duration t_b, will be used by the chain to recover its position at the end of the previous forward pulse. The mobility is the same as would be obtained by applying the field forward during a time $(t_a - t_b)$ for each period of duration $(t_a + t_b)$. Things are more subtle for a trapped chain: In the mechanism of trapping by "threading" described in Fig. 9, the trap only acts one-way, i.e., a chain trapped during a forward pulse is free to reptate back when the field is reverted (*see* e.g., Fig. 9 D–F). The key point then is whether t_b is large enough for the chain to leave totally the trap or not. In the latter case which correponds to long chains or short reverse pulses (Fig. 9E), when the field is reverted again at the end of a backward pulse, the chain will immediately return into its old trap and it will have chances to leave this trap only by the (very rare) thermal process. Smaller chains (Fig. 9F) are detrapped at the end of the backward pulse and free to move until they encounter another trap. Using a Fokker-Plank equation that takes into account velocity fluctuations, one can calculate the fraction of electrically detrapped chains at the end of a backward pulse,

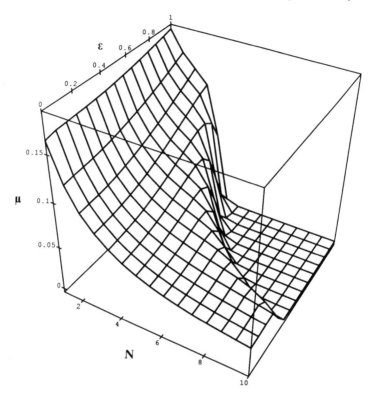

Fig. 11. Mobility during constant-field electrophoresis vs chain length N (x axis) and reduced field (y axis). Other parameters are $p = 1000$ and $\alpha = 1$.

as a function of N and t_b. The crossover N^{retrap} roughly scales linearly with t_b. The fraction of untrapped chains $\rho_0(t)$ varies continuously during each pulse. The self-consistent calculation for the mobility as a function of the chain size is rather lengthy but straightforward, and it can be done for any values of t_a and t_b (93).

4.3.2. Discussion of Results

Figure 12 displays a few examples of the evolution of the mobility when varing t_r at fixed t_a, for two values of t_a. In the general case, one now obtains two distinct regions of exponentially decreasing mobility. The first one, around N^{trap}, corresponds to the onset of trapping in constant-field, i.e., to the size at which thermal motions become ineffective for detrapping. Its position depends on the field strength, but it is independent of pulse times, except for very short and unpractical times. The mobility, however, levels off to an intermediate plateau,

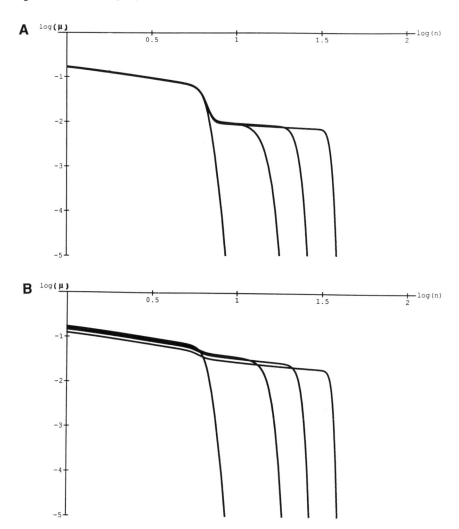

Fig. 12. **A.** Log(mobility) vs Log(N) for a proteinated chain during FIGE dynamic. Fixed parameters are ε = .5, p = 1000, α = 1, t_a = 50,000. The reverse pulse time t_b is respectively 0, 200, 400, 800, from left to right curves. (t_a and t_b are expressed in units of $\xi\, a^2/(2\, kT)$). **B.** Same as **A**, with a shorter forward pulse time, t_a = 5000.

instead of continuously decreasing to zero as in the case of constant-field. The height of this plateau is essentially controlled by the duration of the forward pulse, t_a, as compared with the reptation time between traps t_r. The qualitative reason is as follows: After the field is switched to the forward direction for the first time, chains in this range of size get trapped after a typical time t_r, and most of them remain trapped during the remainder of the pulse. They are detrapped by the backward pulse, however, so that they can start the cycle again by moving during a time t_r at the beginning of the next forward pulse: The "efficiency" of their motion is of order $(t_r - 2t_b)/(t_a + t_b)$, which roughly corresponds to the observed plateau. At fixed t_b, a larger t_a leads to a lower plateau, and to a smaller mobility. This plateau ends up around N^{retrap}, where the mobility drops down exponentially to zero. Chains larger than this cannot detrap even with the help of the reverse pulse. The position of this second exponential domain is controlled mainly by t_b. More quantitatively, we predict that, for chains of order 1 kb and fields of order 50 V/cm, N^{detrap} should scale as $(t_b E)^{2/3}$. A second regime with $N^{\text{detrap}} \sim t_b E^2$ is expected for longer chains. In all cases, the longer the reverse pulse, the longer the chains it can detrap.

For separation purposes, the second exponential domain is more interesting than the first one, because its position in size can be tuned to larger chains without decreasing the field strength, i.e., without increasing the experimental time and the diffusion. Therefore, the model suggests that the best way to use pulsed fields in trapping electrophoresis is to keep the plateau as high as possible, and to generate a "custom-made" variation of the mobility with size by a suitable ramp of reverse pulse times. Figure 12B shows that there is a range of t_a in which this plateau is indeed so high that the first exponential region practically disappears. The value of this time is not as crucial as that of t_b for the quality of separation, but an optimized balance between the relative time spent in traps (which increases with increasing t_a), and the relative time spent in reverse motion (which increases when t_a decreases at fixed t_b) will minimize experiment duration. In the absence of more detailed evaluations, we suggest to keep the ratio t_a/t_b at a moderately large value (say, around 5 to 10), during the scan of t_b.

4.3.3. Discussion of Assumptions

It is important at that stage, to distinguish between the general model of trapping electrophoresis, and more specific assumptions that we made for the case of SS DNA labeled with streptavidin in acrylamide.

In the presentation given above, we neglected the stretching of the chain and the internal modes, as in the original biased reptation model. Our evaluations suggest that the experiments in *(91)* (chains around 1 kbp and fields of a few tens of Volts/cm) correspond to the range $N < N^*$ (*see* relation 3). A more detailed solution of the model than the one outlined here, taking stretching modes into account *(93)*, indicates that chain stretching is indeed weak in this case. Fluctuations, however, may play a significant role in the "detrapping path," and possibly change the quantitative predictions. For larger chains, stretching may become very important, and a more complete treatment is necessary *(93)*.

We also assumed trapping by "threading," which seems reasonable for a very flexible gel as acrylamide. Other trapping mechanisms and detrapping paths, however, may be imagined. Different expressions for the residence time in a trap (Section 4.2.2.) would result, but the remainder of the theory would remain valid. The same remark also applies to gel flexibility, which would help detrapping and smoothen somewhat the mobility curves.

Direct experimental observation of the trapping mechanism presently seems difficult to imagine, but detailed comparisons of experiments with the predictions of the theory associated to different trapping models may help to tell them apart. Presently available data are compatible with our major predictions, such as the exponentially decaying mobility in constant field, and the effect of pulse times, but they are too scarce for quantitative comparisons. Further systematic investigations will be necessary, in particular to get reliable experimental exponents and to orient further choices of models.

Finally, note that we neglected the fact that, in some of the detrapping processes we assume, the labeled chain has to perform a "head-tail" inversion. This process, however, should not modify the overall mobility as dramatically as the exponential trapping times.

4.4. Trapping as a Very General Mechanism in Electrophoresis

Besides the approach of deliberately labeling DNA to induce trapping, it has become apparent in the last few years that trapping is a very general mechanism in electrophoresis. For instance, Griess et al. *(63)* observed that charged latex spheres electrophoresed in very weak agarose gel trapped within a few minutes, and that FIGE with $t_a/t_b = 3$ is able to restore their migration. Serwer et al. *(94)* also observed trapping of circular DNA in agarose, which occurs only in high fields. In this case, trapping is attributed to a "threading" of the DNA loop by fibers protruding from the gel, but the general model of migration described in Section 4.2.3. and following should remain relevant. Detrapping by FIGE and RGE was also observed *(55,56)*, and this technique is now used to distinguish circular and linear DNA *(55,56,95)*.

Most strikingly, it now becomes obvious that trapping also occurs for linear DNA in agarose gels. This phenomenon which, to our knowledge, was suggested by Carle and Olson *(96)* for the first time in 1987, is probably the major hindrance to further progress in the field *(44,67,96–98)*. The trapping of linear chains, which spreads the bands and ultimately prevents DNA from leaving the loading wells, occurs rather sharply above a critical size that decreases with increasing field. Experimental linear or exponential dependences of N^{trap} with the inverse of the field have been reported, and the precise evolution as well as the molecular mechanism of this trapping remain obscure. We recently observed that, in contrast with the trapping of streptavidin-labeled DNA, latex spheres or circular DNA, that of linear DNA is irreversible *(97)*: When Mbp DNA are electrophoresed in low field, then trapped in a higher field, they are unable to start their migration again when the initial electrophoresis conditions are restored. Interestingly, Turmel et al. (*67, see also* Chapter 7, this vol.) showed that the trapping of linear chains can be significantly reduced if the chains are "shaken" by a high-frequency modulation of the field. This suggests that the origin of trapping is very local *(67)*.

5. Conclusions

At present, four mechanisms for the electrophoretic migration of macromolecules in gels have been identified: sieving, reptation, growth of "hernias," and trapping. Sieving is relevant to particles with an overall size of the order of the pore size of the gel, and it leads to a velocity that is linear as a function of field strength. Therefore, it presents no intereset for pulsed-field electrophoresis, except for the very particular case when it can help to separate DNA with different topologies that fortuitously migrate at the same velocity.

The large majority of applications of pulsed electrophoresis deal with linear DNA with an extended length much larger than the pore size of the gel. In that case, two mechanisms identified by theory and computer simulation, and later confirmed by fluorescence videomicroscopy, are responsible of the migration: "head-on" reptation and growth of hernias. These two mechanisms appeal to the idea of a "tube," i.e., a sequence of pores of the gel, which is defined by a "head" and followed by the remainder of the chain. In the case of conventional reptation, the "head" can only be an extremity of the chain, so that the tube is linear. When hernias develop, the tip of the hernia plays the role of the head. This can occur anywhere along a linear DNA (or along a DNA with a different topology), and most often several hernias develop at the same time, so that the "tube" becomes ramified. According to the statistical theory of flexible chains, entropy, i.e., thermal motions, favor head-on reptation in a linear configuration, whereas high fields favor the growth of hernias. The latter should become more and more important as the field is increased. This is indeed what is observed. Theory also predicts that hernias should form more easily in gels with larger pores, i.e., at lower concentrations. A second theoretical prediction, common to head-on reptation and hernias, is that the tube presents in the direction of the field an orientation that increases with field strength. This orientation increases the velocity of chains, and renders the mobility field-dependent. The combination of this field-dependent mobility owing to orientation, and of the time taken by a chain to reach this orientation that increases with chain length, qualitatively explains why pulsed fields can separate chains by size.

Regarding the theory, reptation and growth of hernias have a very different status: Hernias are extremely difficult to fit into a simple theoretical frame, and their investigation was essentially by computer simulations or "numerical experiments" difficult to generalize. Reptation is much less resistant to analytical calculations, though it is important to distinguish further between two different levels of approximation: conventional biased reptation, in which the chain is considered as an inextensible "rope" with fixed length, and generalized biased reptation, in which the longitudinal elasticity of the chain is taken into account. The first model has been treated rather extensively on a theoretical ground, and many results were achieved completely analytically. In particular, it seems to provide a good description of homogeneous crossed-field electrophoresis (CHEF, RGE, PACE, and so on). Using a few parameters measured in standard conditions, the evolution of the mobility as a function of field strength, pulse times, field angle, and mol wt can be predicted with an accuracy sufficient for many applications (for instance, for extrapolating the variation of band patterns when one changes the field strength or the pulse time, or for determining the optimal conditions for separating DNA in a given range of mol wt). Satisfying predictions are obtained for field inversion too, provided the strengths of the forward and reverse fields are rather different. Methods involving inhomogeneous fields are intrinsically more resistant to quantification, but the qualitative predictions of the biased reptation model are also of some help. Finally, the biased reptation motion provides a handy image of DNA motion, and it can be used to imagine new protocols and to simulate their performance in various conditions before going to actual experiments. This approach, which proved to be successful for the case of field inversion with unequal field strengths in the two directions, for instance, should gain interest in the following years as one goes along with more specific biological problems, more complex field sequences (i.e., still more parameters to vary!) and more accurate apparatus.

In spite of many experimental and theoretical efforts, field-inversion electrophoresis with equal field strengths in the forward and in the reverse direction remains poorly described by theories. It is now well accepted that the peculiar band-inversion observed there arises from the fact that the chain can adopt two very different conformations, either highly stretched or highly collapsed. Different molecular mechanisms were shown to lead to such effects, however, and the one effectively at work in real DNA migration is not really elucidated (notwithstanding the possibility of a synergy between several of them, such as head disorientation and formation of hernias). An important point is that the biased reptation model is qualitatively wrong there, and only its generalization with stretching and collapsing internal modes can approach (although not very accurately) the experimental behavior.

The case of field inversion demonstrates that internal modes can qualitatively alter the biased reptation behavior, in contrast with the established field of reptation without an external field, in which they only intervene as a corrective effect. At the level of understanding we have now reached, in particular thanks to recent videomicroscopy experiments, it is somewhat surprising that the simple biased reptation model, which ignores length fluctuations and hernias obviously occurring in many experimental situations, and qualitatively fails to account for FIGE with equal field strengths, also achieved such a consistent and reasonable description of a large set of data in constant-field, crossed-field, and more recently, trapping electrophoresis. A first argument is that, although hernias do occur, they only do so in competition with (or on top of) head-on reptation, so that the latter process still dominates the macroscopic motion in many cases. In particular, the computer simulations and the videomicroscopy experiments, which put a strong emphasis on hernias most often correspond to rather high values of the field (10 V/cm or more), whereas PFGE is performed in lower fields, especially for the longer chains that interest the technique most. A more general (and more tentative) explanation would be that the presence of hernias, which obviously contradicts the hypothesis of a linear tube, may not ruin the biased reptation as an analytical model. For instance, videomicroscopy suggests that, for DNA above 1 Mbp, hernias develop in several places along the chain. One can thus understand that their effect is averaged out after migration on a long distance, so that it could be cast in a simple renormalization of molecular parameters (for instance an effective tube larger than the real one), without affecting the basic equations. These equations would become qualitatively wrong only in experimental conditions putting a particular weight on hernias.

The last mechanism considered in theories is trapping. The pulsed-field electrophoresis of proteinated single-stranded DNA (following the method recently proposed by Ulanovsky et al.) seems well described by a model combining biased reptation and rate-equation theory. More experiments will be necessary to provide a critical check of the model, however. We expect that this approach will be equally applicable to other particles apparently undergoing trapping, such as circular DNA in agarose or bulky particles at the limit of the sieving process.

The trapping of linear DNA, conversely, is presently poorly understood, and only conjectures on its mechanism can be advanced. This should be an important challenge to theorists in the next years, because trapping is probably responsible for the present limitation of PFGE. Another important challenge is the treatment of hernias, which play an important role in the mechanisms of migration, and more particularly in the reorientation processes occurring in pulsed electrophoresis. There is a reasonable agreement between simulations and videomicroscopy as far as the observation goes, but this presently is of little help in experimental design. Models able to predict without performing expensive computations or experiments, how hernias would be affected by various field conditions, and most importantly how this would in turn affect the mobility, would be highly desirable.

Acknowledgments

We are indebted to F. Caron, T. A. J. Duke, G. Slater, J. Prost, and L. Ulanovsky for fruitful discussion, and for sending to us various manuscripts prior to publication.

Appendix: Outline of Calculations and Important Formulae

Section 2.1.1.

The reptation motion is a curvilinear one-dimensional diffusion. The square average diffusion distance is $<s^2> = D_{rep} \, t$, where D_{rep}, the reptation diffusion constant, is $kT/N\zeta$ (we ignore numerical constants of the order of unity). Disengagement from the tube requires migration on a distance of the order Na (tube length), so the reptation time is:

$$t_{rept} = \zeta a^2 N^3 / kT = N^3 t_A \tag{A1}$$

where t_A is the equilibration time of a blob, and ζ is the blob friction coefficient.

Section 2.1.2.

In the presence of an electric field, the longitudinal components of the electric forces on the chain induce a curvilinear electrophoretic drift obeying the equation:

$$\dot{s} = (q E/\zeta) \, (R_z/Na) \tag{A2}$$

The first factor is the usual balance of electric and frictional forces, and the second arises from the fact that only components of the forces parallel to the tube axis can induce reptation. The macroscopic velocity is then obtained by projecting the tube trajectory (followed at a rate $\dot{s}$) onto the field direction z:

$$v_z = \dot{s} \, (R_z/Na) = (q E/\zeta) \, <\rho^2> \tag{A3}$$

Relation (2) follows.

Section 2.1.3.

The orientational probability distribution function (pdf) of the charged head in the presence of an electric field is given by the Boltzmann weight:

$$Q(\theta,\varepsilon) = \cos\theta \, \exp(\varepsilon \cos \theta) \, / \, 2 \, \text{sh} \, (\cos\theta) \tag{A4}$$

This leads, for a sequence of N segments with pdf Q, to:

$$<\rho^2> = <\cos^2\theta>/N + <\cos\theta>^2 \, (N-1)/N \qquad (A5)$$

with

$$<\cos\theta> = 1/(\text{th } \epsilon) - 1/\epsilon \qquad \sim\epsilon/3 \text{ for } \epsilon <<1$$
$$<\cos^2\theta> = 1 - 2/(\epsilon \text{ th } \epsilon) + 2/\epsilon^2 \qquad \sim 1/3 \text{ for } \epsilon <<1$$

Relations (3) and (4) by supposing that the tube is followed at uniform velocity; for a more accurate treatment, *see*, e.g., refs. *6,13,16,17*.

Section 2.2.1.

The steady-state velocity for a chain completely aligned along, say E_A, is given by relation A3:

$$V_\infty \cong E_A \, \mu_1 \, <\rho^2(E_A)> \qquad (A6)$$

The same relation is valid for pulse B. The velocity along the diagonal is obtained by projecting the "zigzag" path onto z, i.e.:

$$V_z \, (t_{or} << \tau) \approx V_\infty \cos(\phi/2) \qquad (A7)$$

where is the angle between the two fields.

Longer chains with $t_{or} >> \tau$, which do not have the time to completely reorient in a single pulse, "feel" an effective field along z given by the vector sum:

$$E_{\text{eff}} = (E_A + E_B)/2 \qquad (A8)$$

The amplitude of this effective field is $E_A \cos(\phi/2)$, so that the effective orientation of long chains along z is

$$<\rho^2(E_{\text{eff}})> = <\rho^2(E_A)> \cos^2 (\phi/2) \qquad (A9)$$

The net velocity is obtained by replacing E_A by E_{eff} in relation A6, i.e.,

$$V_z \, (t_{or} >> \tau) \approx V_\infty \cos^3(\phi/2) \qquad (A10)$$

The complete expressions for the intermediate case can be found in ref. *33*.

Section 2.3.

If two opposed fields of amplitude E_1 and E_2 are alternatively applied during times τ_1 and τ_2, respectively, chains with an orientation time much smaller than τ_1 and τ_2 spend most of the time migrating with an orientation that corresponds to the applied field: The macroscopic migration velocity is given by the weighted average of the permanent-regime velocities corresponding to fields E_1 and E_2, respectively. Following relations A3 and A5,

$$V_z(t_{or} \ll \tau_{1,2}) \approx V_\infty (E_1) \; [\tau_1 - (E_2/E_1)^3 \tau_2] / (\tau_1 + \tau_2) \qquad (A11)$$

Chains with an orientation time much longer than the pulse times, on the other hand, will reptate a curvilinear distance s_1 (much smaller than the tube length Na) during τ_1 and s_2 (smaller than s_1) during τ_1. Once oriented, they destroy at the beginning of each forward pulse the tube section created in the last backward pulse, so that the macroscopic orientation is essentially that corresponding to E_1. These chains are fully oriented and migrate at a velocity:

$$V_z \, (t_{or} \gg \tau_{1,2}) \approx \lambda \; V_\infty \, (E_1) \qquad (A12)$$

where λ is an "effective field factor" defined by:

$$[\tau_1 E_1 - \tau_2 E_2] / (\tau_1 + \tau_2) = \lambda E_1$$

Finally, for very short pulse times, chains should only feel the effective field λE_1 on a molecular level, leading to a mobility:

$$V_z \, (\tau_{1,2^-} < t_A) \approx \lambda^3 \; V_\infty \, (E_1) \qquad (A13)$$

Section 4.2.1.

The mobility of a proteinated chain can be calculated using A2 and A3, taking into account the extra friction owing to the protein:

$$\dot{s} = [NqE/\zeta(N + \alpha)] \; R_z/Na$$

or

$$<V_z> = E \mu_u = \mu_1 E \frac{N}{(N + \alpha)} \rho^2 \qquad (A14)$$

where we define the mobility in a gel without traps as μ_u and the "projection factor" ρ was defined in relation (A5).

Section 4.2.2.

Once the detrapping path corresponding to Fig. 8 A–C is assumed, the Kramers theory of activated processes gives the average residence time in one trap:

$$t_d = t_A[(N+\alpha)/(N\epsilon)]\,[\pi\rho\,/(2\epsilon)]^{1/2}\exp\,[(\epsilon\,\rho\,N^2)/2] \qquad (A15)$$

The important point to notice about relation (A15) is that the time for escaping the trap is dominated by the last factor: It increases exponentially with chain size and electric field.

Section 4.2.3.

In permanent regime, ρ_o is given by the balance between the detrapping rate $w_d = 1/t_d$ (relation A15), and the trapping rate w_t. According to our model, the proteinated chain trapped when it threads a pore of size smaller than r. The occurrence of trapping events depends on the fraction of such pores, $1/p$, and on their distribution in space. The present knowledge of gels is not sufficient to accurately predict $1/p$, and we keep it as an unknown parameter in the model. Assuming that small pores are distributed at random in the gel, one can calculate the mean time spent in reptation between two traps, t_r:

$$t_r = t_A\,p\,(N+\alpha)\,/\,(N\epsilon\,\rho\,) \qquad (A16)$$

w_t, the trapping probability by time unit is the inverse of t_r. $\rho_o(t)$ the fraction of untrapped chains at time t, starts at 1 and reaches a steady-state value $w_d/(w_d + w_t)$. Therefore, the mobility in the presence of traps is:

$$\mu(t) = \mu_u\rho_o(t) = \mu_u\,w_d/(w_d + w_t) \qquad (A17)$$

and the crossover N^{trap} occurs when $w_d/w_t \approx \exp[-(\epsilon\,\rho\,N^2)/2]\,(\epsilon/\rho^3)^{1/2}$ is of order 1. The calculations leading to Fig. 11 are given in ref. *(34)*.

References

1. Lerman, L. S. and Frisch, H. L. (1982) Why does the electrophoretic mobility of DNA in gels vary with the length of the molecule? *Biopolymers* **21**, 995–997.
2. Lumpkin, O. J. and Zimm, B. H. (1982) Mobility of DNA in gel electrophoresis. *Biopolymers* **21**, 2315,2316.
3. Schwartz, D. C. and Cantor, C. R. (1984) Separation of yeast chromosome-sized DNAs by pulsed field gradient gel electrophoresis. *Cell* **37**, 67–75.
4. Lumpkin, O. J., Déjardin, P., and Zimm, B. H. (1985) Theory of gel electrophoresis of DNA. *Biopolymers* **24**, 1573–1593.
5a. Zimm, B. H. (1988) Size fluctuations can explain anomalous mobility in field inversion electrophoresis of DNA. *Phys. Rev. Lett.* **61**, 2965.

5b. Zimm, B. H. (1991) "Lakes-straits" model of field-inversion electrophoresis of DNA. *J. Chem. Phys.* **94**, 2187–2206.

6. Déjardin, P. (1989) Expression of the electrophoretic mobility of polyelectrolytes through gels. *Phys. Rev.* A **40**, 4752–4755.

7. Levene, S. D. and Zimm, B. H. (1989) Understanding the anomalous electrophoresis of bent DNA molecules: A reptation model. *Science* **245**, 396.

8. Lumpkin, O., Levene, S. D., and Zimm, B. H. (1989) Exactly solvable reptation model. *Phys. Rev.* A**39**, 6557–6566.

9. Slater, G. W. and Noolandi, J. (1985) New biased-reptation model for charged polymers. *Phys. Rev. Lett.* **55**, 1579–1582.

10. Slater, G. W. and Noolandi, J. (1986) On the reptation theory of gel electrophoresis *Biopolymers* **25**, 431–454.

11. Slater, G. W., Rousseau, J., and Noolandi, J. (1987) On the stretching of DNA in the reptation theories of gel electrophoresis. *Biopolymers* **26**, 863–872.

12. Slater, G. W., Rousseau, J., Noolandi, J., Turmel, C., and Lalande, M. (1988) Quantitative analysis of the three regimes of DNA electrophoresis in agarose gels. *Biopolymers* **27**, 509–524.

13. Noolandi, J., Rousseau, J., Slater, G. W., Turmel, C., and Lalande, M. (1987) Self-trapping and anomalous dispersion of DNA in electrophoresis. *Phys. Rev. Lett.* **58**, 2428–2431.

14. Slater, G. W. and Noolandi, J. (1989) The biased reptation model of gel electrophoresis. *Proceedings of the International Symposium on New Trends in Physics and Physical Chemistry of Polymers*, L. H. Lee, ed., Plenum, New York, pp. 567–600.

15. Olvera de la Cruz, M., Deutsch, J. M., and Edwards, S. F. (1986) Electrophoresis in strong fields. *Phys. Rev.* A **33**, 2047–2055.

16. Viovy, J. L. (1988) Anomalous dispersion in gel electrophoresis: A solvable biased reptation model. *Europhy. Lett.* **7**, 657–661.

17. Doi, M., Kobayashi, T., Makino, Y., Ogawa, M., Slater, G. W., and Noolandi, J. (1988) Band inversion in gel electrophoresis of DNA. *Phys. Rev. Lett.* **61**, 1893–1896.

18. Kremer, K. (1988) DNA electrophoresis: A Monte Carlo simulation. *Polymer Comm.* **29**, 292–294.

19. Batoulis, J., Pistoor, N., Kremer, K., and Frisch, H. L. (1989) Monte Carlo simulation of DNA electrophoresis. *Electrophoresis* **10**, 442–446.

20. Melenkevitz, J. and Muthukumar, M. (1989) Electrophoresis of a polyelectrolyte in a random medium: Monte Carlo simulations. *Science at the John Von Neumann National Supercomputer Center, Annual Report* (University of Massachusetts, Amherst, MA) pp. 145–152.

21. Deutsch. J. M. (1987) Dynamics of pulsed-field electrophoresis. *Phys. Rev. Lett.* **59**, 1255–1258.

22. Deutsch, J. M. (1989) Explanation of anomalous mobility and birefriengence measurements found in pulsed field electrophoresis. *J. Chem. Phys.* **90**, 7436–7441.

23. Deutsch, J. M. (1988) The motion of DNA during gel electrophoresis. *Science* **240**, 222.

24a. Deutsch, J. M. and Madden, T. L. (1989) Theoretical studies of DNA during gel electrophoresis. *J. Chem. Phys.* **90**, 2476.

24b. Madden, T. L. and Deutsch, J. M. (1991) Theoretical studies of DNA during orthogonal field alternation gel electrophoresis. *J. Chem. Phys.* **94**, 1584–1591.

25. Lalande, M., Noolandi, J., Turmel, C., Rousseau, J., and Slater, G. W. (1987) Pulsed-field electrophoresis: Application of a computer model to the separation of large DNA molecules. *Proc. Natl. Acad. Sci. USA* **84**, 8011–8015.

26. Slater, G. W., and Noolandi, J. (1989) Effect of non-parallel alternating fields on the mobility of DNA in biased reptation model of gel electrophoresis. *Electrophoresis* **10**, 413–428.

27. Noolandi, J., Slater, G. W., Lim, H. A., and Viovy, J. L. (1989) Generalized tube model of biased reptation for gel electrophoresis of DNA. *Science* **243**, 1456–1458.

28. Lalande, M., Noolandi, J., Turmel, C., Rousseau, R., Rousseau, J., and Slater, G. W. (1988) Scrambling of bands in gel electrophoresis of DNA. *Nucleic Acids Res.* **16**, 5427–5437.

29. Jamil, T., Frisch, H. L., and Lerman, L. S. (1989) Relaxation effects in gel electrophoresis of DNA in intermittent fields. *Biopolymers* **28**, 1413–1427.

30. Viovy, J. L. (1987) Pulsed electrophoresis: Some implications of reptation theories. *Biopolymers* **26**, 1929–1940.

31. Viovy, J. L. (1987) Vers une théorie de l'électrophorèse résonante. *C.R. Acad. Sci. Paris* **305**, 181–184.

32. Viovy, J. L. (1988) Molecular mechanism of field inversion electrophoresis. *Phys. Rev. Lett.* **60**, 855–858.

33. Viovy, J. L. (1989) Reptation-breathing theory of pulsed electrophoresis: Dynamic regimes, antiresonance and symmetry breakdown effects. *Electrophoresis* **10**, 429–441.

34. Défontaines, A. D. and Viovy, J. L. (1991) Theory of trapping electrophoresis. *Proceedings of the First International Conference on Electrophoresis, Supercomputing and the Human Genome* (Cantor, C. R. and Lim, H. A., eds.), World Scientific, Singapore, pp. 286–313.

35. Lumpkin, O. (1989) One-dimensional translational motion of a two-spring chin with tron nonlinear drag: A possible model for time-dependent DNA gel electrophoresis. *Phys. Rev. A* **40**, 2634–2642.

36. Duke, T. A. J. (1989) Tube model of field inversion electrophoresis. *Phys. Rev. Lett.* **62**, 2877–2880.

37a. Duke, T. A. J. (1990) Monte Carlo reptation model of gel electrophoresis: Steady state behavior. *J. Chem. Phys.* **93**, 9049–9054.

37b. Duke, T. A. J. (1990) Monte Carlo reptation model of gel electrophoresis: Response to field pulses. *J. Chem. Phys.* **93**, 9055–9061.

38. Southern, E., Anand, R., Brown, W. R. A., and Fletcher, D. S. (1987) A model for the separation of large DNA molecules by crossed field gel electrophoresis. *Nucleic Acids Res.* **15,** 5925–5943.

39. Serwer, P. (1988) The mechanism of DNA's fractionation during pulsed-field agarose gel electrophoresis: A hypothesis. *Appl. Theor. Electrophoresis* **1,** 19–22

40. Carle, G. F. and Olson, M. V. (1984) Separation of chromosomal DNA molecules from yeast by orthogonal-field-alternation gel electrophoresis. *Nucleic Acids Res.* **12,** 5647–5666.

41. Smith, C. L. and Cantor, C. R. (1986) Pulsed-field gel electrophoresis of large DNA molecules. *Nature* **319,** 701,702.

42. Sor, F. (1987) A computer program allows the separation of a wide range of chromosome sizes by pulsed field gel electrophoresis. *Nucleic Acids Res.* **15,** 4853–4863.

43. Mathew, M. K., Smith, C. L., Cantor, C. R., Gaal, A., and Hui, C. F. (1988) High resolution separation and accurate size determination in pulse-field gel electrophoresis of DNA. *Biochemistry* **27,** 9204–9226.

44. Vollrath, D. and Davis, R. W. (1987) Resolution of DNA molecules greater than 5 megabases by contour-clamped homogeneous electric fields. *Nucleic Acids Res.* **15,** 765–7876.

45. Chu, G., Vollrath, D., and Davis, R. W. (1986) Separation of large DNA molecules by contour clamped homogeneous electric fields. *Science* **234,** 1582–1585.

46. Chu, G. (1989) Pulsed leld electrophoresis in contour clamped homogeneous electric fields. *Electrophoresis* **10,** 290–295.

47. Clark, S., Lai, E., Birren, B., and Hood, L. (1988) A novel instrument for separatin large DNA molecules with pulsed homogeneous electric fields. *Science* **241,** 1203–1205.

48. Birren, B., Lai, E., Clark, S., Hood, L., and Simon, M. (1988) Optirnized conditions for pulsed-field gel electrophoresis of DNA. *Nucleic Acids Res.* **16,** 7563–7582.

49. Birren, B. W., Hood, L., and Lai, E. (1989) Pulsed field gel electrophoresis: Studies of DNA migration made with the programmable autonomously controlled electrode electrophoresis system. *Electrophoresis* **10,** 302–309.

50. Gemmil, R. M., Coyle-Morris, J. F., McPeek, F. D., Ware-Uribe, L. F., and Hecht, F. (1987) Construction of long range restriction maps in human DNA using pulsed field gel electrophoresis. *Gene Anal. Techn.* **4,** 119–131.

51. Anand, R. (1986) Pulsed field gel electrophoresis: A technique for fractionating large DNA molecules. *Trends Genet.* **2,** 278–283.

52. Southern, E. M., Anand, R., Brown, W. R. A., and Fletcher, D. S. (1987) A model for the separation of large DNA molecules by crossed field gel electrophoresis. *Nucleic Acids Res.* **15,** 5925–5944.

53. Serwer, P. (1987) Gel electrophoresis with discontinuous rotation of the gel: An alternative to gel electrophoresis with changing direction of the electrical field. *Electrophoresis* **8**, 301–304.

54. Serwer, P. (1989) Sieving of double-stranded DNA during agarose gel electrophoresis. *Electrophoresis* **10**, 327–331.

55. Serwer, P. and Hayes, J. H. (1989) A new mode of rotating gel electrophoresis for fractionating linear and circular duplex DNA: The effects of electrophoresis during the gel's rotation. *Appl. Theor. Electrophoresis* **1**, 95–98.

56. Serwer, P. and Hayes, S. J. (1989) A typical sieving of open circular DNA during pulsed field agarose gel electrophoresis. *Biochemistry* **28**, 5827–5831.

57. Louie, D. and Serwer, P. (1989) A hybrid mode of rotating gel electrophoresis for separating linear and circular duplex DNA. *Appl. and Theor. Electrophoresis* **1**, 169–173.

58. Gardiner, K., Laas, W., and Patterson, D. (1986) *Somatic Cell Mol. Genet.* **12**, 185–195.

59. Gardiner, K. and Patterson, D. (1989) Transverse alternating field electrophoresis and application to mammalian genome mapping. *Electrophoresis* **10**, 296–301.

60. Bancroft, I. and Wolk, C. P. (1988) Pulsed homogeneous orthogonal field gel electrophoresis. *Nucleic Acids Res.* **16**, 7405–7418.

61. Carle, G. F., Frank, M., and Olson, M. V. (1986) Electrophoretic separations of large DNA molecules by periodic inversion of the electric field. *Science* **232**, 65–68.

62. Heller, C. and Pohl, F. M. (1989) A systematic study of field inversion gel electrophoresis. *Nucleic Acids Res.* **17**, 5989–6003.

63. Griess, G. A. and Serwer, P. (1990) Gel electrohoresis of micron-sized particles: A problem and a solution. *Biopolymers* **29**, 1863–1866.

64. Linsley, J. (1991) Mobility models and experimental data for lambda phage concatemers during FIGE. *Proceedings of the First International Conference on Electrophoresis, Supercomputing and the Human Genome* (Cantor, C. R. and Lim, F. A., eds.), World Scientific, Singapore, pp. 123–156.

65. Lai, E., Davi, N. A., and Hood, L. E. (1989) Effect of electric field switching on the mobility of single stranded DNA in polyacrylamide gels. *Electrophoresis* **10**, 65–67.

66a. Crater, G. D., Gregg, M. C., and Holzwarth, G. (1989) Mobility surfaces for field inversion gel electrophoresis of DNA. *Electrophoresis* **10**, 310–314.

66b. Kobayashi, T., Doi, M., Makino, Y., and Ogawa, M. (1990) Mobility minima in field inversion gel electrophoresis. *Macromolecules* **23**, 4480, 4481.

67. Turmel, C., Brassard, E., Slater, G. W., and Noolandi, J. (1990) Molecular detrapping and band narrowing with high frequency modulation of pulsed field electrophoresis. *Nucleic Acids Res.* **18**, 569–575.

68. Turmel, C., and Lalande, M. (1988) Resolution of Schizosaccharomyces pombe chromosomes by field inversion gel electrophoresis. *Nucleic Acids Res.* **16,** 4727.

69. Fesjian, S., Frisch, H. L., and Jamil, T. (1986) Diffusion of DNA in a gel under an intermittent electric field. *Biopolymers* **25,** 1179–1184.

70. Lai, E., Birren, B. W., Clark, S. M., and Hood, L. (1988) Relaxation intervals alter the mobility of large DNA molecules in pulsed field gel electrophoresis. *Nucleic Acids Res.* **16,** 10376.

71. Suterland, J. C., Monteleone, D. C., Mugavero, D. H., and Trunk, J. (1987) Unidirectional pulsed-field electrophoresis of single and double stranded DNA in agarose gels. *Anal. Biochem.* **162,** 511–520.

72a. Smith, C. L., Klco, S. R., and Cantor, C. R. (1988) Pulsed-field gel electrophoresis and the technology of large DNA molecules. *Genome Analysis: A Practical Approach* (Davies, K. E., ed.), IRL, London, pp. 41–72.

72b. Smith, C. L., Matsumoto, T., Niwa, O., Klco, S., Fan, J. B., Yanagida, M., and Cantor, C. R. (1987b) An electrophoretic karyotype for Schizosaccharomyces pombe by pulsed field gel electrophoresis. *Nucleic Acids Res.* **11,** 4481–4489.

73. Barlow, D. P. and Lehrach, H. (1987) Genetics by gel electrophoresis: The impact of pulsed field gel electrophoresis on mammalian genetics. *TIG* **3,** 167–171.

74. Carle, G. F., and Olson, M. V. (1987) Orthogonal field alternation gel electrophoresis. *Methods Enzymol.* **155,** 468–482.

75. Smith, S. B., Aldridge, P. K., and Callis, J. B. (1989) Observation of individual DTA molecules undergoing gel electrophoresis. *Science* **243,** 203–206.

76. Schwartz, D. C. and Koval, M. (1989) Conformational dynamics of individual DNA molecules during gel electrophoresis. *Nature* **338,** 520.

77. Gurrieri, S., Rizzarelli, E., Beach, D., and Bustamante, C. (1990) Imaging of kinked configurations of DNA molecules undergoing orthogonal field alternating gel electrophoresis by fluorescence microscopy. *Biochemistry* **29,** 3396–3401.

78. Ogston, A. G. (1958) The spaces in a unifom random suspension of fibres. *Trans. Faraday Soc.* **54,** 1754–1757.

79. Rodbard, D. and Chrambach, A. (1971) Estimation of molecular radius, free mobility and valence using polyacrylamide gel electrophoresis. *Anal. Biochem.* **40,** 95–134.

80. Rodbard, D. and Chrambach, A. (1970) Unified theory of gel electrophoresis and gel filtration. *Proc. Natl. Acad. Sci. USA* **4,** 970–977.

81. Lunney, J., Chrambach, A., and Rodbard, D. (1971) Factors affecting resolution, band width, number of theoretical plates and apparent diffusion coefficients in polyacrylamide gel electrophoresis. *Anal. Biochem.* **40,** 158–173.

82. Ferguson, K. A. (1964) Starch-gel electrophoresis: Application to the classification of pituary proteins and polypeptides. *Metabolism* **13,** 985–1002.

83. Flint, D. H. and Harrington, R. E. (1972) Gel electrophoresis of deoxyribonucleic acid. *Biochemistry* 11, 4858–4863.
84. McDonnell, M. W., Simon, M. N., and Studier, F. W. (1977) Analysis of restriction fragments of T7 DNA and determination of molecular weights by electrophoresis in neutral and alkaline gels. *J. Mol. Biol.* 110, 119–146.
85. Hervet, H. and Bean, C. P. (1987) Electrophoretic mobility of lambda phage Hind III and Hae III DNA fragments in agarose gels: A detailed study. *Biopolymers* 26, 77–742.
86. de Gennes, P. G. (1971) Reptation of a polymer chain in the presence of fixed obstacles. *J. Chem. Phys.* 55, 572–579.
87. de Gennes, P. G. (1979) *Scaling Concepts in Polymer Physics*, Cornell University Press, Ithaca, NY, pp. 223–240.
88. Doi, M. and Edwards, S. F. (1986) *The Theory of Polymer Dynamics*, Clarendon, Oxford, UK, pp. 188–216.
89. Sturm, J., and Weill, G. (1989) Direct observation of DNA chain orientation and relaxation by electric birefringence: Implications for the mechanism of separation during pulsed field gel electrophoresis. *Phys. Rev. Lett.* 62, 1484–1487.
90. Heiger, D. N., Cohen, A. S., and Karger, B. L. (1990) Separation of DNA restriction fragments by high performance capillary electrophoresis with low and zero crosslined polyacrylamide using continuous and pulsed electric fields. *J. Chromatogr.* 516, 3 3–8.
91. Ulanovsky, L., Drouin, G., and Gilbert, W. (1990) DNA trapping electrophoresis. *Nature* 343, 190.
92. Rousseau, J. (1988) A generalized reptation model of DNA gel electrophoresis. *M. Sc. Thesis*, Waterloo University, Ontario, Canada.
93. Défontaines, A. D. and Viovy, J. L. Theory of trapping electrophoresis. *To be published.*
94. Serwer, P. and Hayes, S. J. (1987) A voltage-gradient induced arrest of circular DNA during agarose gel electrophoresis. *Electrophoresis* 8, 244–246.
95. BioRad™ (1990) *Chef-Mapper™ Documentation.*
96. Olson, M. V. (1989) *Genetic Engineering*, Setlow, J. K., ed., Plenum, New York, pp. 183–227.
97. Viovy, J. L., Miomandre, F., Miguel, M. C., Caron, F., and Sor, F. Irreversible trapping in crossed-fields pulsed electrophoresis. *Electrophoresis* in press.
98. Hurley, I. (1986) DNA orientation during gel electrophoresis and its relation to electrophoretic mobility. *Biopolymers* 25, 539–554.
99. Akerman, B., Jonsson, M., and Norden, B. (1985) Electric orientation of DNA detected by linear dichroism spectroscopy. *J. Chem. Soc. Chem. Commun.* 422,423.
100. Akerman, B., Jonsson, M., Norden, B., and Lalande, M. (1989) Orientational dynamics of T2 DNA during agarose gel electrophoresis. Influence of gel concentration and electric field strength. *Biopolymers* 28, 1541.

101. Jonsson, M., Akerman, B., and Norden, B. (1988) Orientation of DNA during gel ekctrophoresis studied with linear dichroism spectroscopy. *Biopolymers* **27**, 381.
102. Holzwarth, G., McKee, C. B., Steiger, S., and Crater, G. (1987) Transient orientation of linear DNA molecules during pulsed field gel electrophoresis. *Nucleic Acids Res.* **15**, 10031–10044.
103. Chu, B., Xu, R., and Wang, Z. (1988) Low field transient electric birefringence of DNA in agarose gels. *Biopolymers* **27**, 2005.
104. Wang, Z. and Chu, B. (1989) Electrophoretic mobility and deformation of large DNA during gel electrophoresis. *Phys. Rev. Lett.* **63**, 2528–2531.
105. Chu, B., Wang, Z., Xu, R., and Lalande, M. (accepted for publication) Study of large DNA fragments in agarose gels by transient electric birefringence.
106. Stellwagen, N. C. and Stellwagen, J. (1989) Orientation of DNA and the agarose gel matrix in pulsed electric fields. *Electrophoresis* **10**, 332–344.
107a. Chu, B., Wang, Z., and Wu, C. (1989) Measurement of electrophoretic mobility of dye-labeled large DNA fragments in agarose gels by movement of fluorescence pattern after photobleaching. *Biopolymers* **28**, 1491.
107b. Wu, C., Wang, Z., and Chu, B. (1990) Electrophoresis and movements of fluorescence pattern after photobleaching of large DNA fragments in agarose gels. *Biopolymers* **29**, 491–500.
108. Holzwarth, G., Platt, K. J., McKee, C. B., Whitcomb, R. W., and Crater, G. D. (1989) The acceleration of linear DNA during pulsed-field gel electrophoresis. *Biopolymers* **28**, 1043–1058.
109. Doi, M. (1983) Explanation for the 3.4 power law for viscosity of polymeric liquids on the basis of the tube model. *J. Polym. Sic. Pol. Phys. Ed.* **21**, 667–684.
110. Lim, H. A., Slater, G. W., and Noolandi, J. (1990) A model of the DNA transient orientation overshoot during gel electrophoresis. *J. Chem. Phys.* **92**, 709–720.
111. Rubinstein, M. (1987) Repton model of entangled polymers. *Phys. Rev. Lett.* **59**, 1946.

CHAPTER 27

Elastic Bag Model of One-Dimensional Pulsed-Field Gel Electrophoresis (ODPFGE)

Jaan Noolandi and Chantal Turmel

1. Introduction

Gel electrophoresis is one of the most common techniques used in molecular biology for the separation of DNA molecules. Conventional gel electrophoresis (using a static electric field) does not permit separation of DNA fragments larger than 30–50 kbp *(1)* as shown in Fig. 1A of Chapter 7. This is a surprising result as one would think that larger molecules would suffer a larger retardation, and separation over any size range would be possible. The inability to separate is related to the molecular conformation of a polyelectrolyte, such as DNA, migrating in a disordered medium, such as a gel, under the influence of a static electric field. During continuous field electrophoresis, the larger DNA fragments tend to orient and stretch in the field direction because they migrate in a one-dimensional fashion between the gel fibers *(2–4)*. When this orientation is negligible, e.g., for smaller molecules or for very low field intensities, they maintain a three-dimensional random-walk conformation intertwined with the gel fibers during migration, and experience a retardation that is proportional to the mol size. However, when the orientation becomes large, the molecules become

From: *Methods in Molecular Biology, Vol. 12: Pulsed-Field Gel Electrophoresis*
Edited by: M. Burmeister and L. Ulanovsky
Copyright © 1992 The Humana Press, Inc., Totowa, NJ

stretched and migrate essentially linearly along the field direction
(5). The electrophoretic mobility then becomes independent of
the mol size and no separation of molecules of different sizes is
possible (Fig. 1A of Chapter 7). Physically, this is a consequence of
the fact that for long molecules stretched and oriented in the field
direction, both the electrical force on the molecule and the average friction opposing the forward motion are proportional to the
length. It follows that the velocity, which is the ratio of these two
quantities, depends only on the force per unit length and is independent of the actual mol length. This explains why a plateau of
length-independent mobility is reached in a continuous electric
field (Fig. 2 of Chapter 7).

Since 1983, many different pulsed-field gel electrophoresis
techniques have been developed, based on an empirical approach
to find the optimum electrode geometry, which coupled to the
right choice of pulses would give good separations *(6–13).* This is
characteristic of a field where the applications of a technique are
important *(7),* but the basic principles are poorly understood. Separation of DNA molecules in orthogonal field alternation gel
electrophoresis (OFAGE; *8),* field inversion gel electrophoresis
(FIGE; *9),* transverse alternating field electrophoresis (TAFE; *10),*
clamped homogeneous electric field (CHEF; *11),* programmable
autonomously controlled electrode gel electrophoresis (PACE; *12),*
pulsed homogeneous orthogonal field gel electrophoresis (PHOGE;
13), zero-integrated field electrophoresis (ZIFE; *14),* and one-
dimensional pulsed-field electrophoresis (ODPFGE; *15)* all depend
in some way on the size-dependent retardation/reorientation of
large molecules.

Current practice is dominated by two main electric field geometries. In one case, transverse fields are alternately applied in two
directions and the direction of net migration depends on the angle
between them (CHEF, PACE, PHOGE, OFAGE, and TAFE). In the
other cases (FIGE, ZIFE, and ODPFGE, *see later discussion)* the electric field geometry involves inversion of the direction of the field
in one dimension and (with the exception of ZIFE and ODPFGE)
the change of the *magnitude* of the electric field. It is now known
that for obtaining good separations the electrode geometry is an
irrelevant parameter—the separation of large molecules in pulsed
fields takes place because a basic physical symmetry (the identical

scaling of friction and electrical force for large stretched molecules) is broken, and this can be accomplished by electrical, mechanical, or other means; however, the separation technique is useful only if the symmetry-breaking is highly size-specific. The deeper reason for the effectiveness of pulsed fields in separating large DNAs is thus related to the symmetry of space, and this being so, we choose the simplest subspace, which is one-dimensional.

We have developed the pulsed-field technique using one-dimensional electric field geometry, and a variant of this method called ZIFE *(14)*. We make a distinction between ZIFE, ODPFGE on the one hand and FIGE on the other, because we not only invert the direction of the electric field, but also change the ratios of the voltages in the forward and reverse directions to selectively take advantage of the different reorientation and forced relaxation times of molecules of different sizes. In this technique a particular sequence of pulses provides the code for the migration of a molecule of a given size to a designated position in the gel, keeping the correct order according to size, which is not always possible with FIGE. A complex pulse train can then be predetermined by a computer algorithm for the separation of a large number of DNA fragments. Figures 1B and 2 of Chapter 7, show examples of logarithmic distributions covering approx 6000 kbp on a single gel. The logarithmic distribution is chosen so that a large range of sizes can be separated on a single gel. This technique also allows the use of conventional gel trays, and gel lane comparisons are simplified since the bands migrate in straight lines (unlike in some crossed-field systems). This approach to the separation of DNA based on the dynamics of molecular motion is based on a considerable amount of experimental and theoretical work on the effects of field intensity, time and field pulse ratios, and gel concentrations for different sizes of DNA, during which over 4000 gels were run.

As we show later, the field and time pulse ratios ($R_t = t_+/t_-$, $R_E = E_+/E_-$) in one-dimensional electrophoresis are the most important parameters to be determined for an optimal separation. These parameters (R_t and R_E) are chosen depending on the range of separation desired. We have demonstrated previously that near zero-integrated field electrophoresis (ZIFE) conditions allow an optimal separation for molecules ranging from 10–2000 kbp *(14)*. *See also* Chapter 7 for a discussion of pulse strategies. One-dimensional pulsed-field electrophoresis (ODPFGE) with a forward-bias *(15)* can be used for efficient separations above 2 Mbp. The ZIFE pulse strategy greatly increases the power of separation, taking advantage of the fact that the migration of large molecules can be kept frozen whereas smaller molecules (for a narrow range of sizes) migrate quickly. With this pulse scheme, band inversion is minimized and separation in a given size range is maximized.

2. Elastic Bag Model

The goal of this section is to explain how the mobilities of DNA molecules of different sizes are affected by different pulsing conditions. The most important practical result is that the region of band inversion can always be moved out of the size range of interest by choosing the right pulses. This is only possible if different voltages, as well as times, are used for the forward and reverse pulses. With the solution of the band inversion problem, the one-dimensional system has an advantage over more complicated electrode arrangements because of its simplicity. Another result in this section is that the molecular mobilities can be related to recent measurements of velocity changes during the pulse cycle (transient velocity), and not to the myriad details of individual conformations, as has been previously thought *(17)*. A similar conclusion has been arrived at for two-dimensional systems by Chu *(18)*.

The most important studies of the conformational dynamics of DNA during gel electrophoresis are the elegant experiments by the Swedish group *(19–22)* using linear dichroism spectroscopy, the transient velocity measurements of Holzwarth and coworkers *(23,24)*, and the electric birefringence studies of Sturm and Weill *(25)* and Chu et al. *(26)*. The theory of extensible model chains has been used to discuss the results of these groups. For our purposes we focus mainly on the transient velocity measurements. Platt and Holzwarth *(23)* have carried out measurements of the average velocity of T4 DNA (170 kbp) and λDNA (48.5 kbp) during field inversion gel electrophoresis, as shown in Fig. 1. They observed that the transient velocity immediately after field inversion has the general shape shown in Fig. 2, with an initial increase in the velocity, followed by a rapid drop in the velocity (deceleration phase), and a slower increase (acceleration phase) with some oscillations before reaching a steady-state velocity.

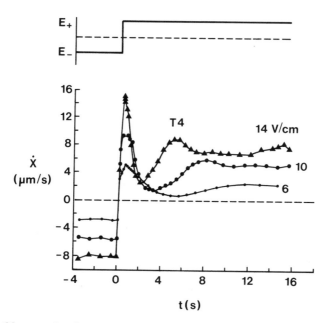

Fig. 1. Observed velocity of DNA during field inversion *(24)*. The velocity of T4 DNA in 1% agarose is shown for E = 6, 10, and 14 V/cm.

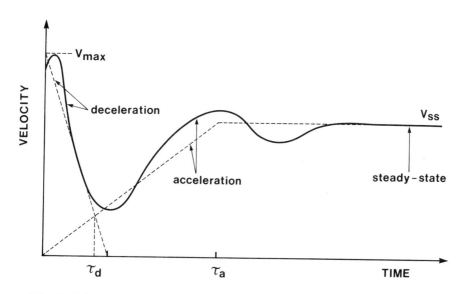

Fig. 2. Schematic diagram showing transient velocity measurements of Platt and Holzwarth *(24)* for field inversion, and an idealized curve to show the main features with parameters as described in the text.

There is an enormous number of conformations experienced by a molecule subjected to pulsed fields. In the following, we discuss only a couple of representative examples. Both the generalized reptation model *(27,28)*, giving results shown in Fig. 3, and the fluctuating bond model *(29)*, giving results shown in Fig. 4, predict results qualitatively similar to those of Platt and Holzwarth *(23)*. The sharp maximum in the initial velocity after field reversal is caused by the different parts of the molecules moving into spaces where there are no obstacles, before they realize they are connected and start pulling against each other in response to the electric field. In Fig. 3 the deceleration phase results from both ends of the molecule growing out of an initial collapsed configuration while the middle part is hooked around gel fibers, thereby stretching the chain (i.e., decreasing the average amount of DNA per gel pore) until the longer arm starts to grow at the expense of the smaller one (acceleration phase) and the trailing end is released, after which the chain contracts and moves forward. In Fig. 4 the deceleration phase is related to the development of a large "hernia" while the free end moves forward in the field direction and the acceleration phase corresponds to the hernia pulling the free end backward through the region of herniation *(29)*. Although the Platt and Holzwarth experiment has not been carried out for molecules in the megabase size range, the general results derived from Fig. 2 *(see below)* as well as extensive computer simulation of DNA motion in gels *(28,29)* lead us to believe that this velocity profile is similar for molecules of all sizes in one-dimensional pulsed-field gel electrophoresis.

To simplify the following analysis, we idealize the experimental curve by a series of straight lines. The extrapolated deceleration and acceleration times are denoted by τ_d and τ_a, neglecting the small secondary oscillations. The initial maximum velocity is V_{max} and V_{ss} is the steady state velocity. A more detailed mathematical description will be given later. The main result of this simple picture is that it gives an explanation of the phenomenon of band inversion, in which intermediate-sized DNA molecules migrate more slowly than both smaller and larger ones *(14)*. The first theoretical explanation of this effect for small molecules was obtained from the biased reptation model, in which both the 5' and 3' ends of the DNA move forward at the same time (for single stranded DNA),

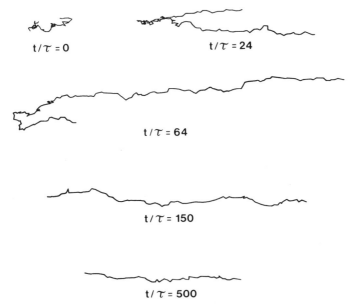

Fig. 3. Computer simulation of the motion of a DNA molecule as it moves to the right out of a collapsed configuration in response to an electric field. The initial stretching of the contour length (deceleration phase) of the bead-spring "molecule" corresponds to the decrease of the amount of DNA per gel pore when one end is hooked around gel fibers (not shown). The basis of the simulation is the generalized reptation model *(27,28)*, and the time is given in terms of a convenient time constant.

leading to temporarily immobilized loop-like conformations, in which the electrical force on both arms is the same. This model however neglected longitudinal and transverse fluctuations during migration *(30)*.

The control of band inversion is the single most important goal for making one-dimensional pulsed-field electrophoresis a useful tool for separating large molecules. Before proceeding with a discussion of band inversion, based on the velocity profile shown in Fig. 2, it is important to note that V_{max} is reduced in amplitude, and the deceleration time is decreased, if the molecules do not have time to reorient completely during the preceding pulse of opposite polarity. Thus there is a strong hysteresis effect on the velocity profile, which can be described mathematically by a scaling law, given later.

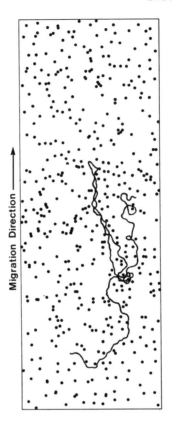

Fig. 4. Computer simulation of a DNA molecule showing a giant hernia, which leads to a transient slowing down of the molecule (deceleration phase). The computer simulation was carried out using a Monte Carlo algorithm *(29)*.

Figure 5 shows a series of forward and reverse pulses for one-dimensional pulsed-field electrophoresis. Experimentally it was found that both τ_d and τ_a are approximately proportional to M/E, where M is the mol mass and E is the field intensity. In Carle, Frank, and Olson's FIGE technique *(9)*, the forward pulse duration t_+ is typically three times the reverse pulse duration t_- and $E_+ = E_-$ in magnitude.

For small mol sizes, both forward and reverse pulses are much longer than the deceleration (τ_d) and acceleration (τ_a) times, and Carle, Frank, and Olson's method gives a net velocity that is proportional to a superposition of the displacements during the two parts of the cycle,

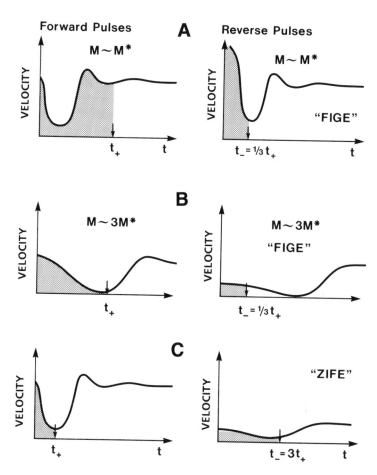

Fig. 5. Schematic diagrams explaining the source of the band inversion observed in one-dimensional pulsed-field electrophoresis. The curves for FIGE in **(A)** correspond roughly to the critical mol size M^* showing minimum mobility (*see* text) and in **(B)** to $M \sim 3M^*$. The curve for **(C)** shows the transient velocity for ZIFE. In both **(A)** and **(C)** the deceleration phase is much reduced in the forward direction. In FIGE **(A)**, with $M \sim M^*$, the forward pulse integrates only the large minimum in velocity, whereas only the large deceleration peak is effective in the reverse direction. **(B)** for larger molecules, $M \sim 3M^*$, the forward pulse integrates the large deceleration peak, whereas the smaller deceleration peak in the reverse direction is not very effective in displacing the molecule backward, with the result that the larger molecules in **(B)** move faster than the smaller molecules in **(A)**. **(C)** in ZIFE, both deceleration phases are reduced in amplitude, and both pulses integrate over the same number of oscillations, i.e., they are are in phase.

$$V_{FIGE} = V_{SS} \left(\frac{t_+}{t_+ + t_-} \right) - V_{SS} \left(\frac{t_-}{t_+ + t_-} \right) \qquad (1)$$

where V_{SS} is the steady state velocity in a continuous field of intensity $E_+ = E_-$. For small M, Eq. (1) reduces to $V_{FIGE} = 1/2 \, V_{SS}$ for $t_+ = 3t_-$. At a critical size, denoted by M^*, the average mol velocity in the reverse direction actually increases (Fig. 5, panel A), because the initial transient velocity peak for the reverse direction becomes larger than for the forward direction. The displacements of the molecules are shown in Fig. 5 as the integrated (shaded) areas under the transient velocity time curves. This occurs because complete molecular reorientation cannot take place during the short reverse pulse for Carle, Frank, and Olson's FIGE conditions. The net result is that the average velocity decreases at the critical size, and this leads to large separations for sizes smaller than but approaching the critical size. However, for larger sizes (Fig. 5, panel B) the same effect eventually leads to a decrease of the velocity maximum for the reverse pulse as well (since in this case complete reorientation cannot take place for sufficiently large molecules during the forward pulse) resulting in an increase of the velocity of the larger molecules compared to the velocity around the critical size. The shaded area for the forward pulse is now bigger than for the reverse pulse in panel B, unlike in panel A where the shaded areas are comparable. In other words the reverse deceleration phase starts vanishing only for molecules about three times larger, $M \sim 3M^*$. For these large molecules the velocity in both directions is given by only the deceleration phase, with the result that $V_{FIGE} (M \sim 3M^*) > V_{FIGE} (M^*)$, giving rise to band inversion. This difficulty can be avoided in the ZIFE and forward-biased techniques as described in Chapter 7. The limitation of Carle, Frank, and Olson's technique is that the forward and reverse pulses are *empirically* "tuned" to the reorientation times of the molecules for the same electric field strengths in the forward and reverse directions, and there is no systematic way of controlling the band inversion as a function of both pulse times and electric field intensity ratios.

If a lower field $E_- < E_+$ is selected in the reverse direction (Fig. 5, lower panel C), the reverse velocities are lower, and the reverse velocity oscillations are less important. Hence the reduction in V_{max} for the forward pulse has less effect on the average forward velocity than in Carle, Frank, and Olson's technique, and the band inversion is greatly reduced. To minimize the effect of the velocity oscillations we choose pulse conditions for which the oscillations are approximately in phase in the forward and reverse directions. Since τ_d, $\tau_a \sim M/E$, this leads to the requirement that $E_+ t_+ = E_- t_-$, which is consistent with the results of Platt and Holzwarth *(23)*, where the secondary velocity peak moves to longer times for lower fields.

Matching the velocity oscillations in the forward and reverse directions is equivalent to having a field that vanishes when averaged over one cycle, and we call this technique zero-integrated field electrophoresis, or ZIFE. In practice we often use the condition $E_+ t_+ \geq E_- t_-$ to speed up the separation (the average electric field over one cycle is still smaller than in other techniques). The price paid for the violation of the exact ZIFE condition is that band inversion then reappears at the high end of the size range. However, a computer algorithm can be used to keep track of where this occurs and to assemble a sequence of pulses with no ramping, which pushes the band inversion region out of the range specified by the user. This is possible because an experimental study of different pulse times and pulse time ratios as well as electric fields and electric field ratios in the forward and reverse directions, coupled with molecular modeling techniques, has led us to an understanding of the trends in the mobility curves that are required for designing the best pulse conditions for the needs of the user. Some of these trends are seen in the experimental data shown in Figs. 4, 5, 7, and 8, Chapter 7.

Figure 4, Chapter 7 shows a series of experiments for near-ZIFE conditions, where the pulse times have been increased to move the sharp drop in mobility systematically along the size axis while keeping the same pulse time ratio as well as the same field ratio. For $t_+ = 160$ s, band inversion appears between 2 and 3 Mbp. For similar pulse time and electric field ratios, Fig. 5, Chapter 7, shows the results for a higher electric field, with an initial velocity which is double that shown in Fig. 4, Chapter 7. In this case, substantial band inversion appears between 1 and 2 Mbp for $t_+ = 65$ s. Note that the largest velocities of the smallest fragments in Fig. 5, Chapter 7, are greater than 3 mm/h, and that the smallest velocities of the largest fragments are close to zero, demonstrating the good resolving power of the process in a narrow size range where the DNA fragments have large differential velocities. Figure 7,

Chapter 7, shows the velocity curves for a series of pulses with a strong forward bias. This pulse strategy has poor resolving power for the smaller times ($t_+ \leq 300$ s) but works well for longer times and larger DNA sizes. A good pulse strategy for a wide range of sizes is to use ZIFE-like conditions for sizes up to approx 2 Mbp, and to use the forward bias conditions above 2 Mbp. Figure 8, Chapter 7, shows how the band inversion can be pushed toward larger sizes by increasing the duration of the pulses. The pulses with $t_+ = 2000$ s separate the two smaller chromosomes of *S. pombe*, whereas the $t_+ = 4000$ s pulses separate the largest chromosome from the smaller ones. A combination of both types of pulses is effective in separating all three chromosomes.

Figures 6 and 7 show exmples of the calculated velocity vs size curves for a range of pulse time and electric field ratios according to the transient velocity–time model presented earlier. A complete theoretical analysis of the data base is given elsewhere *(31)*. These figures show that the theory can predict the shift in the velocity vs size curves as the value of t_+ is increased for constant electric field and pulse time ratios. The curves were generated by integrating the velocity–time profile shown in Fig. 2 with

$$V^{\pm}_{max} = C_{V_{max}} E_{\pm} \; min \, [\,(t_{\pm}/\tau_a^{\pm})^{E_{Vmax}}, 1\,] \tag{2}$$

$$V^{\pm}_{ss} = C_{V_{ss}} \, (E_{\pm})^{E} v_{ss} \tag{3}$$

$$\tau_a^{\pm} = C_{\tau a} \, (M/E_{\pm}) \tag{4}$$

$$\tau_d^{\pm} = C_{\tau d} \, (M/E_{\pm}) \; min \, [\,(t\mp/\tau_a^{\mp})^{E_{\tau d}}, 1\,] \tag{5}$$

where M is in kbp, E_+ in V/cm, and t_+, $\tau_a^{\pm}$ in s. The numerical values of the constants used to generate Figs. 6 and 7 are $C_{\tau a} = 1/3$, $C_{\tau d} = 1/6$, $C_{V_{ss}} = 1/3.6$, $C_{V_{max}} = 1.0$, $E_{V_{ss}} = 4/3$, $E_{V_{max}} = 1/5$, $E_{\tau d} = 2/5$. Note that $\tau_d^{\pm}$ also depends on the parameters for the previous pulse, $t\mp$, τ_a^+, which is important for understanding the cumulative effect of the pulses on the average mobility of the DNA fragments. Equations (2)–(5) form the mathematical description of the experimental data base, and can be used to estimate pulse parameters for molecules larger than *S. pombe*.

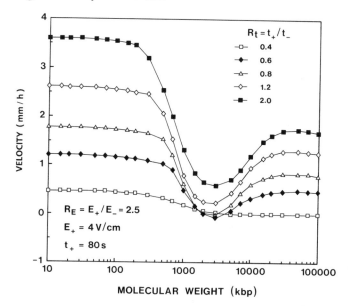

Fig. 6. Calculated velocity–size curves for indicated pulse parameters and t_+ = 80 s using elastic bag model.

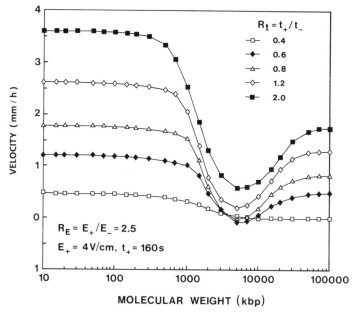

Fig. 7. Calculated velocity–size curves for indicated pulse parameters and t_+ = 160 s using elastic bag model.

For the separation of large molecules, the low frequency pulse strategy discussed in this section has to be augmented by the addition of a high frequency component to obtain useful separations and to control band broadening. As discussed in Chapter 7, the optimum conditions for the combination of high and low frequency pulses are currently under investigation. It is interesting to note that although one of the principal reasons for using a gel to separate biological molecules was to reduce convection in the fluid, a certain amount of local disturbance (at least on the order of a few pore sizes) is apparently necessary to allow the long molecules to change their conformations efficiently in order to respond to the size-specific pulse signals. The insertion of high frequency pulses into the low frequency pulses is one way to facilitate these conformational changes *(16)*, as discussed in Chapter 7, although this is not the only possibility.

3. Conclusions

This chapter has discussed the surprisingly complicated sequences of electric field pulses ranging from milliseconds to hours that are necessary to separate DNA molecules effectively in the megabase size range. This is a general consequence of the peculiar dynamical behavior of long polyelectrolytes, such as DNA, when forced by an electric field through a random network such as a gel, and (in physics terms) to the symmetry breaking (separation according to size) brought about by the use of pulsed fields. A one-dimensional electrode geometry was chosen, because of its simplicity and because it is possible to maintain a one-to-one size vs migration distance relationship with the proper choice of pulses. Equally important for successful separations are preparation techniques, such as the metaphase blocked protocol used for the chicken microchromosomes (Chapter 10), and the recognition of certain artifacts that appear as bands of separated chromosomes, but actually have a physical origin related to the separation process (Chapter 7).

Although a variety of empirical devices with different electrode geometries will continue to be used by many researchers, it is likely that some role will be played by molecular modeling in the design of the next generation of instruments for the separation of biological molecules. For one thing, this will help in sorting out new biology from the physics of the separation process. An example is the band inversion of nucleic acids that has recently been observed in capillary gel electrophoresis with high electric fields (several hundred V/cm and greater) in polyacrylamide matrices. This is similar to the band inversion (or band scrambling) phenomenon encountered in one-dimensional submarine gel systems using agarose *(30)*. Since the relevant dimensionless parameter for both capillary and submarine gels is qEa/k_BT (where q is an effective charge per unit length of the molecule, E is the electric field, a the gel pore size, k_B Boltzmann's constant, and T the temperature) an increase in the electric field by several orders of magnitude can be compensated by a corresponding decrease in the pore size, resulting in similar physical phenomena taking place in quite different systems. Coupling the pulsed electric field to selected internal modes of molecular motion may also be useful in extending the range of sequencers *(32)*, for separating molecules in solution by capillary electrophoresis, and in protein separation *(33)*.

The role of theory in biology as embodied in computer data bases and computer modeling is not restricted to understanding protein structure/function and coming to grips with the expanding genome data base *(34)*, but can also be useful in the design of more efficient analytical and preparative tools.

References

1. Stellwagen, N. C. (1987) Electrophoresis of DNA in agarose and polyacrylamide gels, in *Advances in Electrophoresis* (Chrambach, A., Dunn, M. J., and Radola, B. J., eds.), VCH, Weinheim, Germany, pp. 177–228.
2. Lerman, L. S. and Frisch, H. L. (1982) Why does the electrophoretic mobility of DNA in gels vary with the length of the molecule? *Biopolymers* **21**, 995–997.
3. Lumpkin, O. J., Dejardin, P., and Zimm, B. H. (1985) Theory of gel electrophoresis of DNA. *Biopolymers* **24**, 1573–1593.
4. Slater, G. W. and Noolandi, J. (1985) Prediction of chain elongation in the reptation theory of DNA gel electrophoresis. *Biopolymers* **24**, 2181–2184.
5. Slater, G. W. and Noolandi, J. (1986) On the reptation theory of gel electrophoresis. *Biopolymers* **25**, 431–454.
6. Schwartz, D. and Cantor, C. R. (1984) Separation of yeast chromosome-sized DNAs by pulsed field gradient gel electrophoresis. *Cell* **37**, 67–75.
7. Barlow, D. P. and Lehrach, H. (1987) Genetics by gel electrophoresis: The impact of pulsed field gel electrophoresis on mammalian genetics. *Trends Genet.* **3**, 167.

8. Carle, G. F. and Olson, M. V. (1984) Separation of chromosomal DNA molecules from yeast by orthogonal-field-alternation gel electrophoresis. *Nucleic Acids Res.* **12,** 5647–5665.

9. Carle, G. F., Frank, M., and Olson, M. V. (1986) Electrophoretic separations of large DNA molecules by periodic inversion of the electric field. *Science* **232,** 65–68.

10. Gardiner, K., Laas, W., and Patterson, D. (1986) Fractionation of large mammalian DNA restriction fragments using vertical pulsed field gradient gel electrophoresis. *Somatic Cell Mol. Genet.* **12,** 185–195.

11. Chu, G., Vollrath, D., and Davis, R. W. (1986) Separation of large DNA molecules by contour-clamped homogeneous electric fields. *Science* **234,** 1582–1585.

12. Clark, S. M., Lai, E., Birren, B. W., and Hood, L. (1988) A novel instrument for separating large DNA molecules with pulsed homogeneous electric fields. *Science* **241,** 1203–1205.

13. Bancroft, I. and Wolk, C. P. (1988) Pulsed homogeneous orthogonal field gel electrophoresis (PHOGE). *Nucleic Acids Res.* **16,** 7405–7418.

14. Turmel, C., Brassard, E., Forsyth, R., Hood, K., Slater, G. W., and Noolandi, J. (1990) High resolution zero integrated field electrophoresis (ZIFE) of DNA, in *Current Communications in Molecular Biology. Electrophoresis of Large DNA Molecules* (Birren, B. and Lai, E., eds.), Cold Spring Harbor Laboratory, Cold Spring Harbor, NY, pp. 101–131.

15. Lalande, M., Noolandi, J., Turmel, C., Rousseau, J., and Slater, G. W. (1987) Pulsed-field electrophoresis: Application of a computer model to the separation of large DNA molecules. *Proc. Natl. Acad. Sci. USA* **84,** 8011–8015.

16. Turmel, C., Brassard, E., Slater, G. W., and Noolandi, J. (1989) Molecular detrapping and band narrowing with high frequency modulation of pulsed field electrophoresis. *Nucleic Acids Res.* **18,** 569–575.

17. Deutsch, J. M. (1990) Theoretical aspects of electrophoresis, in *Current Communications in Molecular Biology. Electrophoresis of Large DNA Molecules* (Birren, B. and Lai, E., eds.), Cold Spring Harbor Laboratory, Cold Spring Harbor, NY, pp. 81–99.

18. Chu, G. (1990) Bag model for DNA migrating in a pulsed electric field. *Proc. Natl. Acad. Sci. USA,* in press.

19. Åkerman, B., Jonsson, M., and Norden, B. (1985) Electrophoretic orientation of DNA detected by linear dichroism spectroscopy. *J. Chem. Soc. Chemical Communications* **7,** 422–423.

20. Åkerman, B., Jonsson, M., and Norden, B. (1989) Orientational dynamics of T2 DNA during agarose gel electrophoresis: Influence of gel concentration and electric field strength. *Biopolymers* **28,** 1541–1571.

21. Åkerman, B. and Jonsson, M. (1990) Reorientational dynamics and mobility of DNA during pulsed field agarose gel electrophoresis. *J. Phys. Chem.* **94,** 3828–3838.

22. Åkerman, B., Jonsson, M., Moore, D., and Schellman, J. (1990) Conformational dynamics of DNA during gel electrophoresis studied by linear

dichroism spectroscopy, in *Current Communications in Molecular Biology. Electrophoresis of Large DNA Molecules* (Birren, B. and Lai, E., eds.), Cold Spring Harbor Laboratory, Cold Spring Harbor, NY, pp. 23–41.

23. Platt, K. J. and Holzwarth, G. (1989) Velocity of DNA in gels during field inversion. *Phys. Rev. A.* **40**, 7292–7300.

24. Holzwarth, G., Whitcomb, R. W., Platt, K. J., Crater, G. D., and Mckee, C. B. (1990) Velocity of linear DNA during pulsed field gel electrophoresis, in *Current Communications in Molecular Biology. Electrophoresis of Large DNA Molecules* (Birren, B. and Lai, E., eds.), Cold Spring Harbor Laboratory, Cold Spring Harbor, NY, pp. 43–53.

25. Sturm, J. and Weill, G. (1989) Direct observation of DNA chain orientation and relaxation by electric birefringence: Implications for the mechanism of separation during pulsed-field gel electrophoresis. *Phys. Rev. Lett.* **62**, 1484–1487.

26. Chu, B., Xu, R., and Wang, Z. (1988) Low-field transient electric birefringence of DNA in agarose gels. *Biopolymers* **27**, 2005–2009.

27. Noolandi, J., Slater, G. W., Lim, H. A., and Viovy, L. (1989) Generalized tube model of biased reptation for gel electrophoresis of DNA. *Science* **243**, 1456–1458.

28. Lim, H. A., Slater, G. W., and Noolandi, J. (1990) A model of the DNA transient orientation overshoot during gel electrophoresis. *J. Chem. Phys.* **92**, 709–721.

29. Schönherr, G. and Noolandi, J. (1991) Fluctuating bond model of DNA gel electrophoresis. *Electrophoresis* **12**, 432–435.

30. Noolandi, J., Rousseau, J., Slater, G. W., Turmel, C., and Lalande, M. (1987) Self-trapping and anomalous dispersion of DNA in electrophoresis. *Phys. Rev. Lett.* **58**, 2428–2431.

31. Hood, K., Slater, G. W., Turmel, C., and Brassard, E. Band inversion in one-dimensional pulsed field gel electrophoresis. To be published 1992.

32. Noolandi, J. (1991) A possible application of zero integrated pulsed field gel electrophoresis (ZIFE) to DNA sequence analysis. *Makromol. Chem. Rapid Commun.* **12**, 31.

33. Brassard, E., Turmel, C., and Noolandi, J. (1991) Observation of orientation and relaxation of protein–sodium dodecyl sulfate complexes during pulsed intermittent field polyacrylamide gel electrophoresis. *Electrophoresis* **12**, 373–375.

34. Gilbert, W. (1991) Towards a paradigm shift in biology. *Nature* **349**, 99.

Index